CLINICAL USE OF ANTIVIRAL DRUGS

Yechiel Becker, Series Editor
Julia Hadar, Managing Editor

DEVELOPMENTS IN VETERINARY VIROLOGY

Payne, L.N. (ed.) *Marek's Disease* (1985)
Burny, A. and Mammerickx, M. (eds.) *Enzootic Bovine Leukosis and Bovine Leukemia Virus* (1987)
Becker, Y. (ed.) *African Swine Fever* (1987)
DeBoer, G.F. (ed.) *Avian Leukosis* (1987)
Liess, B. (ed.) *Classical Swine Fever and Related Viral Infections* (1987)
Darai, G. (ed.) *Virus Diseases in Laboratory and Captive Animals* (1988)

DEVELOPMENTS IN MOLECULAR VIROLOGY

Becker, Y. (ed.) *Herpesvirus DNA* (1981)
Becker, Y. (ed.) *Replication of Viral and Cellular Genomes* (1983)
Becker, Y. (ed.) *Antiviral Drugs and Interferon: The Molecular Basis of Their Activity* (1983)
Kohn, A. and Fuchs, P. (eds.) *Mechanisms of Viral Pathogenesis from Gene to Pathogen* (1983)
Becker, Y. (ed.) *Recombinant DNA Research and Viruses, Cloning and Expression of Viral Genes* (1985)
Feitelson, M. *Molecular Components of Hepatitis B Virus* (1985)
Becker, Y. (ed.) *Viral Messenger RNA: Transcription, Processing, Splicing and Molecular Structure* (1985)
Doerfler, W. (ed.) *Adenovirus DNA: The Viral Genome and Its Expression* (1986)
Aloni, Y. (ed.) *Molecular Aspects of Papovaviruses* (1987)

DEVELOPMENTS IN MEDICAL VIROLOGY

Levine, P.H. (ed.) *Epstein-Barr Virus and Associated Diseases* (1985)
Becker, Y. (ed.) *Virus Infections and Diabetes Mellitus* (1987)
DeClercq, E. (ed.) *Clinical Use of Antiviral Drugs* (1988)

CLINICAL USE OF ANTIVIRAL DRUGS

Edited by

Professor Dr. Erik De Clercq
Rega Institute for Medical Research
Katholieke Universiteit Leuven

Martinus Nijhoff Publishing
a member of the Kluwer Academic Publishers Group
Boston/Dordrecht/Lancaster

Distributors for North America:
Kluwer Academic Publishers
101 Philip Drive
Assinippi Park
Norwell, Massachusetts 02061 USA

Distributors for the UK and Ireland:
Kluwer Academic Publishers
MTP Press Limited
Falcon House, Queen Square
Lancaster LA1 1RN, UNITED KINGDOM

Distributors for all other countries:
Kluwer Academic Publishers Group
Distribution Centre
Post Office Box 322
3300 AH Dordrecht, THE NETHERLANDS

Library of Congress Cataloging-in-Publication Data

Clinical use of antiviral drugs / edited by Erik De Clercq.
p. cm. — (Developments in medical virology)
Includes bibliographies and index.
ISBN-13: 978-1-4612-8966-1 e-ISBN-13: 978-1-4613-1715-9
DOI: 10.1007/978-1-4613-1715-9
1. Antiviral agents. 2. Virus diseases—Chemotherapy. I. De Clercq, Erik. II. Series.
[DNLM: 1. Antiviral Agents—therapeutic use. 2. Virus Diseases—drug therapy. QV 268.5 C641]
RM411.C57 1988
616.9'25061—dc19
DNLM/DLC 87-31405
for Library of Congress CIP

CONTENTS

CONTRIBUTING AUTHORS

1. DR. B.E. JUEL-JENSEN
Nuffield Department of Clinical Medicine
Radcliffe Infirmary
University of Oxford
Oxford OX2 6HE, United Kingdom

2. DR. W.H. PRUSOFF
Department of Pharmacology
Sterling Hall of Medicine
Yale University School of Medicine
333 Cedar Street
P.O. Box 3333
New Haven, Connecticut 06510
U.S.A.

3. DR. H.E. KAUFMAN
Lions Eye Research Laboratories
LSU Eye Center
Louisiana State University Medical Center School of Medicine
New Orleans, Louisiana 70112
U.S.A.

4. DR. P.C. MAUDGAL
Eye Research Laboratory
Ophthalmological Clinic
Academic Hospital St. Rafaël
Katholieke Universiteit Leuven Kapucijnenvoer, 7
B-3000 Leuven, Belgium
Department of Ophthalmology
Academic Hospital Free University
Postbus 7057
1007 MB Amsterdam
The Netherlands

DR. E. DE CLERCQ
Rega Institute for Medical Research
Katholieke Universiteit Leuven
Minderbroedersstraat 10
B-3000 Leuven, Belgium

5. DR. R.J. WHITLEY
Department of Pediatrics and Microbiology
Room 609 CDLD
School of Medicine
The University of Alabama
Birmingham, Alabama 35294
U.S.A.

6. DR. S.L. SPRUANCE
Division of Infectious Diseases
Department of Medicine
University of Utah School of Medicine
Salt Lake City, Utah 84132
U.S.A.
The Centre International de Recherches Dermatologiques (CIRD)
Sophia Antipolis
06565 Valbonne, France

7. DR. S.L. SACKS
Division of Infectious Diseases
Department of Medicine
Health Sciences Centre Hospital
The University of British Columbia
2211 Wesbrook Mall
Vancouver, British Columbia
Canada V6T 1W5

8. DR. R. SARAL
Department of Oncology
The Johns Hopkins University School of Medicine
Baltimore, Maryland 21205, U.S.A.

DR. P. LIETMAN
Division of Clinical Pharmacology
The Johns Hopkins University School of Medicine
Baltimore, Maryland 21205, U.S.A.

9. DR. J. ENGLUND
Department of Pediatrics
Laboratory Medicine and Pathology
University of Minnesota Medical School
Minneapolis, Minnesota 55455
U.S.A.

DR. H.H. BALFOUR, Jr.
Department of Pediatrics
Laboratory Medicine and Pathology
University of Minnesota Medical School
Minneapolis, Minnesota 55455
U.S.A.

10. DR. S. SHIGETA
Department of Bacteriology
Fukushima Medical College
Fukushima 960, Japan

DR. E. DE CLERCQ
Rega Institute for Medical Research
Katholieke Universiteit Leuven
Minderbroedersstraat 10
B-3000 Leuven, Belgium

11. DR. P.R. SKOLNIK
Infectious Disease Unit
Department of Medicine
Massachusetts General Hospital
Boston, Massachusetts 02114
U.S.A.

DR. M.S. HIRSCH
Infectious Disease Unit
Department of Medicine
Massachusetts General Hospital
Boston, Massachusetts 02114
U.S.A.

12. DR. H.G. PRENTICE
Department of Haematology
The Royal Free Hospital
Pond Street
Hampstead, London NW3 2QG
United Kingdom

DR. I.M. HANN
Royal Hospital for Sick Children
Glasgow, United Kingdom

13. DR. C. CRUMPACKER
Division of Infectious Diseases
Beth Israel Hospital
Harvard Medical School
330 Brookline Avenue
Boston, Massachusetts 02215
U.S.A.

14. DR. B. ÖBERG
Department of Antiviral Chemotherapy
Astra Alab AB
S-15185 Sòdertälje, Sweden

DR. S. BEHRNETZ
Department of Clinical Research
Astra Alab AB
S-15185 Södertälje, Sweden

DR. B. ERIKSSON
Department of Clinical Research
Astra Alab AB
S-15185 Sódertalje, Sweden

DR. H. JOZWIAK
Department of Clinical Research
Astra Alab AB
S-15185 Södertälje, Sweden

DR. J.O. LERNESTEDT
Department of Clinical Research
Astra Alab AB
S-15185 Södertälje, Sweden

DR. V. LINDSÖ ÅBERG
Department of Clinical Research
Astra Alab AB
S-15185 Södertälje, Sweden

15. DR. D.A.J. TYRRELL
MRC Common Cold Unit
Harvard Hospital
Coombe Road
Salisbury, Wilts SP2 8BW
United Kingdom

DR. W. AL-NAKIB
MRC Common Cold Unit
Harvard Hospital
Coombe Road
Salisbury, Wilts SP2 8BW
United Kingdom

16. DR. R. DOLIN
Infectious Disease Unit
School of Medicine
University of Rochester
601 Elmwood Avenue
Rochester, New York 14642, U.S.A.

17. DR. V. KNIGHT
Department of Microbiology and Immunology
Baylor College of Medicine
Texas Medical Center
One Baylor Plaza
Houston, Texas 77030, U.S.A.

DR. B.E. GILBERT
Department of Microbiology and Immunology
Baylor College of Medicine
Texas Medical Center
One Baylor Plaza
Houston, Texas 77030, U.S.A.

18. DR. J.B. McCORMICK
Special Pathogens Branch
Division of Viral Diseases
Center for Infectious Diseases
Centers for Disease Control
Atlanta, Georgia 30333, U.S.A.

DR. S.P. FISHER-HOCH
Special Pathogens Branch
Division of Viral Diseases
Center for Infectious Diseases
Centers for Disease Control
Atlanta, Georgia 30333, U.S.A.

19. DR. S.J. SPERBER
Departments of Internal Medicine and Pathology
University of Virginia School of Medicine
Charlottesville, Virginia 22908
U.S.A.

DR. F.G. HAYDEN
Departments of Internal Medicine and Pathology
University of Virginia School of Medicine
Charlottesville, Virginia 22908
U.S.A.

20. DR. L.A. BABIUK
Veterinary Infectious Disease Organization
124 Veterinary Road
Saskatoon, Saskatchewan S7N 0W0, Canada

DR. M.I. SABARA
Veterinary Infectious Disease Organization
124 Veterinary Road
Saskatoon, Saskatchewan S7N 0W0, Canada

DR. P. FRENCHICK
Veterinary Infectious Disease Organization
124 Veterinary Road
Saskatoon, Saskatchewan S7N 0W0, Canada

21. DR. J. BALZARINI
Rega Institute for Medical Research
Katholieke Universiteit Leuven
Minderbroedersstraat 10
B-3000 Leuven, Belgium

DR. S. BRODER
Clinical Oncology Program
National Cancer Institute
National Institutes of Health
Bethesda, Maryland 20892, U.S.A.

22. DR. G.J. GALASSO
Office of the Director
National Institutes of Health
Bethesda, Maryland 20892, U.S.A.

PREFACE

Antiviral chemotherapy has come of age, and, after an initial slow progress, the development of new antiviral agents has proceeded at a more rapid pace and the perspectives for their clinical use have increased considerably. Now, 25 years after the first antiviral assay (idoxuridine) was introduced in the clinic, it is fitting to commemorate the beginning of the antivirals' era.

In its introductory chapter B.E. Juel-Jensen touches on what may be considered as five of the most fundamental requirements of an antiviral drug : efficacy, relative non-toxicity, easy solubility, ready availability and reasonable cost. Surely, the antiviral drugs that have so far been used in the clinic could still be improved upon as one or more of these five essential demands are concerned.

How is all began is narrated by W.H. Prusoff. The first antiviral drugs to be used in humans were methisazone and idoxuridine, the former, which is now of archival interest, in the prevention of smallpox, the latter, which was approved for clinical use in the United States in 1962, for the topical treatment of herpetic keratitis.

In terms of potency, also because of solubility reasons, idoxuridine has been superseded by trifluridine in the topical treatment of herpes simplex epithelial keratitis. H.E. Kaufman did not find trifluridine or acyclovir effective in the treatment of deep stromal keratitis or iritis and he reckons that other antiviral drugs (i.e. bromovinyldeoxyuridine) would not be effective either.

However, P.C. Maudgal and E. De Clercq found that stromal keratitis reacted favorably to topical treatment with bromovinyldeoxyuridine, and so did patients with dendritic and geographic corneal ulcers. Even those patients whose keratitis did not respond to treatment with either idoxuridine, trifluridine, vidarabine or acyclovir healed promptly when treatment was switched to bromovinyldeoxyuridine. Preliminary findings indicate that combined oral and topical bromovinyldeoxyuridine treatment is also efficacious against ophthalmic herpes zoster.

Herpes simplex encephalitis has, because of its severity (mortality in excess of 70 %), since the 1960's been considered as an important target disease for antivirals. Acyclovir is currently the drug of choice for the treatment of biopsy-proven herpes simplex encephalitis, but, as pointed out by R.J. Whitley, therapeutic benefit depends on the age of the patient, the level of consciousness (Glasgow coma score) and the onset of treatment relative to the onset of the disease. Further improvements in therapeutic regimens as well as non-invasive diagnostic procedures are highly desirable.

The development and evaluation of an effective chemotherapy for recurrent herpes simplex labialis in otherwise normal, non-immunocompromised subject has proven to be an arduous task, and S.L. Spruance explains why the treatment of herpes labialis has remained elusive. Clinical benefit may be achieved, however, by improved means of drug delivery and treatment initiated in the earliest lesion stages.

The parameters that should be followed in assessing the efficacy of antiviral drugs in the treatment of genital herpes, either primary (or nonprimary, initial) or recurrent, are addressed by S.L. Sacks. Oral acyclovir is especially useful for the therapy of primary genital herpes and the prophylaxis of frequent recurrences. An acceptable treatment form for recurrent disease is still eagerly waited for.

Controlled double-blind clinical studies have shown that acyclovir, whether administered intravenously, orally or topically, is effective for the treatment of established herpes simplex virus (HSV) infections in immunosuppressed patients. R. Saral and P. Lietman prefer the use of intravenous acyclovir at 250 mg/m^2 every 8 hours, acknowledging that the optimal dose and treatment schedule may not yet have been defined.

For the treatment of varicella and zoster in immunocompromised patients, J. Englund and H.H. Balfour Jr. recommend the use of intravenous acyclovir at 7.5-10 mg/kg every 8 hours for a total duration of 5-10 days. Zoster in immunocompetent hosts may be managed on an outpatient base with very high oral doses (5 x 800 mg daily) of acyclovir. Whether such high dosage regimen could also be recommended for uncomplicated varicella in the normal host needs further investigation. Clearly, there is a need for an orally active drug that could be given at a realistic dose to outpatients with varicella-zoster virus infections.

Amid the compounds which have been described as anti-herpes agents, bromovinyldeoxyuridine ranks as one of the most potent and selective inhibitors of varicella-zoster virus (VZV). Preliminary clinical studies reviewed by S. Shigeta and E. De Clercq point to the efficacy of bromovinyldeoxyuridine in the treatment of varicella (chickenpox) and zoster (shingles) in immunocompromised patients. An additional bonus of bromovinyldeoxyuridine is that it can be given orally at relatively low doses without apparent toxicity for the host. Double-blind clinical studies are needed to further corroborate the great potential of bromovinyldeoxyuridine for the treatment of VZV infections.

One, if not the, major cause of morbidity and mortality in immunocompromised patients, particularly bone marrow transplant (BMT) recipients and patients with the acquired immune deficiency syndrome (AIDS), is cytomegalovirus (CMV). P.R. Skolnick and M.S. Hirsch review the current means available for the therapy and prevention of CMV infections. A significant advance in the therapy of CMV infections is dihydroxypropoxymethylguanine (DHPG) which appears efficacious in certain patient populations. However, needed are new agents with greater activity against CMV which also have enhanced oral bioavailability, thus making long-term therapy and prophylaxis satisfactory.

H.P. Prentice and I.M. Hahn re-emphasize that the prevention of CMV infection in immunocompromised patients is the major challenge facing clinicians. In this situation, acyclovir is of little or no value. However, oral acyclovir or its prodrug, 6-deoxyacyclovir, may be useful in the prophylaxis of HSV recurrences in high-risk patients, and some hope may also be held for 6-deoxyacyclovir in the oral prophylaxis of VZV recurrences in such patients.

C. Crumpacker reviews the mechanisms and clinical importance of resistance of HSV to nucleoside analogues such as acyclovir. This resistance may be based upon a deficiency or alteration in the substrate specificity of the virus-induced thymidine kinase or alteration in the virus-induced DNA polymerase. Acyclovir-resistant HSV variants have occasionally been isolated from immunosuppressed patients, but in immunocompetent persons the emergence of drug-resistant viruses does not appear to be a common problem.

The position of foscarnet in the antiviral armamentarium is still uncertain. According to B. Oberg and his colleagues, it is of little, if any, usefulness in the topical treatment of labial and genital herpes. However, intravenous administration of foscarnet, at doses as high as 6 to 10 g per day for 2 weeks, may have a beneficial effect on CMV infections in renal transplant patients, and, in particular, CMV retinitis in AIDS patients seems to respond favorably to foscarnet treatment. Whether foscarnet has any beneficial effect on the progression of the AIDS disease remains to be investigated.

Considerable progress has been achieved in developing potent antirhinoviral agents, whether interferons (HuIFN-α_2) or synthetic substances. As pointed out by D.A.J. Tyrrell and W. Al-Nakib, HuIFN-α_2 is effective in preventing naturally occurring rhinovirus infection within the family setting, and interferons would have a much wider application as long-term prophylaxis, had it not been for their toxicity. Should interferon prove as synergistic with synthetic antirhinoviral agents in humans as it does in cell culture, it may be possible perhaps to use interferon, in combination with these synthetic drugs, for seasonal prophylaxis at concentrations that are not toxic.

The efficacy of rimantadine in both the prophylaxis and therapy of influenza A virus infections has become increasingly clear. R. Dolin reviews recent studies with rimantadine in young children and the elderly. At commonly employed doses, rimantadine appears similarly efficacious as amantadine but gives rise to fewer central nervous system (CNS) side effects.

Ribavirin has recently been approved for aerosol treatment of respiratory syncytial virus (RSV) infection in infants, and V. Knight and B. Gilbert feel that ribavirin aerosol should also be used in the treatment of influenza A and B, and, possibly, parainfluenza virus infections.

When given systemically (intravenously or perorally), ribavirin is effective in the therapy of Lassa fever, and, as pointed out by J.B. McCormick and S.P. Fisher-Hoch, it would now seem mandatory to examine whether other hemorrhagic fever virus infections, such as those caused by the Argentinian and Bolivian hemorrhagic fever viruses, Rift valley fever virus and Crimean-Congo hemorrhagic fever virus, are also amenable to antiviral therapy.

Despite the successes obtained with interferon, amantadine, rimantadine and ribavirin in the prophylaxis and therapy of certain respiratory virus infections, S.J. Sperber and F.G. Hayden infer that there is still a need for more effective agents, better delivery systems, more rapid diagnosis and increased knowledge of the epidemiology, transmission and pathogenesis of respiratory virus infections. Virucidal substances might eventually prove useful in reducing transmission of some respiratory viruses.

For the treatment of gastrointestinal tract virus infections no specific antiviral drugs are presently available. L.A. Babiuk, M.I. Sabara and P. Frenchick anticipate that antiviral drug treatment of gastrointestinal virus infections would be much greater a challenge than, let say, the treatment of herpes virus infections, and this because of a number of reasons, i.e. the variety of causative virus agents involved, the need to interfere at the very early stage of infection and the problems concerning delivery and maintenance of the antiviral drugs at the site of infection.

One of the major targets, if not the most challenging target, for current chemotherapeutic attempts is AIDS, and, as asserted by J. Balzarini and S. Broder, there are grounds for optimism that these efforts may eventually prove successful. The 2',3'-dideoxynucleoside analogues, i.e. azidothymidine and dideoxycytidine, show special promise in this regard, and, taking advantage of the pharmacological principles which govern the metabolism and mode of action of these compounds, their efficacy could undoubtedly be improved upon.

It thus appears that antiviral agents have many promises to keep, the most challenging one being their presumptive activity against AIDS. According to G.J. Galasso, the ideal antiviral agent should be effective, non-toxic (or minimally toxic), highly viral-specific, non-inhibitory to the immune system, able to reach its target organ, unable to lead to virus-drug resistance, and preferably ingenious enough to overcome viral latency. This may sound utopian, but, looking back at what has been accomplished during the past few years and considering the various new antiviral agents that have recently been developed, selective antiviral chemotherapy no longer falls beyond the realm of reality.

E. De Clercq

Acknowledgment

The Editor is particularly obliged to Christiane Callebaut for her dedicated editorial assistance.

Prof. E. De Clercq

CLINICAL USE OF ANTIVIRAL DRUGS

1

CLINICAL USE OF ANTIVIRAL DRUGS

B.E. JUEL-JENSEN

Nuffield Department of Clinical Medicine, University of Oxford, Oxford, United Kingdom

ABSTRACT

This introductory chapter touches on some general problems of the clinical use of antiviral agents, as illustrated by drugs so far used in poxvirus infections, in the herpesvirus group and in some RNA virus infections.

INTRODUCTION

Some fifty years ago saw the beginning of the antimicrobial chemotherapy era. Domagk (1) had discovered prontosil in 1935, and clinicians, for example Colebrook and Kenny (2) in 1936 had first used the drug in the treatment of human puerperal infection. Revolutionary though the sulphonamides were in the treatment of meningococcal, pneumococcal and streptococcal infection, the range of activity was limited and toxic side-effects were considerable.

A whole new era began when Florey and his Group in 1941 first purified penicillin, and first used it systemically in man (3). The drug had been studied for its antimicrobial effect in mice, and its action in patients was little short of miraculous. A surviving member of the team said in passing to the writer not long ago: "We were lucky there was no Safety of Drugs Committee". Perhaps it should read mankind was fortunate that there was no Watchdog Committee which had queried its use in man before more extensive animal experiments had gone on, for instance in the rabbit, which would have put paid to the most valuable antimicrobial discovered to date. A wide variety of antibacterial antimicrobials has since been discovered and found to be very useful in clinical practice.

However, bacteria differ fundamentally from viruses, insofar that they often contain components which are not present in the host.

Muramic acid is an example. It is, therefore, at least theoretically possible, and often in practice feasible to attack the bacterium without doing the host significant harm. In all treatment, the clinician must weigh the possible benefit of the drug against the possible harm done by the drug. With penicillin the decision is usually easy, the antimicrobial is harmless even in very high doses, and the benefit conveyed may be considerable. With some drugs, e.g., vancomycin, extreme caution must be used.

Fungal infections have on the whole proved much more difficult to attack than bacteria. Most of our few active drugs are relatively toxic, such as amphotericin and flucytosine.

Malaria is the most important protozoal disease. The last thirty years has been one continuous tale of disappointments,as P. falciparum became resistant to chloroquine, to Fansidar, to quinine and quinidine, and some prophylactics such as Maloprim (pyrimethamine and dapsone) showed myelotoxicity. Among the anthelmintics those against the important tropical and subtropical infestations schistosomiasis and filariasis are many of them very toxic. Only the recent advent of praziquantel, active against all forms of schistosomiasis, has caused a revolution in the treatment of this ubiquitous infestation. Onchocerciasis is still difficult to treat when such drugs as suramin have to be used against the adult worm.

However, the difficulties encountered in the treatment of other infections and infestations fade into insignificance when it comes to the clinical treatment of viral infections. The fundamental problem is that the virus shares so many of its metabolic processes with the host, and has no components peculiar to itself. Therefore, treatment of virus diseases must be a compromise between, on the one hand arresting replication of the virus, preferably killing it, and on the other not harming the host or at worst doing only minimal damage which could be justified by the possible gain. In the following some groups of viruses and antiviral drugs which have been used in clinical practice are mentioned briefly to give an overview. The following chapters will treat individual problems in depth and give recent developments.

3

POXVIRUSES

Smallpox

Bauer in 1955 had suggested that methisazone might be effective prophylactically in people who had been exposed to smallpox infection (4). A trial was carried out in Madras in 1963 on a group of contacts where people exposed to infection were assigned to treatment or no treatment alternately. Methisazone was given by mouth as a 20% suspension in sucrose syrup according to four dosage schedules. Some or all of the drug issued was taken by 2,610 contacts; 18 developed smallpox during the observation period and 4 died. In the untreated group of 2,710 contacts, 111 developed smallpox and 21 died, or in this trial the case incidence was 0.69% in the treated contacts, and 4.17% in untreated contacts. There were flaws in the trial but on balance there can be little doubt that methisazone worked as a prophylactic drug (5). A therapeutic trial of methisazone in established smallpox was carried out by Rao and others in 1964-5 (6). There was no significant difference between the outcome in patients treated with 12 g at once and 3 g six hourly to a total of 12 doses and those given placebo. The drug was given to 208 patients and placebo to 215. The disappearance of smallpox has made the problem of dealing with that disease obsolete, but a small trial of cytarabine by Dennis and others (7) failed to show any benefit from the drug. The numbers entered into this double-blind controlled trial were small: in 18 Ethiopian patients a single intravenous dose of 100 mg/sq.m of body surface on each of four consecutive days did not alter the natural course of the disease. Criticism of the trial is twofold. The numbers (18) were small, and the disease in Ethiopians was very mild. Only about 5% of people admitted to hospital with the disease used to die. Second, the dose was very modest. Should poxviruses again become a problem, one would surely use a considerably higher amount of drug. Hossain *et al*, had in 1972 shown that cytarabine is active against variola in man provided the drug is given early enough in the disease (8). Anecdotal evidence showed in the past that systemic cytarabine was effective in eczema vaccinatum. The writer saved one tour of the All Blacks when they visited England many years ago. One of their number had misbehaved and was sent back and a replacement was sent having been vaccinated without reference to the fact that he had atopic eczema. On arrival he had widespread eczema vaccinatum and our systemic cytarabine cleared up the lesions within four days, enough time to allow him to play in an important fixture with the English rugby side. There was a need then for a drug effective against vaccinia, for doctors could prick themselves on their fingers and get vaccinia whitlows. Idoxuridine 35% in DMSO applied on dressings would also stop the spread of infection. Vaccination on any large scale has now ceased. However, in experimental work in academic units, the use of vaccinia virus as a carrier for genetic material has become fashionable, and there is no doubt that vaccination has started again among people who work with vaccinia virus for that purpose. It is therefore useful and necessary to have in reserve drugs that will cope with the infection when things go wrong.

HERPESVIRUSES

Idoxuridine

This is perhaps the most important group of viruses where effective antiviral chemotherapy is needed. Over the years often quite sophisticated chemotherapy and radiotherapy has been developed in the battle against malignancies, particularly various forms of lymphoproliferative disease. Transplantation, in particular of kidneys, has become widespread. Most transplanted patients must of necessity take immunosuppressant drugs to prevent rejection. This carries with it a much increased risk of reactivation of the herpesviruses, and regrettably patients are still lost because of generalized infection with herpes simplex or varicella zoster virus or cytomegalovirus. The role of Epstein-Barr virus, the fourth herpesvirus is less certain. This group of viruses was one of the earliest for which there was a glimmer of hope that practical clinical treatment might be possible. The first antiviral drug of real importance, idoxuridine, was synthesized by Prusoff (9) in 1959, and Kaufman treated herpetic keratitis successfully in 1962, first in rabbits and then in man without significant toxicity to the host (10). But the eye is a favoured organ when there are superficial infections with herpes simplex, for even fairly insoluble substances can reach the virus. Double-blind controlled trials of 0.5%

idoxuridine in an inactive water soluble cream base compared with inactive base as placebo in patients with recent recurrent cold sores when the cream was applied six hourly showed no benefit from the active substance. Such trials were carried out by Burnett and Katz (11) and by Juel-Jensen and MacCallum (12) and subsequently by Kubrick and Katz (13). Though recurrent cold sores are trivial complaints, the lesson to the clinician was an important one. However excellent an antiviral drug may be in vitro, it is vital that it should be possible to get it to the lesion where the virus is. We tried spraying a weak watery solution through the skin. It showed significant effect in a double-blind controlled trial, but it was impracticable to use spray guns in ordinary work. We showed (14) that idoxuridine could be dissolved in dimethyl sulphoxide and that that would successfully penetrate the skin and successfully shorten the duration of and the period of virus shedding in recurrent herpetic lesions. It became clear that idoxuridine, useful though it is not only in recurrent herpes simplex, but also in the treatment of uncomplicated segmental zoster (15), was far too toxic when given systemically in doses large enough possibly to influence herpes simplex encephalitis. The Boston Interhospital Virus Study Group clearly showed this in 1975 (16). Hopefully that once and for all put a stop to systemic use of a drug of considerable toxicity. It had been used on repeated occasions on an anecdotal basis where there had been no real evidence of efficacy.

Quite apart from its toxic properties, idoxuridine showed one characteristic which unfortunately is shared by many of the antiviral drugs at present in use. It is very insoluble at ordinary pH. In the early days when it was used intravenously, it was necessary to dissolve it in a solution at pH 11, when as much as 0.5% would go into solution. Unfortunately many other drugs active against herpes simplex virus are also very insoluble and the question of solubility must enter into consideration when new drugs are planned. If they are insoluble much thought should be given to producing analogues that are soluble and therefore easy to give to the patient.

Cytarabine

The need in severe herpes simplex infection as well as in varicella zoster and cytomegalovirus infection for a reliable systemic drug was obvious. Cytarabine, which had been found by Bauer to be about five times as active as idoxuridine against herpes simplex virus (17), is one of the very few drugs that is easily soluble. Kaufman and Maloney had found that a 5% solution of cytarabine was active in the treatment of herpetic keratitis (18), but repeated instillations of 0.5% or 0.1% solutions into the eye caused reversible speckling of the cornea due to distortion of the corneal epithelium, and its use in the eye was stopped. Give a dog a bad name...... Reports on toxicity in man have, however, mainly been of experience in patients with malignant disease that in itself often affects the bone marrow. In patients with normal marrow one only sees a transient macrocytosis, usually in the second week after the institution of treatment, and the drug has low toxicity. Provided it is given in a bolus rather than continuously, a therapeutic level of drug is achieved and the effect on the bone marrow is minimal, whereas slow administration will give maximum adverse effects on the bone marrow and probably no effect on the virus. Cytarabine was used successfully by Hall et al in 1969 to treat patients with zoster (20), and in our hands the drug was effective in treatment of severe systemic herpes simplex virus infections. It was found to be an invaluable adjunct in a series of 15 consecutive patients of which 14 had proven herpes simplex encephalitis. All survived, 13 to go on to lead useful lives. The fifteenth patient turned out in the event to have been infected with Simian herpesvirus and died (21). A national trial, however, had to be stopped, because patients both in placebo and active drug groups had a survival rate lower than that of the untreated average for Britain. Though the multicentre trial was carefully constructed, we failed to take into account the wide variation in accessibility and sophistication of intensive care units.

Vidarabine

Although it turned out to be of limited value as a potential anticancer agent, vidarabine, synthesized by Lee et al (22) was found to be active against herpes simplex, vaccinia and varicella zoster virus in vitro by Miller et al 1969 (23). In clinical practice it was as active in the 3.3% ointment as 1% drops in herpetic keratitis (Coster et al (24)) but once again the problem of insolubility came up. The drug is even less soluble than idoxuridine in ordinary water. This presents a problem when the drug is given systemically and large amounts have to be administered, not least when the patient is on the verge of heart failure. The problem may be overcome with simultaneous use of a diuretic. Given systemically vidarabine is at present the drug of choice in complicated zoster such as motor zoster, zoster of the trigeminal nerve, zoster of S2 and below which often involves bladder and bowel and in immunosuppressed patients, unless they have profound marrow depression. The efficacy of the drug has been proved by do ble-blind controlled trials by Whitley _et al_ (25). We have found a limitation in the elderly, that is in patients over about 65. A limitation caused by an unfortunate tendency of the patients to develop extrapyramidal signs like Parkinsonism. These signs disappear when the drug is withdrawn, but unless there is an absolute contraindication to the use of acyclovir that drug should be preferred despite its shortcomings. Vidarabine was used in herpes simplex virus encephalitis (26), but with disappointing results. It has already been mentioned above how antiviral chemotherapy in that condition can play a minor role only and strains of herpes simplex virus are much less ($<$10 times) sensitive to vidarabine than to idoxuridine _in vitro_.

The clinician must regret that soluble derivatives of vidarabine, such as Ara-AMP (27), were never produced commercially. Had it been generally available a major clinical disadvantage would have been overcome. Although the half-life of vidarabine is short, perhaps half an hour, the metabolite hypoxanthine arabinoside, which is also active against varicella zoster although less so, has a half-life of 24 to 28 hours. In zoster constant exposure to the drug is essential and a long acting substance is therefore of great value. It may also be that such a substance is needed to deal with the virus that causes multiple papillomata of the larynx in children.

Acyclovir

About ten times more active than idoxuridine against some strains of herpes simplex virus _in vitro_, acyclovir compares with cytarabine in potency. Varicella zoster virus is not so sensitive (28), probably because the thymidine kinases of this virus have far less avidity for the drug than does that of herpes simplex virus. Unfortunately, acyclovir is practically insoluble in water, and less than 10% solution can be achieved in dimethyl sulphoxide. It is relatively non-toxic when given systemically, but to get any amount into solution in a small volume it is necessary as with idoxuridine to present the drug at pH 11. It cannot be given intravenously in a bolus in the dose used for the treatment of zoster (10 mg/kg eight hourly), for there is transient impairment of the renal function probably due to crystalluria. Unless it is infused slowly in a large volume there may be serious interference with the kidney. The drug has been proven to be of value in controlled trials in the prophylaxis of herpes simplex infections in immunosuppressed patients (29) and (30) and in primary genital herpes (31). Double-blind controlled trials have also shown that intravenously infused acyclovir was of undoubted value in herpes simplex encephalitis when compared with vidarabine (32). In 53 confirmed cases the mortality in the group treated with acyclovir (10 mg/kg eight hourly for ten days) was 19% as against a mortality of 50% in the group treated with vidarabine (15 mg/kg daily for ten days). What is not said in the published paper, but what is of vital importance is that the supporting treatment of these patients was first class.

As could have been anticipated, namely that acyclovir might be less effective in varicella zoster than in herpes simplex infection turned out to be so. In three double-blind controlled trials, where doses of 5mg/kg eight hourly (33) and ten mg/kg eight hourly (34) and (35) had been used,there was no statistically significant effect on the most dreaded of all sequelae of zoster, namely postherpetic neuralgia. There was a suggestion of a probable effect on the severity of late pain in the studies by Bean and colleagues and by the Oxford Group. But using the most convenient tool, the retrospectoscope, it is easy to see where the weakness of all trials in zoster lies. Much too little attention has been paid to what constitutes the onset of the illness. The prodromal period before a vesicular rash appears may vary widely, insofar as one can measure this period. The only recordable index is preliminary pain. This can be as much as three weeks. It is hard to imagine that any drug, however good, would have as striking an effect on a lesion that had been symptomatic for three weeks as it would in a lesion that appeared overnight. Bean and colleagues had trouble with transient impairment of renal function. This is almost certainly because patients had been treated on an outpatient basis and the drug was given too quickly. We did not find this problem, but it is of course expensive to have people in hospital for frequent intravenous slow infusions. McKendrick et al (36) found that huge doses of oral acyclovir (800 mg five times a day) significantly shortened the period to healing and reduced short term pain as compared with placebo in a double-blind controlled trial. Whether there will be any effect on postherpetic neuralgia remains to be seen.

Oral acyclvir in doses of 200 mg five times a day undoubtedly has a good effect on herpes simplex lesions and it may well be that some of the second generation drugs may be more readily absorbed than acyclovir. In the writer's view it is a mistake to apply any drug which has a potential intravenously in preparations to be applied to the skin. Quite apart from the fact that, except with the aid of solvents such as dimethyl sulphoxide, it is extremely difficult to get a near insoluble drug into the skin, there is a very real risk of sensitizing more readily by contact with the skin. It may make it impossible subsequently to use the drug when there is a life-threatening infection. Sensitization has throughout been a problem in the use of anti-herpes simplex drugs in the eye, where treatment of herpetic keratitis demands long-term use.

None of these and similar drugs active against herpes simplex and less active against varicella zoster, has so far turned out to be very effective against cytomegalovirus. Cytarabine will arrest cytomegalovirus excretion in immunosuppressed patients and so will some of the second generation acyclovir compounds, but the moment the drug is withdrawn the infection becomes active again. Nothing can be more distressing than seeing a patient, grossly immunosuppressed because of a myeloproliferative disorder and because of aggressive cytoxic treatment, rapidly losing his sight because of cytomegalovirus infection of the retina. One arrests the lesion but has to withdraw the drugs mentioned because of profound thrombocytopenia. We desperately need a drug that is sufficiently non-toxic and yet capable of inhibiting cytomegalovirus. A drug, in other words, that is not dependent on the presence of thymidine kinase in the virus.

There is little doubt that acyclovir will inhibit the replication of Epstein-Barr virus. There is in addition growing evidence that the virus may persist in the nasopharynx in patients who have prolonged symptoms in infectious mononucleosis. It would therefore seem reasonable to explore in greater depth the possibility of cutting short the course of persistent infectious mononucleosis, particularly in those who show a continuously elevated level of Epstein-Barr virus early antigen antibody.

Other Drugs

Bromvinyldeoxyuridine (BVDU) has not been generally available. Highly active against herpes simplex virus type 1 but less so against type 2, it is of particular interest because it is very active against varicella-zoster *in vitro*. Unfortunately it has not until the present been readily available, but it may well be that one is able to salvage grossly immunosuppressed patients with widespread varicella zoster disease if it were freely available. This and other interesting compounds have been extensively discussed by De Clercq (37).

For the whole of the herpesvirus group the situation seen from the clinician's point of view currently is this: we have an admirable drug in the shape of acyclovir against most strains of herpes simplex. One of its few drawbacks is its insolubility. If a patient is on the brink of heart failure or is in heart failure one may still have to use the equally potent, but more toxic substance, cytarabine. Varicella-zoster can be treated after a fashion, but none of the drugs available are ideal, for if a sufficiently high level of antiviral drug is to be attained, there is a risk of side-effects as mentioned above. Manufacturers of antiviral drugs should always ask as one of the first questions of any potentially useful substance: is it easily soluble? We should not lose sight either of the fact that any level we may measure in the blood may not represent the level that is present where the virus is found in the host. An easily soluble drug would probably more readily get to the affected area.

RNA VIRUSES

Ribavirin

This drug, which has a considerable range of activity against RNA viruses, has in recent years again been subjected to clinical trials which suggest that it may be useful in practice. In a randomized double-blind study Hall et al found that infants receiving ribavirin in an aerosol in respiratory syncytial virus infection had significantly greater improvement in their overall score for severity of illness and lower respiratory tract signs and in arterial oxygen saturation than in the placebo group (38). An aerosol was also found to be effective in a randomized controlled study in the treatment of influenza B virus infection in a group of college students (39). Its role in exotic, and potentially severe disorders such as Congo haemorrhagic fever, Lassa fever, rabies, Marburg virus disease, Ebola virus disease, Dengue and Yellow Fever is still to be determined. Clearly drugs effective against the whole range of RNA virus infections in man is very high on the list of desiderata.

HIV

Currently, one virus disease above all others, namely that caused by human immune deficiency virus (HIV, HIV-2, etc.) must engage maximum attention. There is a little evidence that the virus may undergo drift and shift comparable to that of the influenza virus, and that apart from any other consideration, that may make the production of vaccine difficult. Therefore, an effective antiviral drug is desperately needed. Incidentally, it may be that drugs active against other viruses might be developed in the course of the search of an agent against this plague, which otherwise might not have been developed. The position of interferon in the treatment of virus disease is still unclear. Often, by the time it is possible clinically to diagnose a virus disease, levels of interferon are already so high that it is doubtful whether added external interferon would change the course of the disease. However interesting the effects of drugs on viruses grown in tissue cultures may be, in clinical practice the drug should be effective at the stage when clinical diagnosis is possible.

CONCLUSION

The introductory chapter has not covered anything like all the many needs for good antiviral drugs in clinical medicine. Certain fundamental demands must be made of the drug: 1) It must be effective. 2) It must be relatively non-toxic. 3) It must be easily soluble. However excellent the drug may be *in vitro*, it is useless *in vivo* if it cannot easily be delivered to the patient. 4) The drug must be readily available. 5) The cost of the drug must be such that a large number of people will be able to benefit from it. This last is not least true of drugs with potential action against tropical diseases such as for instance South American haemorrhagic fever, Congo haemorrhagic fever and many of the arboviruses.

Many individual drugs will be discussed in detail in the following chapters, but the reader may find that Sir Charles Stuart-Harris's and John Oxford's book Problems of Antiviral Therapy gives a useful overview of many of the difficulties attached to the whole subject (40).

REFERENCES

1. Domagk, G. Dt. med. Wschr. 61: 250-3 and 829-32, 1935.
2. Colebrook, L. and Kenny, M. Lancet. I: 1279-86, 1936.
3. Florey, H.W. *et al.* Lancet. II: 177-89, 1941.
4. Bauer, D.J. Br. J. Exp. Path. 36: 105-14, 1955.
5. Bauer, D.J. *et al.* Amer. J. Epidem. 90: 130-45, 1969.
6. Rao, A.R., McFazdean, J.A. and Kamalakshi, K. Lancet. I: 1068-72, 1966.
7. Dennis, D.T. *et al.* Lancet. II: 377-9, 1974.
8. Hossain, M.S. *et al.* Lancet II: 1230-2, 1972.
9. Prusoff, W.H. Biochem. Biophys. Acta. 32: 295-6, 1959.
10. Kaufman, H.E. Proc. Soc. Exp. Biol. Med. 109: 251-2, 1962.
11. Burnett, J.W. and Katz, S.L. J. Invest. Derm. 40: 7, 1963.
12. Juel-Jensen, B.E. and MacCallum, F.O. Br. Med. J. II: 987-8, 1964.
13. Kubrick, S. and Katz, A.S. Ann. NY Acad. Sci. 173: 83-9, 1970.
14. MacCallum, F.O. and Juel-Jensen, B.E. Br. Med. J. II: 805-8, 1966.
15. Juel-Jensen, B.E. *et al.* Br. Med. J. IV: 776-80, 1970.
16. Boston Interhospital Virus Study Group. New Eng. J. Med. 292: 599-605, 1975.
17. Bauer, D.J. The Specific Treatment of Virus Disease. MTP Press, Lancaster. 54-5, 1977.
18. Kaufman and Maloney, E.D. Arch. Ophthalmol. 69: 626-9, 1963.

19. Elliot, G.A. and Schut, A.L. Amer. J. Ophthalmol. 60: 1074-8, 1965.
20. Hall, T.C. et al. Trans. Assoc. Amer. Phys. 82: 201-10, 1969.
21. Approaches to Antiviral Agents. Ed. M.E. Harnden. Macmillan. 1985, p 298.
22. Lee, W.W. et al. J. Amer. Chem. Soc. 82: 2648-9, 1960.
23. Miller, F.A. et al. Antimicrob. Agents Chemother. 136-47, 1968.
24. Coster, D.J. et al. Antivirals with Clinical Potential. Univ. Chicago Press. 173-7, 1976.
25. Whitley, R.J. et al. New Eng. J. Med. 294: 1193-9, 1976.
26. Dolin, R. and Reichman, R.C. In Shiota, H., Cheng, Y. and Prusoff, W. (Eds). Herpesviruses, Clinical, Pharmacological and Basic Aspects. Excerpta Medica, Amsterdam, Oxford and Princeton, 1982, pp 129-34.
27. Revankar, G.R. et al. J. Med. Chem. 18: 721-6, 1975.
28. Brigden, D., Fowle, A. and Rosling, A. In Collier, L.H. and Oxford, J. (Eds). Developments in Antiviral Therapy. Academic Press, London. 53-62, 1980.
29. Saral, R. et al. New Eng. J. Med. 305: 65-7, 1981.
30. Hann, I.M. et al. Brit. Med. J. 287: 384-8, 1983.
31. Mindel, A. et al. Lancet I: 697-700, 1982.
32. Sköldenberg, B. et al. Lancet II: 707-11, 1984.
33. Peterslund, N.A. et al. Lancet II: 827-30, 1981.
34. Bean, B., Braum, C. and Balfour, H.H. Lancet II: 118-21. 1982.
35. Juel-Jensen, B.E., Kahn, J.A. and Pasvol, G. J. Infect. 6: Suppl. I, 31-6, 1983.
36. McKendrick, M.W. Brit. Med. J. 293: 1529-32, 1986.
37. De Clercq, E. In Approaches to Antiviral Agents. Ed. M.R. Harnden. Chapter 3. Synthetic Pyrimidine Nucleoside Analogues. Macmillan. 57-99, 1985.
38. Hall, C.B. et al. New Eng. J. Med. 308: 1443-47, 1983.
39 McClung, H.W. et al. J. Amer. Med. Ass. 249: 2671-74, 1983.
40 Stuart-Harris, Sir Charles and Oxford, J. (Eds). Problems of Antiviral Therapy. Academic Press, 1983.

2

IDOXURIDINE OR HOW IT ALL BEGAN

W.H. PRUSOFF

Department of Pharmacology, Yale University, School of Medicine, New Haven, Connecticut 06510, U.S.A.

"In the beginning God created Heaven and the Earth and the Earth was without form and void and darkness was upon the face of the deep ... Then God said : Let there be Light and there was Light" (1).

The earth is believed to have been formed about 4.6 billion years ago from a mass of gas or dust which travelled around the infant sun. Evidence that life existed more than 3 billion years ago is derived from the discovery of fossils in ancient rock formations. Primitive earth is postulated to have been enveloped in a reducing atmosphere that contained CH_4, NH_3 and H_2O from which organic molecules destined to be the basis of living things were formed. The energy source required for such synthesis may have been ultraviolet radiations present in lightning and sunlight, as well as steam and other hot gases from volcanic eruptions. Support for such reasoning comes from simulation of these primitive conditions in the laboratory which resulted in the creation of amino acids, pentose and hexose sugars, nucleic acid bases and porphyrins. Further evidence exists in the laboratory for the condensation of nucleic acid bases and sugars by ultraviolet irradiations in the presence of phosphates or cyanide. Presumably, nucleotides formed on primitive earth, polymerized randomly, and a rare nucleotide polymer became self-replicating. Similarly, amino acids may have polymerized into protein-like substances (2).

Our present day knowledge of nucleic acids begins with Miescher's discovery in 1869 of "Nuclein" in pus cells derived from discarded surgical bandages. He characterized "Nuclein" as being high in phosphorous, acidic, and soluble in alkali but not in acid. Prior to this time several purine bases (uric acid, xanthine, guanine, hypoxanthine) had been discovered. The term "nucleic acid" was first used by Altman who was a student of Miescher. Kossel and his students identified adenine from beef pancreas, uracil from yeast nucleic acid and both thymine and cytosine from thymus nucleic acid. Levene and colleagues

DeClercq, E. (ed.), *CLINICAL USE OF ANTIVIRAL DRUGS*.

identified D-ribose as a component of yeast nucleic acid and D-2-deoxyribose as a component of thymus nucleic acid. Thus all the purine and pyrimidine bases of nucleic acids were identified (3).

The importance of nucleic acids as genetic determinants is well established today. The concept that nucleic acid plays such a role evolved from the discovery by Avery, MacLeod and McCarty (4) that DNA can induce transformation of pneumococcus type II into pneumococcus type III. Further support for the critical role of nucleic acid came from the investigations of Hershey and Chase (5) who established that phage DNA, but not phage protein, was injected into E.coli.

Analogs of nucleic acid bases and their nucleosides have been synthesized in which extensive variation of the purine, pyrimidine and pentose moieties have been made, and citations may be found in the following references (6-10). Hitchings and his colleagues (11,12) were the first to systematically study analogs of purine and pyrimidine bases as potential inhibitors of nucleic acid synthesis, since the importance of nucleic acids for cellular replication was well appreciated by the 1950s.

The nutritional requirement for purine and pyrimidine bases by several strains of bacteria encouraged study of these analogs as potential inhibitors of nucleic acid metabolism. Purines, pyrimidines, nucleosides, nucleotides and analogs thereof have been useful in the therapy of cancer, metabolic disorders, therapy of various infectious agents as well as in the sorting out of biochemical pathways. However, this discussion will be restricted to their role in antiviral chemotherapy. Welch (13) has discussed the rationale for the use of purine and pyrimidine analogs for therapy of cancer.

Elion and Hitchings (14) have presented an extensive review of purine and pyrimidine analogs as antiviral agents, and references to the original reports may be found therein. Among the purine analogs, 2,6-diaminopurine was the first to show antiviral activity by inhibition of the replication of vaccinia virus (15,16). Among the pyrimidine analogs several 5-substituted uracils (5-bromouracil, 5-hydroxyuracil, 2,4-dithiothymine) were reported to produce a small inhibition of the replication of vaccinia virus (15). Additional references to the early studies of these and other purine and pyrimidine analogs as antiviral agents have been presented in several reviews (9,17-19). A few recent reviews of compounds evaluated for their antiviral activity in cell culture, as well as those in clinical use or trial, have been published (20-38).

Several historical reviews of the discovery of antiviral drugs have been presented (39,40). Perhaps the first report of a compound with antiviral activity was that of Brownlee and Hamre (41) who found p-aminobenzaldehyde thiosemicarbazone inhibited the replication of vaccinia virus when the virus was injected into the yolk sac of fertile eggs. This observation was extended by Thompson and coworkers (42) who replaced the benzene moiety with other ring systems and found isatin-3-thiosemicarbazone protected mice infected with the vaccinia virus. Bauer (43) proposed that this compound might be effective in the prophylaxis of smallpox. Although the 1-ethylisatin-3-thiosemicarbazone was most active, because of solubility problems, 1-methylisatin-3-thiosemicarbazone (Methisazone) was chosen for further evaluation. The effectiveness of this substance in preventing smallpox in man has been established (44). During a smallpox epidemic in Madras, 2,227 persons who had had close contact with established cases of smallpox were vaccinated and, of these, 1,101 received methisazone. The group that did not receive the drug developed 78 cases of smallpox, of whom 12 died, whereas the group on drug therapy had only 3 mild cases of smallpox and no deaths. These studies have been expanded and a report presented by Bauer (45) indicated that among 2,297 treated close contacts, only 6 developed smallpox, of whom 2 died, whereas among 2,842 untreated contacts, 114 developed smallpox and 20 died. Methisazone was reported to be a value also in the therapy of eczema vaccinatum (45) and vaccinia gangrenosa (46). Since smallpox has been eradicated from the earth by the policy of containment and vaccination this antiviral agent is of historical interest only.

The first compound to be approved by the Food and Drug Administration of the United States for clinical use is 5-iodo-2'-deoxyuridine (Idoxuridine, IUdR, IdUrd, IDU). Replacement of the hydrogen on carbon-5 of a pyrimidine by an iodine was first achieved by Johnson and Johns (47) in 1905 in an alkaline reaction mixture. Hitchings et al. (48) were the first to study the effects of pyrimidines on the growth of Lactobacillus casei in 1945. Studies with L.casei and other microorganisms have shown that 5-iodouracil antagonizes the utilization of thymine or thymidine for growth, and that this analog, as well as 5-bromouracil, are incorporated into bacterial DNA (48-52).

These findings stimulated Arnold D. Welch (13), with whom William L. Holmes and I were postdoctoral fellows at the time, to initiate an anticancer program based on the replacement of DNA thymine of neoplastic cells with 5-iodouracil. Would neoplastic cells suffer the same inhibitory fate as microorganisms by such a substitution in this vital macromolecule ?

The first question was whether 5-iodouracil would not only be incorporated into mammalian DNA, but also whether one could achieve preferential localization into the DNA of neoplastic tissues. Bill Holmes and I (53) synthesized radioactive [^{131}I]-5-iodouracil, [^{131}I]-5-iodouridine, and [^{131}I]-5-iodoorotic acid, but no preferential uptake of radioactivity into the plasma of mice was found. Since thymine (54-57) and thymine riboside (57) are very poor precursors of DNA thymine in mammalian systems, it is not surprising in retrospect, to have made such observations with these three iodinated compounds.

With the commercial availability of 2'-deoxyuridine, 5-iodo-2'-deoxyuridine (IdUrd) was synthesized in good yield, and in crystalline form, by iodination of 2'-deoxyuridine in dilute nitric acid (58). Zamenhof _et al_. (52) and Dunn and Smith (49,50) had previously isolated IdUrd from the DNA of microorganisms when grown in media supplemented with 5-iodouracil. Prusoff (60), Mathias _et al_. (59), and Eidinoff and co-workers (61-63) demonstrated incorporation of IdUrd into the DNA of mammalian cells.

That the incorporation of IdUrd into mammalian DNA was a quantitative substitution of DNA thymidine by this analog was shown by Mathias _et al_. (59). They incubated L5178 cells with radioactive IdUrd and isolated the DNA after 28 hours of incubation. The DNA was hydrolyzed enzymatically to the deoxyribonucleoside monophosphate and separated by ion-exchange chromatography. Each nucleotide was quantitated and the amount of IdUMP (0.33 µmole) plus dTMP (0.65 µmole) very closely equaled that of dAMP (0.94 µmole). Similarly, the amount of dGMP (0.99 µmole) was similar to that of dCMP (0.92 µmole).

The substitution of IdUrd for thymidine is not precluded on steric considerations, even though there is an increase in the bulk volume of the halogen substituent on carbon-5 of the pyrimidine moiety. The methyl group has a van der Waals' atomic radius of 2.00 Å, whereas that of the iodine atom is 2.15 Å. Nevertheless, there is no hindrance in the capacity of IdUrd to substituted for thymidine in the metabolic pathway leading to the formation of the mono-, di- and triphosphate derivative with subsequent incorporation into DNA. The Km of IdUrd for thymidine kinase as well as the Km of IdUTP for DNA polymerase is essentially equivalent to that of thymidine and dTTP, respectively. The biological consequence of incorporation of IdUrd into the DNA of mammalian cells is an inhibition of replication.

Studies on the mechanism of action of IdUrd were reported by Prusoff (64) in 1960 in which it was shown that IdUrd inhibited the utilization of radioactive orotic acid, formate and thymidine for the biosynthesis of DNA-thymine, but not of orotic acid for the biosynthesis of DNA-cytosine or of RNA-pyrimidines by mouse Ehrlich ascites tumor cells in vitro. IdURd was also shown to be a competitive antagonist of the utilization of thymidine.

These observations were the basis for two subsequent independent and simultaneous investigations which culminated in establishing IdUrd as an effective antiviral agent. Herrmann Jr. (65) and Rada et al. (66) independently made a very important contribution to antiviral chemotherapy by their development of a plaque-inhibition test for evaluation of potential antiviral agents. Rada et al. (66) described the inhibitory effect of 6-azauracil riboside on the multiplication of vaccinia virus. Herrmann Jr. (65) screened a wide variety of compounds to detect specific inhibitors of DNA containing viruses. Among the compounds evaluated he identified IdUrd to be a potent inhibitor of vaccinia virus and herpes simplex virus. Prior to this report, Thompson et al. (15) observed 5-bromouracil significantly inhibited the replication of vaccinia virus, and Smith et al. (67) the inhibition of polyoma virus production by BrdUrd.

The other critical observation was contributed by Kaufman (68) who, as a research ophthalmologist, was engaged in a program concerned with finding a therapy for herpes keratitis using as the experimental model rabbits whose eyes were inoculated with the herpes simplex virus. He reasoned that since the herpes simplex virus was a DNA-containing virus, compounds which inhibited the synthesis of DNA or were incorporated into DNA might be efficacious in the therapy of herpes keratitis.

Kaufman did indeed find that instillation of a solution of IdUrd into the conjunctival sac of the herpesvirus infected rabbit eye would result in a cure of experimental herpes keratitis in rabbits. The antiviral activity of IdUrd in rabbits was confirmed by Perkins et al. (69). Of great importance, thanks to the liberal governmental regulations that existed at that time, was the demonstration by Kaufman that this analog of thymidine, when applied topically, produced clinical cures in man suffering from herpes keratitis, which is a major cause of blindness in the United States (70-72). This finding by Kaufman and co-workers (70-72) that a drug can attack successfully an established virus infection in man and produce a cure represents a milestone in antiviral chemotherapy.

The ability to cure herpetic keratitis in man without concomitant systemic toxicity is related to the following factors :

(a) One can obtain a high concentration of IdUrd in the infected cell by application of a solution or ointment directly to the eye.

(b) If absorbed into the blood stream, IdUrd is rapidly cleaved by phosphorolysis to 5-iodouracil, an innocuous substance. The extremely rapid rate of degradation of IdUrd in man had been previously established by Calabresi _et al_. (73).

(c) Infection with herpes simplex virus results in a marked increase in viral encoded thymidine kinase. Normal resting cells have a very low cellular thymidine kinase activity. Hence, during the relatively brief period of time, that the solution of IdUrd is in the conjunctival sac, it is transported into both the viral infected and uninfected cell. However, it is trapped in the infected cell by phosphorylation to the monophosphate (IdUMP) by thymidine kinase, is subsequently phosphorylated to the triphosphate (IdUTP) and then incorporated into the viral DNA. There is a direct relation between the incorporation of IdUrd into the herpesvirus DNA and loss of infectivity (74).

The consequence of incorporation of IdUrd into the herpes simplex virus DNA has a profound effect on subsequent RNA and protein synthesis (75-77). The biological and physical consequences of incorporation of idoxuridine and other halogenated uracil derivatives into mammalian and viral DNA have been reviewed by Prusoff _et al_. (78 and references cited therein). Among the physical consequences of incorporation of idoxuridine into viral DNA are (a) increased density, (b) increased lability to stress, (c) increased stacking energy, (d) increased temperature of helix-coil transition, (e) increased binding of proteins, (f) slight increased pH required for separation of DNA strands, and (g) increased sensitivity to X-ray and UV-radiations. Among the biological effects are (a) increased rate of mutation, (b) inhibition of the formation of polyadenylated RNA, (c) inhibition of the formation of specific proteins, (d) a small increase in the formation of herpesvirus encoded thymidine kinase, (e) inhibition of cellular and viral replication, (f) inhibition of the expression of differentiated cellular or viral functions (79-81) and (g) virus induction.

IdUrd cannot be administered systemically for therapy of viral infections. Calabresi _et al_. (73) have shown that the systemic administration of IdUrd to man may produce toxicities such as bone marrow depression, stomatitis and alopecia which are dose-dependent. Unfortunately, the investigators (82) involved in the evaluation of systemic IdUrd for therapy of herpes encephalitis were apparently not aware of the report of Calabresi _et al_. (73). Hence, the toxicities reported earlier by Calabresi and co-workers (73) were confirmed by the Cooperative Study Group (82).

Although IdUrd is approved in the United States only for the therapy of herpetic keratitis, it has been used for therapy of other viral infections. At present, the use of idoxuridine for therapy of herpetic keratitis is being replaced by trifluridine and vidarabine in the United States, and in addition acyclovir is in clinical trial in Great Britain for therapy of this viral infection. Presumably acyclovir will also be approved for therapy of herpetic keratitis in the United States. Juel-Jensen (83,84) has reveiwed the clinical utility of IdUrd and, in addition to the therapy of herpes keratitis, beneficial results have been reported in the therapy of herpetic whitlow, genital herpes, herpes zoster, vaccinia lesions and vaccinia whitlow when a concentrated solution of this drug in dimethyl sulfoxide (DMSO) is used.

At present there are a total of 6 antiviral compounds (Fig. 1) which have been approved by the FDA in the United States for clinical use.

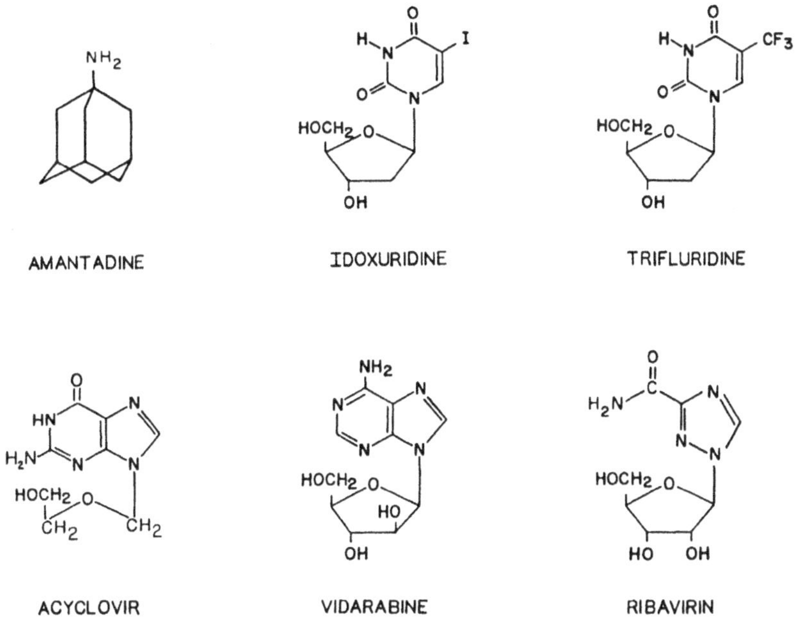

Fig. 1. Structures of FDA-approved antiviral agents.

These compounds and their year of approval are as follows :

1. 5-Iodo-2'-deoxyuridine (Idoxuridine; IdUrd) : 1962.
2. 5-Trifluoromethyl-2'-deoxyuridine (Trifluridine, CF_3-dUrd; TFT) : 1964.
3. Amantadine (Symmetrel) : 1966.
4. 9-β-D-Arabinofuranosyl)adenine (Vidarabine, ara-A) : 1972.
5. 9-(2-Hydroxyethoxymethyl)guanine (Acyclovir, ACV) : 1982.
6. 1-β-D-Ribofuranosyl-1,2,4-triazole-3-carboxamide (Virazole, Ribavirin) : 1985.

In summary we have discussed how the primordial origins of nucleic acids, the discovery of their importance in replication, and the directed synthesis of analogs of purines and pyrimidines for interference with the formation or function of viral nucleic acids. Although, the number of available antiviral drugs are few as of today, the expectation in the near future for the development of more potent and non-toxic agents is very encouraging.

ACKNOWLEDGEMENT

Supported by US Public Health Service Grant CA-05262 from the National Cancer Institute.

REFERENCES

1. "Genesis. The First Book of Moses", p. 1.
2. Encyclopaedia Britannica 10: 893-911, 1981.
3. Potter, V.R. Nucleic Acid Outlines 1 : 3-22, 1960.
4. Avery, O.T., MacLeod, C.M. and McCarty, M. J. Exp. Med. 79 : 137-157, 1944.
5. Hershey, A.D. and Chase, M. J. Gen. Physiol. 36 : 39-56, 1953.
6. Michelson, A.M. The Chemistry of Nucleosides and Nucleotides, Academic Press, New York, 1963, pp. 54-58.
7. Brown, D.J. The Pyrimidines, Interscience, 1962.
8. Brown, D.J. The Pyrimidine, Suppl. 1, Interscience, 1970.
9. Elion, G.B. and Hitchings, G.H. In: Advances in Chemotherapy (Eds. A. Goldin, F. Hawking and R.J. Schnitzer), Academic Press, New York, vol. 2, 1965, pp. 91-177.
10. Prusoff, W.H., Cheng, Y.-C. and Neenan, J. In: Progress in Chemotherapy 2 (Ed. G.K. Daikos), Hellenic Soc. Chemother., Athens, Greece, 1974, pp. 881-898.
11. Hitchings, G.H., Elion, G.B., Falco, E.A., Russell, P.B., Sherwood, M.B. and Vanderwerff, H. J. Biol. Chem. 183: 1-9, 1950.
12. Hitchings, G.H., Elion, G.B., Falco, E.A. and Vanderwerff, H.V. Ann. N.Y. Acad. Sci. 52: 1318-1335, 1950.
13. Welch, A.D. Cancer Res. 21: 1475-1490, 1961.
14. Elion, G.B. and Hitchings, G.H. In: Advances in Chemotherapy (Eds. A. Goldin, F. Hawking and R.J. Schnitzer), Academic Press, New York, vol. 2, 1965, pp. 130-137.
15. Thompson, R.L., Wilkin, M.L., Hitchings, G.H., Elion, G.B., Falco, E.A. and Russell, P.B. Science 110: 454, 1949.

16. Thomson, R.L., Price, M.L., Minton, S.A. Jr., Elion, G.B. and Hitchings, G.H. J. Immunol. 65: 529-534, 1950.
17. Herrmann, E.C. Jr. (Ed.) Ann. N.Y. Acad. Sci. 130: 1-482, 1965; 173: 1-844, 1970; 284: 1-720, 1977.
18. Matthews, R.E.F. Pharmacol. Rev. 10: 359-406, 1958.
19. Sidwell, R.W. and Witkowski, J.T. In: Burger's Medicinal Chemotherapy (Ed. M.E. Wolf), John Wiley & Sons Inc., 4th Ed., Part II, 1979, pp. 543-592.
20. Rapp, F. (Ed.) Herpesvirus. UCLA Symposia on Molecular and Cellular Biology, New Series. Alan R. Liss Inc., vol. 21, 1984.
21. De Clercq, E. and Walker, R.T. (Eds.). Targets for the Design of Antiviral Agents. Plenum Press, New York, 1984.
22. Galasso, G.J., Merigan, T.C. and Buchanan, R.A. (Eds.). Antiviral Agents and Viral Diseases of Man. Raven Press, New York, 1984.
23. Becker, Y. (Ed.). Antiviral Drugs and Interferon: The Molecular Basis of Their Activity. Martinus Nijhoff, Boston, 1984.
24. Shugar, D. (ed.). Viral Chemotherapy. Vol. 1. Pergamon Press, Oxford, 1984.
25. Larder, B.A. and Darby G. Antiviral Res. 4: 1-42, 1984.
26. Darby, G. and Field, H.J. Pharmacol. Ther. 23: 217-251, 1984.
27. De Clercq, E. Biochem. Pharmacol. 33: 2159-2169, 1984.
28. Brown, F. Ann. Rev. Microbiol. 38: 221-235, 1984.
29. Whitley, R.J. J. Antimicrob. Chemother. 14, Suppl. A: 57-74, 1984.
30. Watson, R.J. and Enquist, L.W. Prog. Med. Virol. 31: 84-108, 1984.
31. Prusoff, W.H., Zucker, M., Mancini, W.R., Otto, M.J., Lin, T.-S. and Lee, J.J. Antiviral Res. Suppl. 1: 1-10, 1985.
32. Dolin, R. Science 227: 1296-1303, 1985.
33. Oxford, J.S. and Öberg, B. (Eds.) Conquest of Viral Diseases - A Topical Review of Drugs and Vaccines. Elsevier Sci. Publ., 1985.
34. Mitsuya, H., Weinhold, K.J., Furman, P.A. St. Clair, M.H., Nusinoff Lehrman, S., Gallo, R.C., Bolognesi, D., Barry, D.W. and Broder, S. Proc. Natl. Acad. Sci. USA 82: 7096-7100, 1985.
35. Pratt, W.B. and Fekety, R. (Eds.), The Antimicrobial Drugs. Oxford University Press, Oxford, 1986.
36. Robins, R.K.. Chemical and Engineering News, Jan. 27: 28-40, 1986.
37. De Clercq, E. J. Med. Chem. 29: 1561-1569, 1986.
38. Prusoff, W.H., Lin, T.-S. and Zucker, M. Antiviral Res. 6: 311-328, 1986.
39. Bauer, D.J. Brit. Med. Bull. 41: 309-314, 1985.
40. Prusoff, W.H. Pharmacol. Rev. 19: 209-250, 1967.
41. Brownlee, K.A. and Hamre, D.A. J. Bacteriol. 61: 127-134, 1951.
42. Thompson, R.L., Minton, S.A. Jr., Officer, J.E. and Hitchings, G.H. J. Immunol. 70: 229-234, 1953.
43. Bauer, D.J. Brit. J. Exp. Pathol. 36: 105-114, 1955.
44. Bauer, D.J., St. Vincent, L.S., Kempe, C.H. and Downie, A.W. Lancet ii: 494-496, 1963.
45. Bauer, D.J. Ann. N.Y. Acad. Sci. 130: 110-117, 1965.
46. Hansson, O., Johansson, S.G.O. and Vahlquist, B. Acta Paediatr. Scand. 55: 264-272, 1966.
47. Johnson, T.B. and Johns, C.O. J. Biol. Chem. 1: 305-318, 1905-06.
48. Hitchings, G.H., Falco, E.A. and Sherwood, M.B. Science 102: 251-252, 1945.
49. Dunn, D.B. and Smith, J.D. Biochem. J. 67: 494-506, 1957.
50. Dunn, D.B., Smith, J.D., Zamenhof, S. and Griboff, G. Nature 174: 305-307, 1954.
51. Luzzati, D. Compt. Rend. Acad. Sci. 245: 1466-1468, 1957.

52. Zamenhof, S. and Griboff, G. Nature 174: 307-308, 1954.
53. Prusoff, W.H., Holmes, W.L. and Welch, A.D. Cancer Res. 13: 221-225, 1953.
54. Plentle, A.A. and Schoenheimer, R. J. Biol. Chem. 153: 203-217, 1944.
55. Holmes, W.L., Prusoff, W.H. and Welch, A.D. J. Biol. Chem. 209: 503-509, 1954.
56. Brown, G.B., Roll, P.M. and Weinfield, H. In: Phosphorous Metabolism (Eds. W.A. McElroy and B. Glass), Johns Hopkins Press, Baltimore, 1952, p. 388.
57. Reichard, R. Acta Chem. Scand. 9: 1275-1285, 1956.
58. Prusoff, W.H. Biochim. Biophys. Acta 32: 295-296, 1959.
59. Mathias, A.P. Fischer, G.A. and Prusoff, W.H. Biochim. Biophys. Acta 36: 560-561, 1959.
60. Prusoff, W.H. Fed. Proc. 18: 305, 1959.
61. Eidinoff, M.L., Cheong, L., Gambetta Gurpide, E., Benua, R.S. and Ellison, R.R. Nature 183: 1686-1687, 1959.
62. Eidinoff, M.L., Cheong, L. and Rich, M.A. Science 129: 1550-1551, 1959.
63. Eidinoff, M.L., Cheong, L. and Rich, M.A. Fed. Proc. 18: 220, 1959.
64. Prusoff, W.H. Cancer Res. 20: 92-95, 1960.
65. Herrmann, E.C. Jr. Proc. Soc. Exp. Biol. 107: 142-145, 1961.
66. Rada, B., Blaskovic, D., Sorm, F. and Skoda, J. Experientia 16: 487-488, 1960.
67. Smith, J.D., Freeman, G., Vogt, M. and Dulbecco, R. Virology 12: 185-196, 1960.
68. Kaufman, H.E. Personal communication.
69. Perkins, E.S., Wood, R.M., Sears, M.L., Prusoff, W.H. and Welch, A.D. Nature 194: 985-986, 1962.
70. Kaufman, H.E. Proc. Soc. Exp. Biol. Med. 109: 251-252, 1962.
71. Kaufman, H.E., Martola, E.-L. and Dohlman, C. Arch. Ophthalmol. 68: 235-239, 1962.
72. Kaufman, H.E., Nesburn, A.B. and Maloney, E.D. Arch. Ophthalmol. 67: 583-591, 1962.
73. Calabresi, P., Cardoso, S.S., Finch, S.C., Kligerman, M.M., von Essen, C.F., Chu, M.Y. and Welch, A.D. Cancer Res. 21: 550-559, 1961.
74. Fischer, P.H., Chen, M.S. and Prusoff, W.H. Biochim. Biophys. Acta 606: 236-245, 1980.
75. Otto, M.J., Lee, J.J. and Prusoff, W.H. Antiviral Res. 2: 267-281, 1982.
76. Otto, M.J., Goz, B. and Prusoff, W.H. In: Antiviral Drugs and Interferon: The Molecular Basis of Their Activities (Ed. J. Becker), Martinus Nijhoff, Boston, 1984, pp. 11-38.
77. Zucker, M.L., Mancini, W.R., Otto, M.J., Shim Lee, J.-J. and Prusoff, W.H. Antiviral Res. 6: 69-81, 1986.
78. Prusoff, W.H., Mancini, W.R., Lin, T.-S., Lee, J.J., Siegel, S.A. and Otto, M.J. Antiviral Res. 4: 303-315, 1984.
79. Goz, B. and Prusoff, W.H. J. Biol. Chem. 243: 4750-4756, 1968.
80. Goz, B. and Prusoff, W.H. Ann. N.Y. Acad. Sci. 173: 379-389, 1970.
81. Goz, B. Pharmacol. Rev. 29: 249-272, 1978.
82. Boston Interhospital Virus Study Group and the NIAID-Sponsored Cooperative Antiviral Clinical Study. New Engl. J. Med. 292: 599-603, 1975.
83. Juel-Jensen, B.E. Brit. Med. J. i: 406-410, 1973.
84. Juel-Jensen, B.E. Practioner 213: 508-518, 1974.

3

THE TREATMENT OF HERPETIC EYE INFECTIONS WITH TRIFLURIDINE AND OTHER ANTIVIRALS*

H.E. KAUFMAN

Lions Eye Research Laboratories, LSU Eye Center, Louisiana State University Medical Center School of Medicine, New Orleans, Louisiana, U.S.A.

ABSTRACT

In the treatment of epithelial herpes, trifluridine seems approximately as effective as any of the thymidine kinase selective drugs, such as acyclovir and bromovinyldeoxyuridine. The toxicity of trifluridine might seem to be avoidable by the use of truly thymidine kinase selective agents. In fact, perhaps because treatment is of such short duration, there is no good evidence to date that the thymidine kinase selective agents are less toxic for topical use than trifluridine. At this time, the treatment of choice for disciform edema is long-term corticosteroid therapy to reduce inflammation and prevent neovascularization and scarring, with concurrent antiviral coverage (trifluridine) to antagonize the corticosteroid effect and minimize the severity of epithelial recurrences. During long-term treatment combined with corticosteroids, trifluridine is usually used only two or at most three times a day, and toxicity is rarely seen with this reduced frequency.

Trifluridine is not effective in the treatment of clinical cases of deep stromal disease or iritis, although it appears to have some effect in experimental animal models. To date, there is no good evidence that either acyclovir or other antiviral drugs are effective in treating these syndromes in man in the absence of corticosteroid treatment. The potential problem of creating widespread drug resistant virus strains seems to outweigh potential advantages that may or may not accrue from the use of these thymidine kinase selective agents.

INTRODUCTION

It is estimated that there are approximately 500,000 cases of herpes simplex keratitis in the United States each year. Herpes keratitis is a very

---

*Supported in part by PHS grants EY02672, EY02389, and EY02377 from the National Eye Institute, National Institutes of Health, Bethesda, Maryland.

significant cause of morbidity, and with the damage caused by recurrent disease, a major cause of blindness. In terms of human disease and misery, it is the most common specific corneal infection encountered in ophthalmology, and was the first human disease shown to be treatable by antiviral agents (1).

PATHOGENESIS OF OCULAR HERPETIC DISEASE

One of the problems in treating herpetic keratitis is understanding the pathogenesis of the multiple manifestations of disease, particularly in relation to the presence or absence of actively multiplying virus in the various anatomically distinct layers of the cornea (Fig. 1). Much work in this area has been accomplished in animal models. Some of the advantages of studying herpes simplex keratitis in an animal model include the fact that the animal disease mimics the human disease in some of its stages, the lesions are easily observed and quantitated, and the availability of two eyes per animal offers the potential for unique controls and observations.

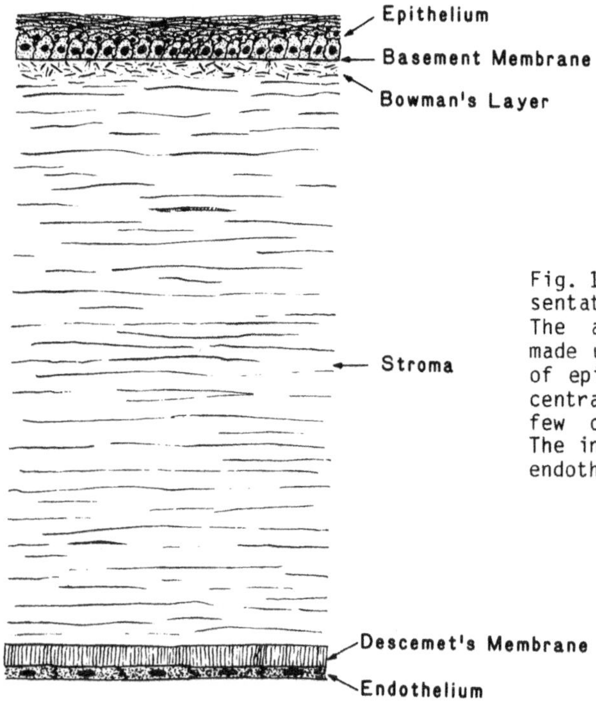

Fig. 1. Diagramatic representation of the cornea. The anterior surface is made up of several layers of epithelial cells. The central stroma contains few cells (keratocytes). The innermost layer is the endothelium.

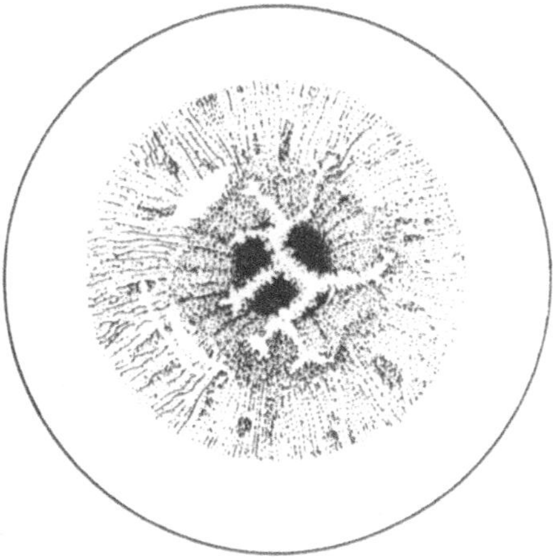

Fig. 2. Dendritic keratitis is the most common form of herpetic corneal infection. The ulcer is confined to the epithelial layers of the cornea, and the typical tree-shaped lesion is pathognomonic for herpesvirus infection.

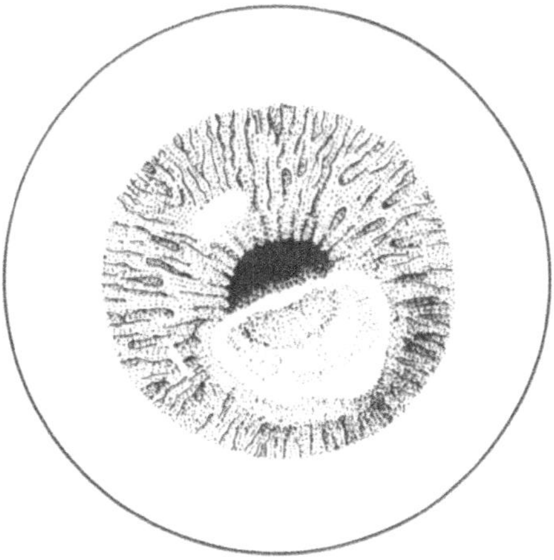

Fig. 3. As a dendrite progresses, the lesion can remain in the epithelial layers and assume a geographic pattern, or progress to the deeper layers of the stroma, where necrosis takes the form of a dense white corneal infiltrate.

Epithelial keratitis.

Whether the epithelial disease is primary or secondary, the typical branching dendritic ulcer that forms on the surface of the cornea is caused by multiplying virus (Fig. 2). Virus is easily cultured, and is seen in the cells with fluorescent antibody techniques.

Necrotizing stromal disease.

Necrotizing stromal disease (Fig. 3) is more obscure in its pathogenesis. There is good reason to believe that a significant amount of the stromal necrosis, if not most of it, is caused by a hypersensitivity reaction to the virus and viral products, and the relative importance of virus multiplication and hypersensitivity remains undetermined. For this reason, laboratory models that emphasize rapid virus multiplication and its treatment may not be relevant to the human disease. Although models of epithelial keratitis have been shown to be extremely predictive in the treatment of human infection, this has not been the case for necrotizing stromal disease or iritis.

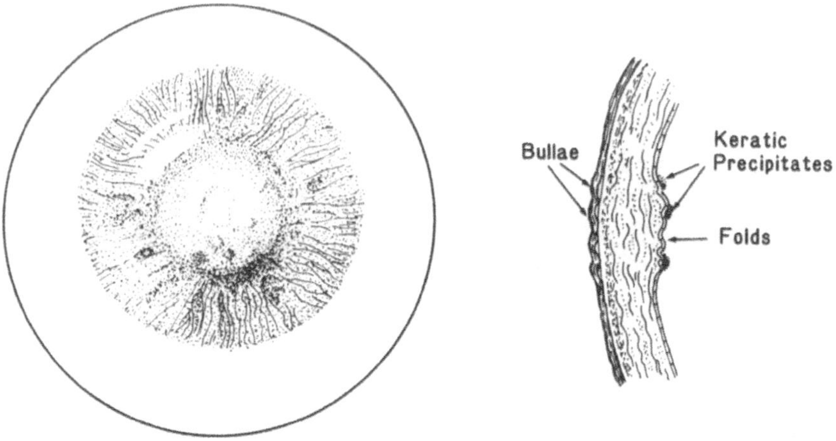

Fig. 4. Disciform keratitis is a disc shaped area of edema, usually in the central cornea, which is surrounded by normal appearing, clear cornea. The cornea swells posteriorly, causing wrinkling in Descemet's membrane. Often there are inflammatory cells adherent to the endothelial surface (keratic precipitates).

Disciform edema.

Disciform edema of the cornea (Fig. 4) is a round patch of edema typically occurring after the epithelial herptic infection has resolved, and appears to be almost entirely the result of some hypersensitivity phenomenon (2,3). Whether this is a typical delayed hypersensitivity or some toxic phenomenon caused by antigen-antibody complexes in the anterior chamber is unclear. It is clear, however, that topical corticosteroids can ameliorate the symptoms of this disease (4,5). Although the corticosteroids do not actually provide a cure, they do suppress inflammation, which seems to reduce scarring and vascularization, and with it, the permanent damage that can follow. If the corticosteroids are continued for a period of three months or more, they can usually be withdrawn with no recurrence of the discomfort and blindness due to the distortions and fogginess of corneal edema from disciform herpes. Because corticosteroids may exacerbate epithelial herpes and cause recurrent epithelial ulcers, I originally recommended treating disciform edema with corticosteroids and an antiviral. Unfortunately, to this day, this remains the treatment of choice. However, specific antiviral agents alone have not been shown to be beneficial in this syndrome.

Metaherpetic keratitis.

Metaherpetic keratitis is a condition in which epithelial ulcers occur over damaged basement membrane or necrotic patches of stroma. This is a problem of previous tissue damage, in that the epithelial cells of the cornea attach to the basement membrane by specialized hemidesmosomes, and if these desmosomal attachments are damaged, adhesion between the epithelium and the basement membrane is deficient. As a result, epithelial defects occur in the absence of multiplying virus (6). With this pathogenesis, antiviral drugs are not of value; in fact, frequent topical applications may be harmful. Therapy must be directed toward retaining the epithelial cells in place until they can form new desmosomal attachments to secreted basement membrane. If there is some question about whether virus is present, a therapeutic soft contact lens (7,8) may be placed on the surface of the eye to hold the epithelium in place, and still permit treatment with antiviral drugs.

Herpes simplex virus (HSV) types 1 and 2.

Further complicating the problem of therapy and its evaluation is the type of virus infection. Although most ocular virus infections are caused by HSV type 1, type 2 virus has been cultured from the eye. Thus, there is a danger in using therapeutic agents that might be effective against type 1 but not against type 2. Also, we have shown that the shape, extent, and virulence of corneal infections are determined by the specific DNA composition of the strain or isolate of virus (9,10). Viral genetics determine the size and shape of the epithelial ulcer, as well as the tendency for viruses to recur and the frequency of such recurrences, independent of immune factors (9-11).

ANTIVIRAL DRUGS AND THEIR PROPERTIES

Idoxuridine.

I first utilized idoxuridine to treat herpetic keratitis in rabbits and man in 1962 (1,12), and ample studies have confirmed its effectiveness. Idoxuridine, however, is relatively insoluble, and in part because of this, is not as potent a viral inhibitor as some of the newer drugs. In addition, apparently because of the iodine moiety, the continued use of idoxuridine induces a degree of toxicity and allergy to the compound over time; although this is not common, it is frequent enough to be undesirable. The primary shortcoming of idoxuridine, however, is its lack of potency.

Trifluridine, vidarabine, and acyclovir.

Trifluridine was first synthesized by Heidelberger (13) as an anti-cancer drug and introduced for the treatment of herpetic keratitis in 1964 (14). The superiority of this agent was shown in our (15) and other (16,17) subsequent studies.

In general, although antimetabolites have multiple mechanisms of action, all of them must be phosphorylated to the triphosphate to be optimally active. With trifluridine, vidarabine, and idoxuridine, both cellular and viral thymidine kinase can phosphorylate the drugs, and, therefore, these drugs can have some inhibitory effect on cellular multiplication. Although all of the antiviral drugs have multiple sites of activity, and all of them may inhibit the action of the DNA polymerase to some extent, the five substituted uridines tend to be incorporated into DNA, providing a DNA that cannot replicate appropriate viral enzymes and proteins and does not reproduce properly. The drugs with abnormal sugars, such as the arabinosides and probably the acyclic compounds, tend to interfere with the polymerase and to be chain terminators; thus the normal 3'-5'-deoxyribose-phosphate bonds that make up the sugar backbone of the DNA cannot form, and the sugar backbone is terminated or distorted after linking to one of these abnormal sugars. Multiple other sites of action are also possible. For example, trifluridine also inhibits the function of thymidylic synthetase and may reduce the size of the thymidine pool.

Trifluridine and acyclovir have been shown to penetrate the eye and appear in the aqueous, in both rabbits and man. Especially in patients with corneas damaged from herpes, significant amounts of both drugs have been found in the aqueous and the stroma (18,19). When multiplication of virus is induced in the stroma in the rabbit model, drug penetration can produce a reduction in iritis and prevent stromal disease, but penetration into the stroma and aqueous has not been shown to be of value in the treatment of human disease to date, perhaps because virus multiplication is not rapid, and hypersensitivity is a major component of the clinical disease.

Treatment of epithelial keratitis.

In experimental herpes simplex keratitis, idoxuridine can be shown to be active; animals treated with idoxuridine heal more rapidly than untreated controls. Vidarabine treatment has approximately the same results as treatment with idoxuridine in terms of efficacy, perhaps in part because vidarabine is also relatively insoluble and must be administered in ointment form. Solubility limits the effectiveness of both idoxuridine and vidarabine. Trifluridine and acyclovir are more active than idoxuridine; trifluridine, acyclovir, and bromovinyldeoxyuridine are all approximately equally effective in the treatment of experimental herpetic keratitis (20).

In man, it is clear that trifluridine is significantly better than idoxuridine in the treatment of epithelial keratitis. In most studies (15-17,21), approximately 97% of patients with epithelial herpes treated with trifluridine are cured within two weeks, compared with 76.5% treated with idoxuridine and 86% treated with vidarabine (Table 1). It is clear, then, that trifluridine and acyclovir are a generation ahead of the earlier antivirals in terms of the treatment of epithelial herpes.

Toxicity. There is an overwhelming temptation to assume automatically that drugs that are more selective in inhibiting cellular metabolism will be less toxic in the treatment of human disease. To date, there is no evidence that this is correct for the topical treatment of herpes. Vidarabine, for example, can be shown in some experimental studies to have less inhibitory effect on corneal epithelial cells. In the studies done by Parke-Davis (22), however, which were randomized, double-blind controlled studies comparing vidarabine with idoxuridine, no significant reduction in toxicity was observed with vidarabine. Long-term toxicity and allergies were not measured in the study, but significant short-term toxicity of vidarabine was documented. Similarly, in controlled studies comparing acyclovir with idoxuridine, acyclovir did not cause less epithelial toxicity. There have been no adequately

Table 1. Summary of historically controlled* clinical trial experience with trifluridine

Study	Healed/treated	% Healed	Mean days to heal
McGill et al., 1974	24/24	100.0	7.1
Laibson et al, 1977	12/12	100.0	6.0
Pavan-Langston and Foster, 1977	13/15	86.6	6.3
Unpublished data, Burroughs Wellcome Co., USA	63/77	82.0	-

*Open, unmasked evaluation of trifluridine in patients unresponsive to available forms of antiviral chemotherapy.

Reprinted with permission from: Heidelberger, C. and King, D.H. Pharmacol. Ther. 6:427-442, 1979.

Table 2. Adverse experiences related to therapy

Adverse experiences	Acyclovir (N = 30) TOT	PCT	Idoxuridine (N = 34) TOT	PCT
Superficial punctate keratopathy*	3	11.1	14	42.4
Burning or stinging	1	3.3	1	2.9
Punctal occlusion	1	3.3	0	0.0
Follicular conjunctivitis	0	0.0	1	2.9
Lid edema	0	0.0	1	2.9

Reprinted with permission from: McCulley, J.P., Binder, P.S., Kaufman, H.E., O'Day, D.M. and Poirier, R.H. Ophthalmology 89:1195-1200, 1982.

controlled, randomized, double-blind studies of the other thymidine kinase selective agents, such as bromovinyldeoxyuridine, to determine their epithelial toxicity during the actual treatment of patients. On the basis of previous studies with acyclovir and idoxuridine (Table 2), there is no reason to assume that these agents will be any less toxic than other antivirals (23).

Treatment of geographic ulcers.

As an infected ulcer becomes larger, it may lose the shape of a branching tree and assume an irregular geographic shape. These geographic ulcers have not only the problem of rapid virus multiplication, but also the problem of epithelial healing over a much larger area, with a possibly damaged basement membrane. This means that the clinical response requires not only the elimination of the multiplying virus, but also healing on the part of the host, which is more variable and less predictable than the eradication of the virus by antiviral therapy. In this form of ocular herpes, also, trifluridine is significantly better than idoxuridine or vidarabine, and comparable to acyclovir in efficacy.

Treatment of necrotizing stromal keratitis.

In the rabbit model, trifluridine, acyclovir, and other compounds have been shown to prevent the development of necrotizing stromal keratitis when given relatively soon after the injection of live virus into the stroma (24,25). When live virus is injected into the anterior chamber, a severe iritis supervenes, and these drugs can be shown to have a beneficial effect in aborting and minimizing the iritis (26). In experimental keratitis, these effects are clear cut and beyond dispute.

Unfortunately, however, the experimental data have not been transferable to the treatment of human stromal keratitis and iritis. Although in some cases of stromal keratitis, viral particles have been seen in the stroma (27-29), the amount of visible virus was small, and in many specimens, no virus at all was seen. Stromal disease has been mimicked to some extent by the injection of antigen into the stroma. It is likely that even if virus multiplication takes place, the rate of multiplication may be very slow, and hypersensitivity components may be so overwhelming, that antiviral drugs may not be effective in treating this syndrome. The possibility that the human disease and the experimental disease differ is very great, in part because experimental mimicry of the human disease can be obtained by either rapid viral multiplication or hypersensitivity phenomena, and the mechanism predominant in man is unknown.

Human stromal keratitis. Good, controlled, double-blind studies have not been done on the treatment of stromal keratitis in man. However, there is a significant backlog of clinical experience. There is general agreement that our early observations suggesting that idoxuridine is ineffective in treating stromal keratitis are correct. This drug is not a primary therapeutic agent for this syndome, despite the fact that it effectively treats epithelial disease. Vidarabine also seems ineffective, and the more soluble congener, adenine arabinoside monophosphate (Ara-AMP), appears to be too toxic for topical administration.

There is, at present, no evidence that trifluridine, acyclovir, or any other drug, is effective in treating deep stromal keratitis in man, although some of these drugs do show some effect in animals. In fact, in the initial studies on topical treatment of dendritic keratitis with acyclovir, a small but significant number of patients in the acyclovir-treated group (17.4%) developed stromal keratitis, compared with 29.4% in the idoxuridine-treated group (23). It seemes clear, therefore, that acyclovir does not prevent the development of stromal keratitis even when given during the treatment of acute dendritic disease. In addition, in an uncontrolled non-randomized study done by our group, oral and topical acyclovir were given to a series of patients with stromal keratitis, but did not appear to benefit the disease to a greater extent than topical trifluridine alone (27). The literature on the treatment of stromal disease is muddled, however. The experimental keratitis definitely does respond to antiviral therapy. Furthermore, the addition of even a small amount of corticosteroids appears to dramatically benefit the stromal disease. It may be that the healing of the superficial epithelial ulcer causes an apparent benefit to the stroma without having a direct effect on the underlying disease. The more potent antiviral agents help the epithelium to heal, and they make the use of corticosteroids somewhat safer, but there is no reason to believe, at this time, that they have a direct effect on necrotizing stromal keratitis and the iritis apart from this.

Treatment of disciform edema.

Disciform edema is best treated with corticosteroids, but the tendency of the corticosteroids to reactivate epithelial herpes and cause active ulcers must be counteracted by an antiviral drug. Trifluridine appears to be comparable to acyclovir and bromovinyldeoxyuridine and superior to the other antivirals in antagonizing the corticosteroid effect. The antiviral agents, although not primary therapeutic agents, have made the treatment of this prolonged and debilitating syndrome clinically feasible.

Treatment of recurrent disease.

In a prospective study (30), the rate of recurrence of epithelial herpes was 24.5% within 12 months for patients who had at least two previous recurrences. Age and sex had no effect on the recurrence rate. However, it appeared that the shorter the time between the two previous recurrences, the shorter the time to the next recurrence. At present, there is no known way to prevent recurrences of ocular herpes.

In between active attacks, the herpesvirus appears to become latent in the neural ganglia. It is clear that none of the available antiviral agents, whether given topically or systemically, can eliminate virus from the ganglion. Furthermore, there is no good evidence that any of these agents alters the incidence of recurrences after the drug is stopped, no matter how it was administered or for how long a time.

There is no question that in patients with genital herpes (31,32) and other kinds of herpes simplex infections (33), the administration of oral acyclovir can prevent recurrences, as long as adequate doses of the drug are continually administered. This has not been shown to be the case for ocular disease. One of our experimental studies in rabbits utilizing both oral acyclovir and oral bromovinyldeoxyuridine showed no significant differences in recurrence rates during the administration of these drugs (34). Rabbits, however, metabolize these drugs somewhat differently than man. Even though the serum levels appeared to be adequate in these studies, further investigation seems justified, since in other experimental studies (35), some effect of continued administration on virus shedding in the rabbit eye has been demonstrated.

In man, the continual topical administration of trifluridine does not prevent the appearance of epithelial ulcers on the cornea during recurrences, although lesions that appear during antiviral treatment are very small and disappear quickly.

Drug resistance.

Resistance to antiviral drugs in a clinical setting appears to take different forms: retarded healing, true drug resistance, and thymidine kinase selectivity. Retarded healing involves tissue that has been ravaged by previous virus infection. In such cases, healing may be so slow because of the extensive damage that the disease appears to be resistant, although viral multiplication has halted.

True drug resistance to all antiviral drugs certainly has been observed. However, it is extremely rare with trifluridine, and trifluridine is generally effective against virus that is resistant to other antivirals, including vidarabine and idoxuridine. In a decade of observation, there have been only three trifluridine-resistant strains isolated clinically.

The thymidine kinase selective drugs, i.e., acyclovir and bromovinyldeoxyuridine, tend to show cross resistance. There are, in general, three mechanisms for this resistance: 1) In any population of viruses, a certain proportion are thymidine kinase negative. Since drugs like acyclovir and bromovinyldeoxyuridine depend on viral thymidine kinase for their activation, and are not phosphorylated by cellular thymidine kinase, the loss of thymidine kinase makes the virus drug resistant. Most of the thymidine kinase negative mutants appear to be less virulent, and there is at least some hope that this group of resistant viruses will not be a major clinical problem. 2) Mutation in the portion of the virus genome that specifies the viral thymidine kinase enzyme may make the enzyme more selective. The drug is not phosphorylated, but sufficient thymidine kinase activity remains to foster viral growth. Some of these isolates appear to be significantly more virulent, and the importance of this as a clinical problem is more worrisome, although as yet unclear. 3) Alterations in the DNA polymerase may alter the affinity of this enzyme for the triphosphates of the drug, and make the virus relatively more drug resistant. The clinical importance of this kind of mutation also remains unclear.

In the experimental setting, it is relatively easy to create resistance to the thymidine kinase selective antiviral agents, and virus clinically resistant to acyclovir has also been seen (36-38). Although the magnitude of the clinical problem of drug resistance remains obscure, when a superficial lesion is to be treated, and when topical non-thymidine kinase selective agents are as effective and not significantly more toxic than the thymidine kinase selective agents, it is not clear whether the thymidine kinase selective agents should be used.

Patient compliance.

Patient compliance is a major problem with the topical treatment of viral infections. With idoxuridine, treatment was at first required every hour during the day and every two hours at night (12). Later, ointments simplified this regimen to some extent, but resulted in blurred vision and made the eye functionally blind during treatment. There is no question that compliance is better with drops, such as trifluridine, which do not blur vision. Also, trifluridine drops are given only 5-7 times a day for the short time required for the lesion to heal, and no medication is given at night.

FOR THE FUTURE

There is no clinical evidence that trifluridine or acyclovir are of value in ocular adenovirus infection or established herpes zoster infection in the eye, although further studies are certainly indicated. Much work has been done on the combined use of various agents in an attempt to find synergistic therapeutic effects and to decrease the threat of emergent resistant virus. It is clear that the addition of interferon to trifluridine (39-42) has some beneficial effects, and combinations of antiviral drugs (43,44) can increase efficacy with reduced dosages, but the clinical value of these combinations remains uncertain. It may be that drug combinations or combinations of the thymidine kinase selective drugs with either the unselective agents or interferon could be used to reduce the likelihood of the development of virus resistance while preserving effective therapeutic levels of antiviral activity.

REFERENCES

1. Kaufman, H.E. Proc. Soc. Exp. Biol. Med. 109:251-252, 1962.
2. Williams, L.E., Nesburn, A.B. and Kaufman, H.E. Arch. Ophthalmol. 73:112-114, 1965.
3. Swyers, J.S., Lausch, R.N. and Kaufman, H.E. Br. J. Ophthalmol. 51:843-846, 1967.

4. Kaufman, H.E., Martola, E-L. and Dohlman, C.H. Arch. Ophthalmol. 69:468-472, 1963.
5. Kaufman, H.E., Martola, E-L. and Dohlman, C.H. Trans. Am. Acad. Ophthalmol. Otolaryngol. 67:695-701, 1963.
6. Kaufman H.E. Am. J. Ophthalmol. 57:983-987, 1964.
7. Gasset, A.R. and Kaufman, H.E. Am. J. Ophthalmol. 69:252-259, 1970.
8. Kaufman, H.E., Uotila, M.H., Gasset, A.R., Wood, T.O. and Ellison, E.D.: Trans. Am. Acad. Ophthalmol. Otolaryngol. 75:361-373, 1971.
9. Centifanto-Fitzgerald, Y.M., Yamaguchi, T., Kaufman, H.E., Tognon M. and Roizman, B. J. Exp. Med. 155:475-489, 1982.
10. Wander, A.H., Centifanto, Y.M. and Kaufman, H.E. Arch. Ophthalmol. 98: 1458-1461, 1980.
11. Gerdes, J.C. and Smith, D.S. J. Gen. Virol. 64:2441-2454, 1983.
12. Kaufman, H.E., Nesburn, A.B. and Maloney, E.D. Arch. Ophthalmol. 67: 583-591, 1962.
13. Heidelberger, C., Parsons, D.G. and Remy, D.C. J. Med. Chem. 7:1-5, 1964.
14. Kaufman, H.E. and Heidelberger, C. Science 145:585-586, 1964.
15. Wellings, P.C., Awdry, P.N., Bors, F.H., Jones, B.R., Brown, D.C. and Kaufman, H.E. Am. J. Ophthalmol. 73:932-942, 1972.
16. Pavan-Langston, D. and Foster, C.S. Am. J. Ophthalmol. 84:818-825, 1977.
17. Laibson, P.R., Arentsen, J.J., Mazzanti, W.D. and Eiferman, R.A. Trans. Am. Ophthalmol. Soc. 75:316-324, 1977.
18. O'Brien, W.J. and Edelhauser, H.F. Invest. Ophthalmol. Vis. Sci. 16:1093-1103, 1977.
19. Poirier, R.H., Kingham, J.D., de Miranda, P. and Annel, M. Arch. Ophthalmol. 100:1964-1967, 1982.
20. Kaufman, H.E., Centifanto-Fitzgerald, Y.M. and Varnell, E.D. Ophthalmology 90(6):700-706, 1983.
21. Heidelberger, C. and King, D.H. Pharmacol. Ther. 6:427-442, 1979.
22. Dresner, A.J. and Seamens, M.L. In: Adenine Arabinoside: An Antiviral Agent (Eds. D. Pavan-Langston, R.A. Buchanan, and C.A. Alford, Jr.), Raven Press, New York, 1975, pp. 381-392.
23. McCulley, J.P., Binder, P.S., Kaufman, H.E., O'Day, D.M. and Poirier, R.H. Ophthalmology 89:1195-1200, 1982.
24. McNeill, J.I. and Kaufman, H.E.: Arch. Ophthalmol. 97:727-729, 1979.
25. Varnell, E.D. and Kaufman, H.E. In: Herpetische Augenerkrankungen (Ed. R. Sundmacher), J.F. Bergmann Verlag, Munich, 1981, pp. 303-307.
26. Kaufman, H.E., Ellison, E.D., and Townsend, W.M. Arch. Ophthalmol. 84: 783-787, 1970.
27. Sanitato, J.J., Asbell, P.A., Varnell, E.D., Kissling, G. and Kaufman, H.E. Am. J. Ophthalmol. 98:537-547, 1984.
28. Shimeld, C., Tullo, A.B., Easty, D.L. and Thomsitt, J. Br. J. Ophthalmol. 66:643-647, 1982.
29. Shimeld, C., Tullo, A.B., Hill, T.J., Blyth, W.A. and Easty, D.L. Arch. Virol. 85:175-187, 1985.
30. Shuster, J.J., Kaufman, H.E. and Nesburn, A.B. Am. J. Ophthalmol. 91: 328-331, 1981.
31. Douglas, J.M., Critchlow, C., Benedetti, J., Mertz, G.J., Connor, J.D., Hintz, M.A., Fahnlander, A., Remington, M., Winter, C. and Corey, L. New Eng. J. Med. 310:1551-1556, 1984.
32. Straus, S.E., Takiff, H.E., Seidlin, M., Bachrach, S., Lininger, L., DiGiovanna, J.J., Western, K.A., Smith, H.A., Lehrman, S.N., Creagh-Kirk, T. and Alling, D.W. New Engl. J. Med. 310:1545-1550, 1984.
33. Meyrick Thomas, R.H., Dodd, H.J., Yeo, J.M. and Kirby, J.D.T. Br. J. Dermatol. 113:731-735, 1985.

34. Kaufman, H.E., Varnell, E.D., Centifanto-Fitzgerald, Y.M., De Clercq, E. and Kissling, G.E. Antimicrob. Agents. Chemother. 24:888-891, 1983.
35. Nesburn, A.B., Willey, D.E. and Trousdale, M.D. Proc. Soc. Exp. Biol. Med. 172:316-323, 1983.
36. Sibrack, C.D., Gutman, T., Wilfert, C.M., McLaren, C., St. Clair, M.H., Keller, P.M. and Barry, D.W. J. Infect. Dis. 146:673-682, 1982.
37. Burns, W.H., Saral, R., Santos, G.W., Laskin, O.L., Lietman, P.S., McLaren, C. and Barry, D.W. Lancet 1:421-423, 1982.
38. Crumpacker, C.S., Schnipper, L.E., Marlowe, S.I., Kowalsky, P.N., Hershey, B.J. and Levin, M.J. N. Engl. J. Med. 306:343-346, 1982.
39. Sundmacher, R., Cantell, K. and Neumann-Haefelin, D. Lancet 2:687, 1978.
40. Sundmacher, R., Cantell, K. and Mattes, A. Arch. Ophthalmol. 102:554-555.
41. De Koning, W.J., van Bijsterveld, O.P. and Cantell, K. Br. J. Ophthalmol. 66:509-512, 1982.
42. Colin, J., Chastel, C., Renard, G. and Cantell, K. La Nouvelle Presse Medicale 11:2783, 1982.
43. Colin, J., Chastel, C., Kaufman, H.E. and Kissling, G. In preparation.
44. Collum, L.M.T., Logan, P., McAuliffe-Curtin, D., Hung, S.O., Patterson, A. and Rees, P.J. Br. J. Ophthalmol. 69:847-850, 1985.

4

TREATMENT (BROMOVINYLDEOXYURIDINE) OF HERPETIC EYE INFECTIONS

P.C. MAUDGAL[1*] and E. DE CLERCQ[2]

[1]Eye Research Laboratory, Ophthalmological Clinic and [2]Rega Institute for Medical Research, Katholieke Universiteit Leuven, B-3000 Leuven, Belgium

ABSTRACT

One hundred and fifty-three patients with herpes simplex virus (HSV) keratitis were treated with topical 0.1 % bromovinyldeoxyuridine (BVDU) eyedrops. All patients responded to BVDU treatment, including those who had been treated, albeit unsuccessfully, with other antivirals, i.e. idoxuridine, vidarabine, trifluridine and/or Zovirax, till the time their treatment was switched to BVDU. Seventy-six patients with dendritic keratitis healed within an average time of 8.6 days upon BVDU therapy. Similarly, 35 patients with geographic corneal ulcers healed within an average time of 11.7 days, and 42 patients with stromal keratitis healed within an average time of 30 days upon BVDU therapy. Except for local hypersensitivity reactions to BVDU eyedrops in a small number of patients, no other local or systemic toxic effects were observed. The follow-up period for all patients treated with BVDU has been at an average 52, 48.5 and 38.7 months for the three groups, and in all three groups the number of recurrences after BVDU treatment appeared to be reduced as compared to the number of recurrences before BVDU treatment.

An additional 15 patients with ophthalmic herpes zoster were treated with oral BVDU capsules at 375 mg/day for 5 days, combined with topical 0.1 % BVDU eyedrops. The skin eruption and the eye disease responded briskly to this treatment. Routine blood and urine tests, as well as urea, creatinine, platelets, electrolytes and liver enzyme (SGPT, SGOT, γGT) estimations before, during and after oral BVDU therapy did not reveal any toxic side effects that could be attributed to the drug.

---

*Present address : Ophthalmology Department, Academic Hospital, Free University, Postbox 7057, 1007 MB Amsterdam, The Netherlands

INTRODUCTION

Bromovinyldeoxyuridine [(_E_)-5-(2-bromovinyl)-2'-deoxyuridine, BVDU] is a 5-substituted analogue of 2'-deoxythymidine (dThd) (1). It is structurally related to idoxuridine (5-iodo-2'-deoxyuridine, IDU) and trifluridine (5-trifluoromethyl-2'-deoxyuridine, TFT). However, _in vitro_ BVDU is more potent and selective in its activity against herpes simplex virus type 1 (HSV-1) than IDU, TFT and several other antiviral compounds, i.e. foscarnet (phosphonoformate), vidarabine (9-β-D-arabinofuranosyladenine, ara-A, Vira-A), acyclovir [9-(2-hydroxyethoxymethyl)guanine, acycloguanosine, ACV, Zovirax] (2-4). BVDU is also effective against varicella-zoster virus (VZV) (5,6), simian varicella virus (7), Epstein-Barr virus and various viruses of veterinary importance, i.e. suid herpesvirus type 1 and bovid herpesvirus type 1 (8).

Very low concentrations (0.001-0.01 μg/mL) of BVDU are required to inhibit the replication of HSV-1 (2-4) and VZV (5,6), whereas cytotoxicity is noted only at a concentration of 50-100 μg/mL. The high antiviral selectivity of BVDU is based upon a specific phosphorylation in virus-infected cells by the HSV-1- and VZV-encoded thymidine kinase (9,10). Once converted to its 5'-triphosphate form, BVDU inhibits viral DNA polymerases in a competitive fashion with regard to 2'-deoxythymidine triphosphate (dTTP). BVDU 5'-triphosphate (BVDUTP) has a greater affinity for HSV-1 DNA polymerase than for the cellular DNA polymerases α, β and γ (11). BVDUTP also serves as an alternate substrate for the DNA polymerase (12), which leads to its incorporation, as BVDU 5'-monophosphate (BVDUMP), into the viral DNA (13). The HSV-2-encoded dThd kinase does not phosphorylate BVDU as efficiently as does the HSV-1-encoded enzyme (10,14), which explains why the drug is less effective against HSV-2. BVDU is not active against dThd kinase-deficient (TK^-) HSV mutants.

In the treatment of herpetic eye infections BVDU is far superior to IDU in both suppressing the development and promoting the healing of epithelial keratitis in rabbits, whether the drugs are applied as eyedrops or ointments (15,16). Similarly, when administered to rabbit eyes as 0.1 % or 0.5 % eyedrops, BVDU exhibited a significantly greater inhibitory effect than 1 % TFT eyedrops on the development of HSV-1 stromal keratitis (17). As 0.5 % eyedrops, BVDU also promoted a significantly greater healing effect than 1 % TFT eyedrops on iritis and endothelitis produced by inoculation of the virus into the anterior chamber (18). These experiments suggest that BVDU must penetrate the cornea in therapeutically sufficient concentrations when applied topically to the rabbit eye. In fact, upon topical administration of [^{125}I]IVDU, a radiolabelled analogue of BVDU, drug concentrations found in the aqueous humor of normal rabbit eyes were 3- to 9-fold higher than those required to inhibit virus replication in cell cultures (19).

Experimental data also indicate that BVDU is effective when given orally in the treatment of HSV-1 keratitis and iritis in rabbits. Upon oral administration to rabbits at 10 mg/kg/day or 100 mg/ kg/day for 4 days, BVDU significantly reduced the severity of iritis and keratitis, produced by inoculation of the virus in the anterior chamber (18).

Our previous reports clearly indicated that BVDU may be a safe and efficacious drug for the treatment of ocular HSV-1 infections (20,21). In this report we describe our observations over a longer follow-up period for a larger group of patients.

DENDRITIC CORNEAL ULCERS

Seventy-six patients with dendritic keratitis were treated with 0.1 % BVDU eyedrops (Table 1). Patients were advised to instill one drop of the drug 9 times a day at 1-hour intervals during the day only. Of these 76 patients, 44 had received other antivirals, i.e. IDU, TFT, Vira-A or Zovirax, albeit unsuccessfully, for at least 10 days before BVDU treatment was started. Associated stromal keratitis was present in 56 patients at the start of BVDU therapy. Thirty-one patients had been treated with topical corticosteroids together with the antiviral drugs. Two patients were on oral corticosteroid treatment after a kidney transplantation. Topical corticosteroids were discontinued in all patients when BVDU treatment was initiated. However, topical corticosteroids had to be reinstituted in 14 patients, as stromal keratitis worsened

Table 1. Topical BVDU treatment in patients with dendritic ulcers

Treatment regimen: BVDU 0.1 % eyedrops 5-9 x per day, up to 3 weeks
Number of patients: 76
Follow-up period: 52 months (4-80 months)
Number of patients with clinical resistance to IDU, TFT, Vira-A and/or Zovirax: 44
Average duration of symptoms before BVDU therapy:
- <1 month : 56 patients
- 1 month-1 year: 18 patients
- >1 year : 2 patients

Average healing time upon BVDU therapy: 8.6 days
Recurrences before BVDU treatment: 34 patients (44.7 %)
Recurrences after BVDU treatment: 21 patients (27.6 %)

despite healing of the dendritic keratitis. These corticosteroid-dependent patients responded very well to combined BVDU and corticosteroid treatment.

Before BVDU eyedrops were prescribed, the duration of acute keratitis was one month or less (average : 10.2 days) in 56 patients, more than one month but less than one year (average : 2.9 months) in 18 patients, and two patients had an history of progressively worsening keratitis for more than 1 year (average : 1.5 years), without being free of symptoms at any time. All patients healed within an average time of 8.6 days upon BVDU therapy. Those patients that had not responded to IDU, TFT and/or Zovirax responded promptly to BVDU treatment. Similarly, in those patients in whom we had to reinstitute topical corticosteroids the stromal disease became quiescent rapidly after combined BVDU and corticosteroid therapy was initiated. However, 10 of these patients remained corticosteroid-dependent. In these patients corticosteroids could be gradually tapered off while under BVDU cover. Other complications observed in this group were bullous keratopathy in 2 patients, dry eye in 4 patients, lower canaliculitis in 2 patients, and local hypersensitivity to topical BVDU eyedrops in 1 patient.

All patients have been followed for an average time of 52 months (4 to 80 months). Dendritic keratitis recurred in 21 patients (27.6 %) after BVDU treatment, whereas 34 patients (44.7 %) had had recurrences before BVDU treatment. Patients with recurrences responded effectively to BVDU treatment.

GEOGRAPHIC CORNEAL ULCERS

Thirty-five patients who presented with geographic corneal ulcers and associated stromal keratitis were treated with 0.1 % BVDU eyedrops (Table 2). Keratic precipitates were present in 18 eyes. Twenty-six patients had been treated, albeit unsuccessfully, with other antiviral drugs, e.g. IDU, TFT, Vira-A and/or Zovirax, when they were switched to BVDU treatment. Twenty-one patients had also been treated with topical corticosteroids. Corticosteroid therapy was discontinued in all patients, but had to be reinstituted in 16 patients as their stromal inflammation worsened. The duration of symptoms was less than 1 month in 20 patients, and more than 1 month in 15 patients at the time BVDU treatment was started. The patients healed within an average time of 11.7 days upon BVDU therapy. However, 14 patients became corticosteroid-dependent, 3 patients developed bullous keratopathy, 2 patients developed aseptic epithelium defects, and dry eye was observed in 8 patients. Contact allergy to BVDU eyedrops was noted in 4 patients.

Table 2. Topical BVDU treatment in patients with geographic corneal ulcers

Treatment regimen: BVDU 0.1 % eyedrops 5-9 x per day, up to 7 weeks
Number of patients: 35
Follow-up period: 48.5 months (3.5-78 months)
Number of patients with clinical resistance to IDU, TFT, Ara-A and/or Zovirax: 26
Average duration of symptoms before BVDU therapy:
<1 month: 20 patients
>1 month: 15 patients
Average healing time upon BVDU therapy: 11.7 days
Recurrences before BVDU treatment: 25 patients (71.4 %)
Recurrences after BVDU treatment: 16 patients (45.7 %)

The patients have been followed for an average period of 48.5 months (3.5 to 78 months). During this period dendritic, geographic or stromal keratitis recurred in 16 patients (45.7 %), whereas 25 patients (71.4 %) had had recurrences before BVDU treatment.

STROMAL KERATITIS

Forty-two patients who presented with stromal keratitis, without any associated epithelial ulceration, were treated with 0.1 % BVDU eyedrops (Table 3). Thirty-two patients had been treated, albeit unsuccessfully, with IDU, TFT and/or Zovirax. Thirty-four patients had also been treated with topical corticosteroids. When BVDU therapy was installed, corticosteroid treatment was discontinued. Upon BVDU treatment, stromal keratitis healed in 13 patients in the absence of corticosteroid administration. In the other patients topical corticosteroid therapy had to be reinstituted, as the stromal disease either did not improve or worsened. Nineteen patients became corticosteroid-dependent. Five patients developed a dry eye, and 2 patients showed local hypersensitivity to BVDU eyedrops. In one patient severe iritis developed one week after the stromal disease had healed and BVDU treatment was stopped. The iritis was again treated successfully with topical BVDU eyedrops, combined with corticosteroids and mydriatics. The duration of symptoms before BVDU treatment was less than 1 month in 11 patients, between 1 month and 1 year in 25 patients, and more than one year in 6 patients. Under topical BVDU therapy stromal disease became quiescent in all eyes within an average time of 30 days.

During the follow-up period (average : 38.7 months) of 6.5 to 78 months, 18 patients (42.8 %) developed a recurrence of herpetic keratitis, while the

Table 3. Topical BVDU treatment in patients with stromal keratitis

Treatment regimen: BVDU 0.1 % eyedrops 5-9 x per day, up to 6 months
Number of patients: 42
Follow-up period: 38.7 months (6.5-78 months)
Number of patients with clinical resistance to IDU, TFT and/or Zovirax: 32
Average duration of symptoms before BVDU therapy:
<1 month : 11 patients
1 month-1 year: 25 patients
>1 year : 6 patients
Average healing time upon BVDU therapy: 30 days
Recurrences before BVDU treatment: 32 patients (76 %)
Recurrences after BVDU treatment: 18 patients (42.8 %)

number of patients in this group who had experienced recurrences before BVDU therapy was 32 (76 %).

OPHTHALMIC HERPES ZOSTER

Fifteen patients with ophthalmic herpes zoster involving both the skin and eye were treated with oral BVDU at 375 mg/kg/day for 5 days, combined with topical 0.1 % BVDU eyedrops. Skin lesions consisted of papules, vesicles, bullae with or without hemorrhage, necrotic bullae, and, occasionally, crust formation on the scalp, forehead, nose, cheek and the temporal and periorbital regions. Ocular lesions consisted of blepharitis, conjunctivitis, punctate keratitis or corneal ulcers, stromal edema or infiltration, aqueous flare and/or keratic precipitates, total ophthalmoplegia and ptosis.

All patients were hospitalized and BVDU 125 mg capsules were administered at 8-hour intervals (375 mg/day) for 5 days. Routine blood and urine tests as well as urea, creatinine, platelets, electrolytes and liver enzyme (SGPT, SGOT, γGT) evaluations were done before, during and after BVDU therapy.

Combined oral and topical BVDU treatment caused a prompt resolution of the skin lesions and ocular inflammation. However, topical corticosteroids had to be administered as corneal edema developed subsequently to endothelial damage. Further details on the response of the ophthalmic zoster patients to BVDU are presented in Table 4. Urine and blood tests did not reveal any abnormality that could be attributed to the drug.

Stromal keratitis and internal rectus palsy developed, apparently as the consequence of a recurrent VZV infection, in one patient 20 days after BVDU treatment was stopped. This recurrence completely resolved within one month upon topical administration of 0.5 % BVDU eyedrops and corticosteroids.

Table 4. Oral BVDU treatment of patients with ophthalmic zoster

Treatment regimen: oral BVDU at 375 mg/day for 5 days, combined with BVDU 0.1 % eyedrops
Number of patients: 15
Average duration of symptoms before BVDU therapy: 5.6 days
Average healing time upon BVDU therapy:
- cessation of new lesion formation: within 1-2 days
- crust formation of skin lesions: within 2-3 days
- complete healing of skin lesions: within 6-12 days
- resolution of keratitis: within 6-12 days
- resolution of conjunctivitis: within 8-20 days

Duration of neuralgia on BVDU therapy: variable (2 days-4 weeks)

DISCUSSION

Two years ago, we reported on the long-term follow-up of 125 patients with HSV keratitis who had been treated with 0.1 % BVDU topical eyedrops (22). This study has now been extended to a total of 153 patients : 76 patients have been followed for an average period of 52 months (dendritic ulcers), 35 patients for an average period of 48.5 months (geographic ulcers), and 42 patients for an average period of 38.7 months (stromal keratitis). Those patients who were lost to follow-up were excluded from this study. The prolonged follow-up of 153 patients confirms the efficacy and safety of topical 0.1 % BVDU eyedrops in the treatment of dendritic, geographic and stromal HSV keratitis. Without any exception all patients responded to topical BVDU treatment, including those patients who had failed to respond to other antivirals, i.e. IDU, TFT, Ara-A and Zovirax (Table 5).

To be effective in the topical treatment of stromal disease and iritis, the antiviral drug should be able to penetrate the cornea in a therapeutically sufficient concentration. TFT penetrates the diseased cornea (23), and is effective in the treatment of experimental keratouveitis. Adequate penetration of topically applied Zovirax, through intact cornea, has been reported (24). Yet, its efficacy in the treatment of stromal disease is a matter of conjecture (25-27). Our own clinial experience is in keeping with experimental data showing dat topically applied Zovirax has some beneficial effect on stromal disease, significantly inferior to that of TFT (28). BVDU shows excellent corneal penetration (19) and can therefore be envisaged for the topical treatment of stromal keratitis and iritis. However, treatment of stromal disease may require concomitant use of topical corticosteroids (29-31), because of the im-

Table 5. Analysis of data on patients clinically resistant to other antiviral drugs

Keratitis (No. of patients)	Antiviral agent used before BVDU treatment	No. of patients	Average healing time upon BVDU therapy (days)
Dendritic ulcers (44)	IDU^r	26	9.2
	TFT^r	18 (5 also IDU^r)	7.4
	$Vira\text{-}A^r$	1 (also IDU^r)	5
	$Zovirax^r$	11 (6 also TFT^r)	6.8
Geographic ulcers (26)	IDU^r	15	10.3
	TFT^r	13 (3 also IDU^r)	10.9
	$Ara\text{-}A^r$	3 (also IDU^r)	11.3
	$Zovirax^r$	7 (1 also TFT^r)	8.7
Stromal keratitis (32)	IDU^r	18	30
	TFT^r	9 (5 also IDU^r)	35.4
	$Zovirax^r$	12 (2 also TFT^r)	26.7

Note: superscript r means clinical resistance to the drug indicated.

mune reaction initiated by the HSV antigens. In the present study, 49 patients with dendritic keratitis had associated stromal disease and 31 of them were using topical corticosteroids along with other antivirals. Upon BVDU treatment we had to reinstall topical corticosteroids only in 14 patients and 10 of them became corticosteroid-dependent. Similarly, in the geographic keratitis group all 35 patients had some degree of stromal disease, and 21 of them also received topical corticosteroids before BVDU therapy was started. We had to reinstitute topical corticosteroids in only 16 patients. All other patients healed on BVDU eyedrops alone. In the stromal keratitis group, again 13 of a total of 42 patients healed upon BVDU treatment in the absence of corticosteroids, while 19 patients showed corticosteroid-dependence. Those patients in whom we had to reinstitute corticosteroids together with BVDU eyedrops had been treated previously with topical corticosteroids in combination with IDU, TFT, Vira-A and/or Zovirax, albeit without any beneficial effect. They healed upon substitution of BVDU treatment for either of the four other antiviral drugs. It is conceivable, therefore, that in these patients BVDU efficiently suppressed the virus replicative cycle in the corneal stroma and thereby prevented the release of virus-specific antigens that would otherwise have initiated or maintained the immunological inflammatory response (29-31).

Upon oral treatment of herpes zoster ophthalmicus with BVDU at 375 mg/day for 5 days, combined with topical 0.1 % BVDU eyedrops, progression of the disease was arrested promptly. New lesions ceased to form within 1-2 days and existing (skin) lesions subsided completely within 6-12 days. An apparent recurrence of VZV eye infection in one patient was treated successfully with 0.5 % BVDU eyedrops combined with topical corticosteroids. Our results on oral BVDU treatment of ophthalmic zoster patients are in agreement with those obtained with BVDU in cancer patients with varicella zoster (32-34), and point to the great potential of BVDU in the systemic treatment of VZV infections.

Complications such as bullous keratopathy, aseptic epithelium defect and dry eye, as noted in some HSV keratitis patients treated with BVDU, are known complications of the disease process itself. Except for local hypersensitivity reaction noted in 7 patients, no toxic side effects of BVDU eyedrops were observed. It is not clear whether this hypersensitivity is due to the drug or other substances present in the eyedrops. In the ophthalmic herpes zoster patients treated with oral BVDU combined with topical BVDU no systemic or local side effects, which could be detected either clinically or through laboratory tests, were observed.

REFERENCES

1. De Clercq, E. and Walker, R.T. Pharmac. Ther. 26 : 1-44, 1984.
2. De Clercq, E., Descamps, J., De Somer, P., Barr, P.J., Jones, A.S. and Walker, R.T. Proc. Natl. Acad. Sci. USA, 76 : 2947-2951, 1979.
3. De Clercq, E., Descamps, J., Maudgal, P.C., Missotten, L., Leyten, R., Verhelst, G., Jones, A.S., Walker, R.T., Busson, R., Vanderhaeghe, H. and De Somer, P. In : Developments in Antiviral Therapy (Eds L.H. Collier and J. Oxford), Academic Press, London, UK, 1980, pp 21-42.
4. De Clercq, E., Descamps, J., Verhelst, G., Walker, R.T., Jones, A.S., Torrence, P.F. and Shugar, D. J. Infect. Dis. 141 : 563-574, 1980.
5. De Clercq, E., Descamps, J., Ogata, M. and Shigeta, S. Antimicrob. Agents Chemother. 21 : 33-38, 1982.
6. Shigeta, S., Yokota, T., Iwabuchi, T., Baba, M., Konno, K., Ogata, M. and De Clercq, E. J. Infect. Dis. 147 : 576-584, 1983.
7. Soike, K.F., Gibson, S. and Gerone, P.J. Antiviral Res. 1 : 325-337, 1981.
8. De Clercq, E. J. Antimicrob. Chemother. 14 : (Suppl. A), 85-95, 1984.
9. Cheng, Y.-C., Dutschman, G., De Clercq, E., Jones, A.S., Rahim, S.G., Verhelst, G. and Walker, R.T. Mol. Pharmacol. 20 : 230-233, 1981.
10. Descamps, J. and De Clercq, E. J. Biol. Chem. 256 : 5973-5976, 1981.
11. Allaudeen, H.S., Kozarich, J.W., Bertino, J.R. and De Clercq, E. Proc. Natl. Acad. Sci. USA 78 : 2698-2702, 1981.
12. Allaudeen, H.S., Chen, M.S., Lee, J.J., De Clercq, E. and Prusoff, W.H. J. Biol. Chem. 257 : 603-606, 1982.

13. Mancini, W.R., De Clercq, E. and Prusoff, W.H. J. Biol. Chem. 258 : 792-795, 1983.
14. Fyfe, J.A. Mol. Pharmacol. 21 : 432-437, 1982.
15. Maudgal, P.C., De Clercq, E., Descamps, J., Missotten, L., De Somer, P., Busson, R., Vanderhaeghe, H., Verhelst, G., Walker, R.T. and Jones, A.S. (E)-5-(2-Bromovinyl)-2'-deoxyuridine in the treatment of experimental herpes simplex keratitis. Antimicrob. Agents Chemother. 17 : 8-12, 1980.
16. Maudgal, P.C., De Clercq, E., Descamps, J. and Missotten L. Bull. Soc. Belge Ophtalmol. 186 : 109-118, 1979.
17. Maudgal, P.C., De Clercq, E., Descamps, J., Missotten, L. and Wijnhoven, J. Arch. Ophthalmol. 100 : 653-656, 1982.
18. Maudgal, P.C., Uyttebroeck, W., De Clercq, E. and Missotten, L. Arch. Ophthalmol. 100 : 1337-1340, 1982.
19. Maudgal, P.C., Verbruggen, A.M., De Clercq, E., Busson, R., Bernaerts, R., de Roo, M., Ameye, C. and Missotten, L. Invest. Ophthalmol. Vis. Sci. 26 : 45-49, 1985.
20. Maudgal, P.C., Missotten, L., De Clercq, E., Descamps, J. and De Meuter, E. Albrecht von Graefes Arch. Klin. Ophthalmol. 216 : 261-268, 1981.
21. Maudgal, P.C., De Clercq, E., Descamps, J. and Missotten, L. In: Herpetische Augenerkrankungen (Ed. R. Sundmacher), J.F. Bergmann Verlag, München, FRG, 1981, pp. 339-341.
22. Maudgal, P.C., Dieltiens, M., De Clercq, E. and Missotten, L. Doc. Ophthalmol. Proc. Series 44 : 247-256, 1985.
23. Pavan-Langston, D. and Nelson, D.J. Am. J. Ophthalmol. 87 : 814-818, 1979.
24. Poirier, R.H., Kingham, J.D., de Miranda, P. and Annel, M. Arch. Ophthalmol. 100 : 1964-1967, 1982.
25. Sanitato, J.J., Asbel, P.A., Varnell, E.D., Kissling, G.E. and Kaufman, H.E. (1984) Acyclovir in the treatment of herpetic stromal disease. Am. J. Ophthalmol. 98 : 537-547.
26. Collum, L.M.T., MacGerrtick, P., Akhtar, J. and Rees, P.J. Doc. Ophthalmol. Proc. Series 44 : 233-240, 1985.
27. McGill, J.I. Doc. Ophthalmol. Proc. Series 44 : 217-225, 1985.
28. Maudgal, P.C., Vrijghem, J.C., Molemans, M. and Missotten, L. Arch. Ophthalmol. 103 : 1389-1392, 1985.
29. Metcalf, J.F. and Kaufman, H.E. Am. J. Ophthalmol. 82 : 827-834, 1976.
30. Meyers, R.L., Chitjian, P.A. and Fiorello, P. In : Proceedings of the 23rd International Congress of Ophthalmology (Eds. K. Shimizu and J.A. Oosterhuis), Excerpta Medica, Amsterdam, The Netherlands, 1979, pp. 1739-1743.
31. Meyers-Elliott, R.H., Pettit, T.H. and Maxwell, W.A. Arch. Ophthalmol. 98 : 897-904, 1980.
32. Wildiers, J. and De Clercq, E. Eur. J. Cancer Clin. Oncol. 20 : 471-476, 1984.
33. Benoit, Y., Laureys, G., Delbeke, M.-J. and De Clercq, E. Eur. J. Pediatr. 143 : 198-202, 1985.
34. Tricot, G., De Clercq, E., Boogaerts, M.A. and Verwilghen, R.L. J. Med. Virol. 18 : 11-20, 1986.

5

DIAGNOSIS AND TREATMENT OF HERPES SIMPLEX ENCEPHALITIS

Richard J. Whitley, M.D.

Department of Pediatrics and Microbiology,
The University of Alabama at Birmingham,
Birmingham, Alabama 35294.

BACKGROUND

Herpes simplex encephalitis, HSE, remains the most common cause of sporadic fatal encephalitis in the Western World.(1-3) Since its recognition in 1941 as a cause of rapidly progressive and diffuse encephalitis in a four week old child which was fatal (4), hundreds of case reports have described various aspects of this disease in the literature. Likely, this first case report represented a child with neonatal herpes simplex virus, HSV, infection of the brain, an entity somewhat different than that which we designate as HSE of older children and adults where the infection is characterized by the focal hemorrhagic necrosis of the brain. Clinical presentation, diagnosis and outcome of patients with HSE had remained confusing until studies performed by the National Institute of Allergy and Infectious Diseases, NIAID, Collaborative Antiviral Study Group. Unequivocal diagnosis often has created both practical dilemmas (3,5,6) and intellectual controversies (7-9). The clarification of many of these issues has become possible because of a uniform diagnostic approach, namely, brain biopsy. This procedure is not routinely employed in therapeutic, natural history or diagnostic investigations performed outside the United States for the prospective evaluation of patients with focal encephalitis. Through the NIAID studies, clinical presentation, the value of brain biopsy for diagnostic purposes, as well as establishing alternative diagnoses which mimic herpes encephalitis, and the definition of effectiveness of treatment and associated factors which influence outcome have all been identified.

Currently, HSE is estimated to occur in approximately one in 250,000 to one in 500,000 individuals per year. At the University of Alabama at Birmingham School of Medicine, a Medical Center which accepts statewide referrals of patients with suspected HSE, we have diagnosed an average of 10 patients per year over the past 10 years by brain biopsy. Alabama has a population of 3.3 million; thus, the incidence of disease is approximately one in 250,000 individuals. The severity of disease, namely its natural history, is best determined by the outcome of patients who have received no therapy or an ineffective antiviral medication such as idoxuridine or cytosine arabinoside. In such situations, mortality is in excess of 70% and only approximately 2.5% of all patients with confirmed disease return to normal function following recovery from their illness (10-14). These data reinforce the severity of HSV infections of the central nervous system. Since brain biopsy with the isolation of HSV from brain tissue was the method of diagnosis in these studies, it is likely that a far broader spectrum of HSV infections of the central nervous system actually exists. In fact, one British study has suggested milder forms of HSE which are associated with lower mortality and improved morbidity (15). Serologic evaluations were the basis for the diagnosis in this latter report, a problem which will be discussed below.

PATHOGENESIS

The pathogenesis of HSE remains unclear. It is apparent that both primary and recurrent HSV infections cause disease of the central nervous system. From studies performed by the NIAID Collaborative Antiviral Study Group, it was learned that approximately one-third of the cases of HSE are the consequence of primary infection. The remaining cases occur in the presence of pre-existing antibodies, often with a history of recurrent herpes simplex labialis and, therefore, are considered to be the consequence of reactivation of HSV (16). The route of access of virus to the central nervous system in primary infection is a subject of debate, especially in humans. While studies performed over two decades ago defined pathways for HSV access to the brain in animals, including both the olfactory and trigeminal nerves among others (17), it is not clear which of these nerve tracts uniformly leads to HSV infection of the central nervous system. The anatomic distribution of nerves from the olfactory tract along the inferiomedial portion of the temporal lobe, the site of onset of HSE in the human brain, suggests that viral access to the central nervous system <u>via</u> this route is a tenable hypothesis. Suggestions in the literature have indicated that,in fact, this has been the case in some individuals with herpes encephalitis (18-21). Similarly, animal model data support the hypothesis that the olfactory tract provides one neurologic avenue for virus to reach the central nervous system and results in localization of the infection in the orbitofrontal region of the brain (22,23). For these studies, virus was either inoculated into the olfactory bulb of the rabbit or administered intranasally at a high concentration. Definitive proof for such progression in humans is lacking. Reactivation of HSV,leading to focal HSE, is similarly confusing from the standpoint of pathogenesis. While it is possible to demonstrate evidence of latent virus infection within infected brain tissue (24), the likelihood of reactivation at that site remains purely hypothetical. Reactivation of virus peripherally, namely in the olfactory bulb or the trigeminal ganglion, has been suggested with subsequent neuronal transmission to the central nervous system (17,23,25,26). Nevertheless, a relevant observation is that with recurrent herpes simplex labialis, HSE is not a common event. Specifically, there are far more cases of recurrent herpes labialis than there are cases of HSE. Furthermore, when patients with biopsy-proven HSE are compared with patients having other diseases which mimic HSE, there is a near equal frequency of fever blisters (6%) and retrieval of virus from the oropharynx (11%) between groups (16). These questions will remain viable subjects for future investigations, particularly with the recent development of appropriate animal models.

Patients with HSE can shed virus from a peripheral site such as the oro- or nasopharynx as noted above (16). Associated with primary infection, HSV accesses the central nervous system in a few patients; the resulting peripheral and central nervous system isolates are identical by DNA restriction endonuclease analysis. However, with disease in a human having pre-existing antibodies, namely recurrent infection, the virus isolated from the peripheral site often is different from that retrieved from the CNS, utilizing the same techniques (27). Thus, the issue of reactivation of virus directly within the CNS, the potential for enhanced neurotropism of certain viruses or, furthermore, the selective reactivation and access of one virus by the trigeminal route or other routes to the CNS remain for further elucidation.

DIAGNOSIS

Initial reports of HSE focused on clinical presentation, pathologic findings at necropsy, and outcome of patients with disease (1-3,5,28-33). Over the past two decades, intense efforts have focused on therapy. In many of the reported studies, diagnosis was not established by a uniform procedure; namely, some patients were diagnosed by isolation of virus from brain biopsy specimens while others were considered to have a diagnosis of HSE predicated upon serologic response in either serum or cerebrospinal fluid or a ratio of serum to cerebrospinal fluid antibodies. As a consequence, patients with other illnesses not caused by HSV likely were included in the groups of patients thought to have "proven" disease. Many of these other diseases have a more benign outcome (7,11,12).

Patients with HSE present at all ages, although the age distribution of cases is essentially biphasic. Approximately 30% of cases occur between the ages of six months and 20 years; 50% of cases occur in individuals over 50 years of age. Caucasians are predominantly afflicted. A comparison of biopsy-proven and biopsy-negative patients is useful. Most patients present with a focal encephalopathic process: focal neurologic findings associated with cerebrospinal fluid pleocytosis and proteinosis, negative evaluations for appropriate bacterial and fungal pathogens, and focal encephalographic, computed tomographic, or technetium brain scan findings. However, only a higher frequency of headache and cerebrospinal fluid pleocytosis occurs in patients with proven HSE; a higher frequency of ataxia occurs in those individuals who have diseases which mimic HSV infection of the central nervous system. Nearly uniformly, patients with HSE present with fever and personality change. Seizures, whether focal or generalized, occur in only approximately two-thirds of all patients with proven disease. Thus, the clinical findings of HSE are non-specific and do not allow for empiric diagnosis of disease predicated solely on clinical presentation.

Non-invasive neurodiagnostic studies have been utilized to support a presumptive diagnosis of HSE. These studies have included the electroencephalogram, and computed tomographic and technetium brain scans. More recently, magnetic resonance imaging has been utilized for diagnostic purposes, although the value of such scans remains to be established in carefully documented and controlled clinical situations. Focality of the electroencephalogram appears the most sensitive of the non-invasive neurodiagnostic procedures (34-38). Characteristic findings on the electroencephalogram include spike and slow wave activity labelled as "pleds" which arise from the temporal lobe. Early in the disease course, electrical activity usually involves one temporal lobe and then spreads to the contralateral temporal lobe as the disease evolves. Sensitivity of the electroencephalogram has been defined as approximately 84%; but, unfortunately a specificity of 32.5% has been demonstrated in the NIAID Collaborative Antiviral Study Group trials. Brain and computed tomographic scans have sensitivities of 60% and 70% but specificities of only 65% and 58%, respectively. The computed tomographic scan initially shows evidence of edema localized to the temporal lobe which will progress to a radiolucent lesion indicative of hemorrhagic necrosis associated with a mid-line shift of the structures of the central nervous system, even including obliteration of lateral ventricals (39,40). Bitemporal disease is common in the absence of therapy, particularly late in the disease course. When these assays are used in combination, the sensitivity can be enhanced; however, the specificity of these diagnostic assays is diminished. At the present time, none of these neurodiagnostic tests is uniformly satisfactory for diagnosing HSE.

The most sensitive and specific means of diagnosis, at least at the present time, remains the isolation of HSV from tissue obtained by brain biopsy. While this diagnostic approach is considered controversial, brain biopsy has not been associated with undue complications either acute or chronic in nature. The frequency of acute complications secondary to brain biopsy is approximately 3% with the most common being poorly controlled cerebral edema at the time of brain biopsy or hemorrhage because of poor visualization of the tissue. While long-term complications have been thought to include seizure disorders, as a consequences of the brain-biopsy, these have been uncommon in the experience of the NIAID Collaborative Antiviral Study Group (41). It is important to recognize that temporal lobe necrosis, as encountered with HSE, is an irreversible process; thus, the tissue must be considered non-viable and not amenable to recovery.

Alternative non-invasive diagnostic procedures are in varying stages of development. A new and promising assay is the demonstration of glycoproteins gB, gD, and gE antigens in cerebrospinal fluid of patients with biopsy-proven disease (42). After the onset of clinical findings of HSE, cerebrospinal fluid has evidence of HSV antigens as early as 5 days after disease onset in 65% to 75% of specimens tested. The assay is nearly 100% specific and increases in sensitivity as the disease duration progresses. It is conceivable that the availability of an antigen detection assay for cerebrospinal fluid will replace the uniform diagnosis of HSE by brain biopsy. It should be remembered, however, that other diseases which mimic HSE can be diagnosed by brain biopsy. These diseases are of significance and often require alternative forms of medical management: cryptococcal infection of the central nervous system, brain abscess, tumor, toxoplasmosis, tuberculosis, and arteriovenous malformations, among others (12). Thus, the future deployment of non-invasive diagnostic procedures must take into consideration alternative diseases which mimic herpes simplex infection of the central nervous system.

THERAPY

The first antiviral evaluated and the first one for which therapeutic usefulness was suggested was idoxuridine, a compound studied in the late 1960's and early 1970's (29,30,33,43-45). In 1972, the NIAID Collaborative Antiviral Study Group, in cooperation with the Boston Interhospital Viral Study Group, demonstrated that idoxuridine was both ineffective and toxic for patients with HSE (10). Toxicity manifested as bone marrow suppression and secondary bacterial infection when patients received purportedly effective dosages of medication; thus, the therapeutic index (ratio of efficacy to toxicity) was unacceptable. Thus, the further utilization of idoxuridine was negated by an unsatisfactory therapeutic index. In the conduct of these studies, it was clear that both uniform case record forms and approaches to management of patients with HSE might lead to clarification of issues which relate to both earlier diagnoses and improved therapy.

Subsequent therapeutic trials defined vidarabine as a useful medication for the management of biopsy-proven HSE (11,12). In the first of a series of controlled studies of HSE, which utilized a double-blind, placebo-controlled study design, vidarabine therapy decreased mortality from 70% to 28% one month after disease onset and to 44% six months later for patients with biopsy proven disease (11). This report was predicated on a study of 28 patients, terminated for ethical reasons, and followed by an open and uncontrolled trial to verify mortality and define long-term morbidity (12). The follow-up study of nearly 100 patients with proven disease defined long-term mortality as 40%. Of importance, variables such as age and level of consciousness, were proven to be major detriments of clinical outcome. Patients less than 30 years of age and with a more normal level of consciousness (lethargic as opposed to comatose) were more likely to return to normal function after HSE than older patients, especially those who were semi-comatose or comatose as displayed in Figure 1.

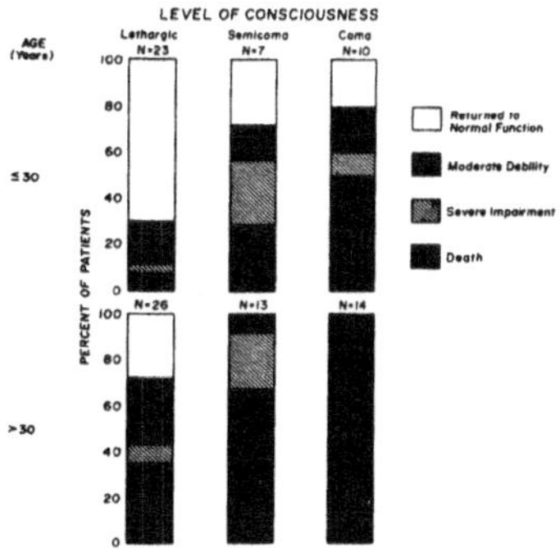

FIGURE 1.
(Reprinted with permission from reference 12).

From these data, it is apparent that older patients (greater than 30 years of age) whether comatose or semi-comatose had mortality rates which approached 70% -- a figure very similar to that encountered in the placebo-studies previously cited. In patients less that 30 years of age, a more acceptable outcome was achieved, as evidenced by a mortality of 25% and 40% returning to normal. Clearly, an important lesson learned from these trials was that if therapy was to be effective, it must be instituted prior to the onset of hemorrhagic necrosis of a dominant temporal lobe as associated with deterioration in the patient's level of consciousness.

More recently, the NIAID Collaborative Antiviral Study Group has demonstrated that acyclovir is superior to vidarabine for the treatment of HSE. Acyclovir, its design and mechanism of action, are elsewhere discussed in this monograph. Criteria for enrollment into this study was isolation of HSV from brain-biopsy tissue, a criteria somewhat different from that diagnostic criteria of a similar study performed in Sweden which also compared these two medications. The NIAID study demonstrated that acyclovir decreased mortality to 19% six months after therapy (46). Importantly, 38% of patients irrespective of age returned to normal function. Scandinavian investigators, led by Dr. Birgit Skoldenberg, defined similar outcome but in a smaller group of patients whose diagnoses were established by a variety of methods (47). Both studies taken together indicate that acyclovir is superior to vidarabine for the treatment of HSE.

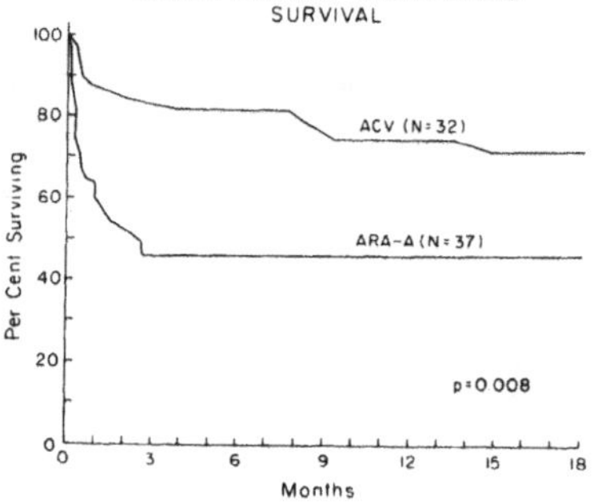

FIGURE 2.
(Reprinted with permission from reference 46).

As displayed in Figure 2, the data from the NIAID Collaborative Antiviral Study Group indicate a mortality of 50% for 37 patients who received vidarabine compared to 19% six months after the onset of treatment and 24% 18 months after the onset of treatment for the acyclovir group. Late deaths in this study were not a consequence of either persistent or reactivated herpes simplex infection of the central nervous system but occurred in patients who were severely impaired as a consequence of their disease.

It should be noted that the mortality following vidarabine therapy in this study was greater than that encountered in the original trials. The reason for enhanced mortality in the vidarabine treated group was because it consisted of older patients who had a a lower level of consciousness--both known to be associated with higher mortality. When patient populations were compared according to specific age and level of consciousness, differences in therapeutic outcome remain significant for long-term mortality when utilizing a two-tail test (P=0.04).

Previous studies indicated that age and level of consciousness influenced long-term outcome. The most recent study defined the Glasgow coma score and disease duration as additional indicators of long-term survival. As shown in Figure 3, a more objective reflection of Glasgow coma score, the level of consciousness which scores patients according to motor, verbal and sensory responses, indicated that scores approaching normal (a score greater than 10) predicted more likely survival, as well as return to normal function, with acyclovir therapy.

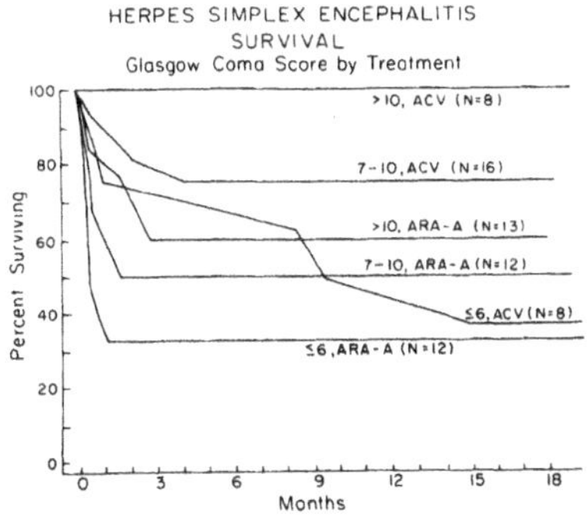

FIGURE 3.

Glasgow coma scores indicative of coma, namely those less than or equal to 6, resulted in a higher mortality with either vidarabine or acyclovir treatment; thus, with lower Glasgow coma scores, survival is no different than in the absence of any form of therapy.

A particularly striking finding in the current trial was the influence of disease duration on survival for acyclovir and vidarabine treated patients. As shown in Figure 4, disease duration less than four days, considering time of onset of fever, disorientation, a seizure or any abnormal clinical symptom as the onset of disease, was associated with the best of all possible outcomes -- a mortality of 12% at 18 months. If disease duration was greater than 4 days, mortality increased from 12% to nearly 35%.

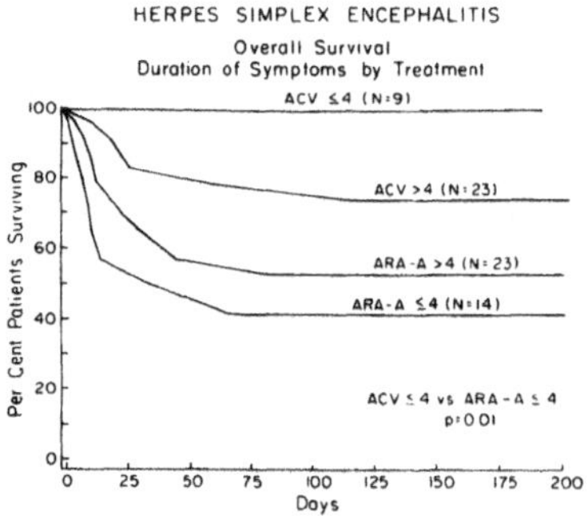

FIGURE 4.

These differences were statistically significant for the acyclovir recipients (P=0.01). Surprisingly, disease duration did not influence outcome when vidarabine was administered.

Long-term morbidity following administration of an antiviral is of particular importance. Historically, the vidarabine therapeutic studies indicated that approximately 15% to 20% of patients overall would develop normally following therapy of HSE on long-term follow-up. The current trial indicated that 13% of vidarabine recipients were left with no or minor sequelae while those with moderate or severe sequelae and dead on follow-up were 22% and 65%, respectively. For acyclovir recipients, 38% of patients were normal or with minor impairment, 9% had moderate sequelae and 53% were left with either severe impairment or dead.

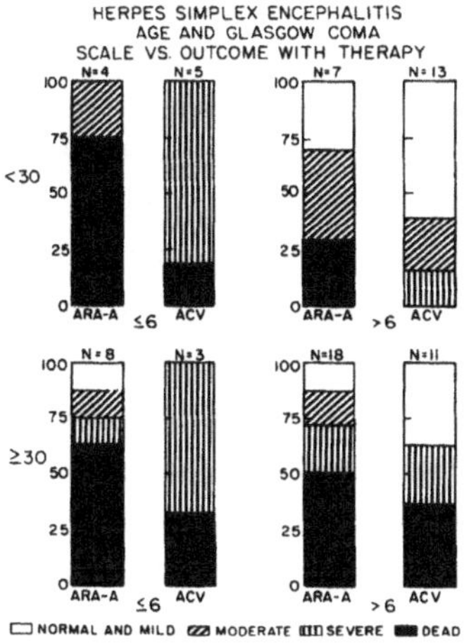

FIGURE 5.

(Reprinted with permission from reference 46).

When Glasgow coma score and age were assessed simultaneously, as displayed in Figure 5, a Glasgow coma score less than or equal to 6 led to a poor therapeutic outcome irrespective of the agent which was administered.

Even in this sub-population of patients, the administration of acyclovir was without significant value. Conversely, for a Glasgow coma score in excess of 6, the most positive therapeutic benefit was achieved in patients less than 30 years of age. Eleven of 13 acyclovir recipients were either normal or left with moderate impairment as compared to 5 or 7 counterpart vidarabine recipients. With advancing age, the number of acyclovir recipients left with no or moderate impairment decreased to 4 of 11 as compared to 5 of 18 counterpart vidarabine recipients. These findings, then directly parallel data derived from the open treatment trial utilizing vidarabine, whereby age and level of consciousness directly influenced long-term outcome. Currently, an open study of acyclovir is underway for biopsy proven disease in order to verify long-term morbidity and mortality.

No patient entered into the current trial suffered a relapse after completion of therapy. Nevertheless, when causes of fever other than HSE were excluded, such as bacterial pneumonia, urinary tract infection, etc., the median duration of an afebrile state was only 3.1 days at the completion of 10 days of treatment. Thus, a longer afebrile state would be considered desirable, extending therapy to a minimum of 14 days. It should be recognized that relapse of HSE has been documented in a few patients following the administration of vidarabine (40,49). Demyelination syndromes have also been identified (50).

In defining the therapeutic index of a compound for the management of HSE, the denominator, or toxicity, becomes an important component of the equation. As shown in Table 1 the vidarabine recipients more likely developed laboratory abnormalities during the course of treatment than the counterpart acyclovir recipients (50% versus 25%, P=0.04).

TABLE 1.

Laboratory Index*	Treatment Group	
	Vidarabine	Acyclovir
	no. of patients (%)	
Platelets (<100,000 cells/mm³)	4 (11)	2 (6)
SGOT (>250 IU/dl)	5 (14)	1 (3)
BUN (>50 mg/dl)	4 (11)	3 (10)
White cells (<2500 cells/mm³)	2 (6)	0 (0)
Total bilirubin (>3 mg/dl)	1 (3)	0 (0)
Creatinine (>3 mg/dl)	0 (0)	2 (6)
Combinations		
SGOT + bilirubin	1 (3)	0 (0)
White cells + platelets	0 (0)	0 (0)
BUN, platelets, + SGOT	1 (3)	0 (0)
Total	18 (49)	8 (25)

*SGOT denotes serum aspartate aminotransferase, and BUN blood urea nitrogen. To convert values for BUN to millimoles per liter, multiply by 0.357; to convert values for bilirubin and creatinine to micromoles per liter, multiply by 17.10 and 88.40, respectively.

(Reprinted with permission from reference 46).

The most significant laboratory abnormalities encountered among vidarabine recipients was a platelet count less than 100,000 (11%), an elevated serum glutamic-oxaloacetic transaminase greater than 250 IU (14%), and an elevated blood urea-nitrogen in excess of 30mg/dl (11%). In contrast, 10% of acyclovir recipients experienced an elevated BUN and 6% developed a creatinine in excess of 2mg/dl. It should be emphasized that the administration of both drugs was not associated with clinical evidence of toxicity and that these findings simply represented laboratory aberations encountered during the course of management of these patients.

These findings taken together indicate the therapeutic of choice for the management for HSE is acyclovir as compared to vidarabine. While not licensed in the United States for the treatment of HSE, a New Drug Application is currently under review by the Food and Drug Administration for utilization of this compound at a dosage of 10mg/kg every 8 hours (total 30 mg/kg/day) for a period of 10 to 14 days. It is conceivable that in certain circumstances a longer period of therapy may be indicated. Clinical response and duration of fever should guide the physician in the deployment of these therapies.

FUTURE THERAPEUTIC DIRECTIONS

The current data indicate that acyclovir is the treatment of choice for biopsy-proven HSE, resulting in significantly improved both morbidity and mortality; however, even in patients with an acceptable Glasgow coma score, mortality and morbidity remain problematic. Thus, alternative therapeutic approaches will need to be developed. The development of sensitive and specific non-invasive diagnostic procedures may contribute to improved outcome by avoiding delays in the onset of therapy while awaiting a biopsy. However, the value of such procedures and their standardization remain to be established.

One theoretical approach to the future therapy of HSE is the utilization of combination chemotherapy as has occurred for the management of malignancy and certain viral infections (51,52). Such approaches have been developed in order to decrease therapeutic failures, minimize potential for resistance to a therapeutic, and potentially decrease dosages of medication to avert toxic effects. The application of combination therapy to the treatment of viral infections has been studied in vitro as well as in animal model systems for several years(53-59). In tissue culture experiments, acyclovir and vidarabine usually have an additive effect for reduction of replication of both HSV-1 and HSV-2 in Vero cells. In studies of this nature, an antagonistic effect has not been demonstrated for these two antiviral compounds (53).

Animal model data indicate that combination chemotherapy, utilizing acyclovir and vidarabine, have at least an additive and, perhaps, a synergistic effect for decreasing mortality, even when therapy is initiated late after inoculation. Similar animal model data have previously predicted the value of both acyclovir and vidarabine alone for treatment of HSE and support the potential utility of combination chemotherapy (58). Studies performed by Dr. Raymond Schinazi indicate synergy between acyclovir and vidarabine at dosages of 150 mg/kg of both medications with delayed therapy. For these studies, combination chemotherapy was initiated 72 hours after intracerebral inoculation with HSV-2 and resulted in mortalities ranging from 20% to 50% as dosages were varied(59). However, when fluoromethylarabinosyl uracil was used alone or in combination with acyclovir or vidarabine, mortality was decreased to 0 to 10%. Because of the existing toxicity profile of this compound, it is unlikely that it will be deployed for therapeutic purposes.

Combination chemotherapy may well have potential for decreasing the development of viral resistance. While resistant viral mutants can be generated easily in tissue culture systems, the appearance of such strains has not been a major problem in experimentally infected animals or humans at the present time (60-63). Combination chemotherapy will be the next area of testing of antiviral therapy.

New compounds with increased activity or increased lipophilicity thereby allowing penetration of drug into the central nervous system, are not currently identified at this time. Thus, future therapeutic efforts, at least initially, will use limited numbers of patients to assess the value of combination chemotherapy.

CONCLUSION

The value of antiviral therapy has been established unequivocally in the management of HSE. Clearly treatment of individuals with this life-threatening and debilitating disease has led to improved outcome and improved quality of life. Nevertheless, the significant mortality and morbidity even in treated patients indicate that improvements in the therapeutic regimens for management of this disease are mandatory. Hopefully, the development of non-invasive diagnostic procedures, or even, the development of vaccines to prevent HSV infection of CNS tissue may be of value in the management and prevention of this disease.

CREDITS

Work performed by the author and cited in this review was supported by a Contract from the NIAID Collaborative Antiviral Study Group (NO1-AI-62554), a grant from the National Cancer Institute (CA-13148) and from DRR (RR-032) and the State of Alabama.

REFERENCES

1. Leider, W., Magoffin, R.L., Lennette, E.H. and Leonards, L.N.R.: N. Engl. J. Med. 273:341-347, 1965.
2. Olson, L.C., Buescher, E.L., Artenstein, M.S. and Parkman, P.D.: N. Engl. J. Med. 277:1271-1277, 1967.
3. Meyer, M.H., Jr., Johnson, R.T., Crawford, I.P., Dascomb, H.E. and Rogers, N.G.: Am. J. Med. 29:334-347, 1960.
4. Smith, M.G., Lennette, E.H. and Reames, H.R.: Am. J. Pathol. 17:55-68, 1941.
5. Rappel, M., Dubois-Dalcq, M., Sprecher, S., Thiry, L., Lowenthal, A., Pelc, S. and Thys, J.P.: J. Neurol. Sci. 12:443-458, 1971.
6. Johnson, R.T., Olson, L.C. and Buescher, E.L.: Arch. Neurol. 18:260-264, 1968.
7. Whitley, R.J. and Alford, C.A., Jr.: JAMA 248:547, 1982.
8. Braun, P.: Am. J. Med. 69:893-902, 1980.
9. Barza, M., Pauker, S.G.: Ann. Intern. Med. 92:641-649, 1980.
10. Boston Interhospital Virus Study Group and the NIAID Sponsored Cooperative Antiviral Clinical Study (Alford, Chien, Whitley, et al.): N. Engl. Med. 292:600-603, 1975.
11. Whitley, R.J., Soong, S-J, Dolin, R., Galasso, G.J., Chien, L.T., Alford, C.A., Jr. and the Collaborative Antiviral Study Group: N. Engl. J. Med. 297:289-294, 1977.
12. Whitley, R.J., Soong, S-J, Hirsch, M.S., Karchmer, A.W., Dolin, R., Galasso, G., Dunnick, J.K., Alford, C.A., Jr. and the NIAID Collaborative Antiviral Study Group: N. Engl. J. Med. 304:313-318, 1981.
13. Longson, M.: Ann. Microbiol. (Paris) 130:5, 1979.
14. Longson, M.M., Bailey, A.S. and Klapper, P.: In: Recent Advances in Clinical Virology. Vol. 2, A.T. Waterson (ed), Churchill Livingstone, Philadelphia, pp. 147-157, 1980.
15. Klapper, P.E., Cleator, G.M. and Longson, M.: J. Neurolog. Neurosurg. Psych. 47:1247-1250, 1984.
16. Nahmias, A.J., Whitley, R.J., Visintine, A.N., Takei, Y., Alford, C.A., Jr. and the NIAID Collaborative Antiviral Study Group: J. Infect. Dis. 145:829-836, 1982.
17. Johnson, R.T.: J. Exp. Med. 119:343-356, 1984.
18. Dinn, J.: Brit. Med. J. 281:1392-1392, 1980.
19. Whitley, R.J.: In: Human Herpesvirus Infections: Pathogenesis, Diagnosis, and Treatment, C. Lopez and B. Roizman (eds), Raven Press, New York, NY, pp. 153-164, 1986.
20. Ojeda, V.J., Archer, M., Robertson, T.A. and Bucens, M.R.: Med. J. Aust. 1:79-81, 1983.
21. Twomey, J.A., Barker, C.M., Robinson, G. and Howell, D.A.: J. Neurol. Neurosurg. Psych. 42:983-987, 1979.
22. Schlitt, M., Lakeman, F.D., Wilson, E.R., To, A., Acoff, R., Harsh, G.R. and Whitley, R.J.: J. Infect. Dis. 153:732-735, 1986.
23. Stroop, W.G. and Schaefer, D.C.: J. Infect. Dis. 153:721-731, 1986.
24. Rock, D.L. and Frasher, N.W.: Nature 302:523-531, 1983.
25. Davis, L.E. and Johnson, R.T.: Ann. Neurol. 5:2-5, 1979.
26. Griffith, J.R., Kibrick, S. Dodge, P.R. and Richardson, E.P.: Electroenceph. Clin. Neurophysiol. 23:263-267, 1967.

27. Whitley, R.J., Lakeman, A.D. Nahmias, A. and Roizman, B.: N. Engl. J. Med. 307:1060-1062, 1982.
28. Johnson, K.P., Rosenthal, M.S. and Lerner, P.I.: Arch. Neurol. 27:103-108, 1972.
29. Sarubbi, F.A., Jr., Sparling, P.F. and Glezen, W.P.: Arch. Neurol. 29:268-273, 1973.
30. Nolan, D.C., Carruthers, M.M. and Lerner, A.M.: N. Engl. J. Med. 282:10-13, 1970.
31. Haymaker, W., Smith, M.G., van Bogaert, L. and de Chenar, C.: In: Viral Encephalitis W.S. Fields, J.R. Blattner (eds), Charles C. Thomas, Springfield, IL, pp.118-162, 1958.
32. Esiri, M.M.: J. Neurol. Sci. 54:209-226, 1982.
33. Rappel, M., Dubois-Dalcq, M., Sprecher, S., Thiry, L., Lowenthal, A., Pelc, S., and Thys, J.P.: J. Neurol. Sci. 12:443-458, 1971.
34. Radermecker, J.: Electroencephalogr. Clin. Neurophysiol. 5(Suppl):239, 1956.
35. Upton, A. and Grumpert, J.: Lancet 1:650-652, 1970.
36. Smith, J.B., Westmoreland, B.F., Reagan, T.J., and Sandok, B.A.: Mayo Clin. Proc. 50:469-474, 1975.
37. Miller, J.H.D. and Coey, A.: Electroencephalogr. Clin. Neurophysiol. 2:582-585, 1959.
38. Ch'ien, L.T., Boehm, R.M., Robinson, H., Liu, C. and Frenkel, L.D.: Arch. Neurol. 34:361-364, 1977.
39. Enzmann, D.R., Ransom, B., Norman, D., and Talberth, E.: Radiology 129:419-425, 1978.
40. Zimmermann, R.D., Russell, E.J., Leeds, N.E. and Kaufman, D.: Amer. J. Roentgenol. 134:61-66, 1980.
41. Soong, S.J., Caddell, G.R., Alford, C.A., Whitley, R.J., and the NIAID Collaborative Antiviral Study Group: 26th Interscience Conference on Antimicrobial Agents and Chemotherapy, New Orleans, LA, 1986.
42. Lakeman, F.D., Koga, J. and Whitley, R.J.: J. Infect. Dis., (In press), 1987.
43. Illis, L.S. and Merry, R.T.G.: J. R. Coll. Physicians (London) 7:34-44, 1972.
44. Breeden, C.J., Hall, T.C. and Tyler, H.R.: Ann. Intern. Med. 65:1050-1056, 1966.
45. Nolan, D.C., Lauter, C.B. and Lerner, A.M.: Ann. Intern. Med. 78:243-246, 1973.
46. Whitley, R.J., Alford, C.A., Jr., Hirsch, M.S., Schooley, R.T., Luby, J.P., Aoki, F.Y., Hanley, D., Nahmias, A.J., Soong, S-J and the NIAID Collaborative Antiviral Study Group: N. Engl. J. Med., 314:144-149, 1986.
47. Skoldenberg, B., Alestig, K., Burman, L., Forkman, A, Lovgren, K., Norby, R., Stiernstedt, G., Forsgren, M., Bergstrom, T., Dahlqvist, E., Fryden, A., Norlin, K. Lancet 8405:707-711, 1984.
48. Davis, L.E. and McLaren, L.C.: Ann. Neurol. 13:192-195, 1983.
49. Dix, R.D., Baringer, J.R., Panitch, H.S., Rosenberg, S.H., Hagedoren, J. and Whaley, J.: Ann. Neurol. 13:196-200, 1983.
50. Koenig, H., Rabinowitz, S.G., Day, E. and Miller, V.: N. Engl. J. Med. 300:1089-1093, 1979.
51. DeVita, V.T., Jr., Young, R.C. and Canellos, G.P.: Cancer 35:98-110, 1975.
52. Rahal, J.J.: Medicine 57:179-195, 1978.

53. Biron, K.K. and Elion, G.B.: Am. J. Med. 73:54-57, 1982.
54. Fischer, P.H., Lee, J.J., Chen, M.S., Lin, T.S. and Prusoff, W.H.: Biochem. Pharmacol. 28:3483-3486, 1979.
55. Ayisi, N.K., Gupta, V.S., Meldrum, J.B., Taneja, A.K. and Babiuk, L.A.: Antimicrob. Agents Chemother. 17(4):558-566, 1980.
56. Wigand, R. and Hassinger, M.: Med. Microbiol. Immunol. 168:179-190, 1980.
57. DeClerq, E., Descamps, J., Verhelst, G., Walker, R.T., Jones, A.S., Torrence, P.F. and Shugar, D.: J. Infect. Dis. 141:563-574, 1980.
58. Schinazi, R.F., Peters, J., Williams, C.C., Chance, D. and Nahmias, A.J.: Antimicrob. Agents Chemother. 22:499-507, 1982.
59. Schinazi, R.F. and Nahmias, A.J.: Am. J. Med. 73:40-48, 1982.
60. Field, J.H.: J. Antimicrob. Chemother 12(Suppl B): 129-135, 1983.
61. Wade, J.C., McLaren, C. and Meyers, J.D.: J. Infect. Dis. 148:1077-1082, 1983.
62. Barry, D.W., Nusinoff-Lehrman, S., Ellis, M.N., Biron, K.K. and Furman, P.A.: Scand. J. Infect. Dis. 47(Suppl):155-164, 1985.
63. Svennerholm, B., Vahlne, A., Lowhagen, G.B., Widell, A. and Lycke, E.: Scand. J. Infect. Dis. 47(Suppl):149-154, 1985.

6

TREATMENT OF HERPES SIMPLEX LABIALIS

S.L. SPRUANCE

Division of Infectious Diseases, Department of Medicine, University of Utah School of Medicine, Salt Lake City, Utah 84103, USA; and the Centre International de Recherches Dermatologiques (CIRD), Sophia Antipolis, 06565 Valbonne, France.

ABSTRACT

The recent experience with new antiviral agents in recurrent herpes labialis in otherwise normal, non-immunocompromised subjects, reveals that small clinical benefits are associated with treatment. These accomplishments have been achieved by improved means of drug delivery and the use of patient-initiated study protocols to begin treatment in the earliest lesion stages.

INTRODUCTION

Herpes simplex labialis is a common and ubiquitous infection of the skin due to herpes simplex virus (HSV).The majority of the population harbours latent infection with HSV type 1, and among approximately half of those latently infected, or roughly 20-40% of the population, reactivation of the virus results in recurrent labial or perioral outbreaks of vesicular lesions (1-4). The frequency of these outbreaks is extremely variable, ranging, in some individuals, from rare episodes every 5-10 years, to monthly outbreaks among a small proportion of subjects. The severity of the illness is most often mild, although uncomfortable and cosmetically disfiguring. Among persons with an underlying immunosuppressing disease, lesions are of longer duration and may spread to cause major morbidity, and recurrent herpes labialis complicated by erythema multiforme reactions can be disabling (5,6).

Major progress has been made in recent years in our understanding of viral infections and in the development of safe and effective antiviral drugs. Nucleoside analogues which inhibit the replication of HSV deoxyribonucleic acid (DNA) are available commercially for the treatment of herpetic keratitis, herpes zoster, herpes genitalis, mucocutaneous herpes simplex in immunosuppressed hosts, neonatal herpes and herpes encephalitis. Despite these advances, an accepted treatment for recurrent herpes labialis in non-immunocompromised patients has remained elusive.

The present review will attempt to define why herpes labialis has been difficult to treat. In order to understand the problem, the natural history of the disease will be described as well as relevant aspects of the pharmaceutics of topical drug therapy. Selected clinical trials will be examined in detail and the utility of existing antiviral preparations will be discussed.

THE NATURAL HISTORY OF HERPES SIMPLEX LABIALIS

Several groups of investigators have studied untreated episodes of herpes labialis, in order to define the disease and to optimally direct efforts at therapy and prevention (4,7,8). Prodromal symptoms are a frequent feature of herpes labialis and potentially could be used to guide the initiation of early treatment. In a study of 80 untreated patients undergoing an episode of herpes labialis, 60% had experienced prodromal symptoms with their present lesion (7).In a prospective study of eight patients with frequent herpes labialis, only 3 of 16 lesions (19%) that developed during the observation period were preceded by a prodrome (9). Moreover, there were twice as many other episodes of prodromata (six occurences) for which the symptoms were not followed by a lesion (false prodromes).

In clinical trials, the utility of prodromal symptoms has depended upon the study design. In two, large clinic-initiated studies, only 5 of a total of 441 (1%) patients were in the prodromal or erythema stage at the time of their first clinic visit (10,11). This likely occurred because the prodrome is of short duration, usually less than 6 hours (7). In self-initiated treatment studies, in which medication was dispensed to the patient for use at the onset of the next episode, 40-57% of participants were able to begin treatment in the prodromal phase (12,13). A high incidence of false prodromata have generally not been a problem for investigators.

The rapid evolution of herpes labialis has been appreciated for many years (14). In a study of untreated patients with the disease, we observed that thirteen of 16 subjects (81%) who had their first clinic visit within 12 hours of lesion onset had already attained the vesicular lesion stage (Fig.1). Median lesion area, the mean lesion pain score and the median lesion virus titers reached their maximum levels within the first 24 hours and declined progressively thereafter (7).

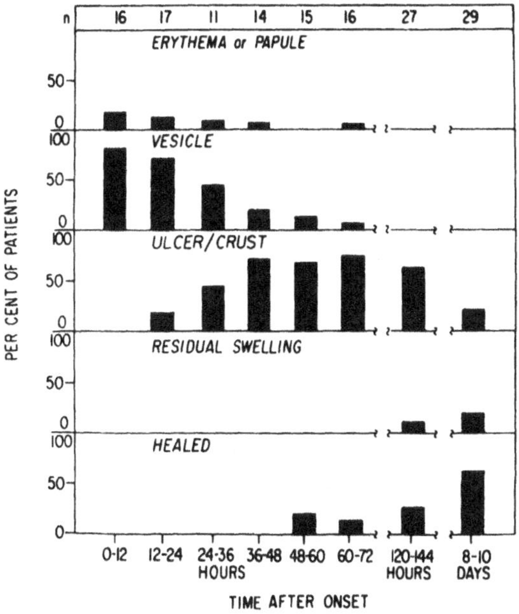

Fig. 1. Frequency of lesion stages at intervals after onset of herpes labialis among 29 untreated patients. (Reprinted with permission from ref. 7).

In order to focus more accurately on the clinical events in the 24 hour period immediately after lesion onset, a second report from our center (15) utilized a different method of data analysis which circumvented the problem of variability in disease severity between patients. A cohort of untreated or placebo-treated subjects were selected from the populations of three large Utah studies. All patients had a complete evaluation in the Herpes Clinic within 24 hours and again one day later, and their time of lesion onset was known. To determine the early course of the disease, each patient was compared with himself for the two visits, and the lesion was determined to be increased in severity, unchanged, or better. Patients were further divided into three subgroups based on the age of the lesion at the first visit (0-8, 9-16, 17-24 hrs old). The results of these determinations on 122 subjects is shown schematically in Fig. 2. The data reveal that a majority of the patients in all three lesion age groups had no change in lesion severity between visits, or improved, using either lesion area, pain, or virus titer as the measure of severity.

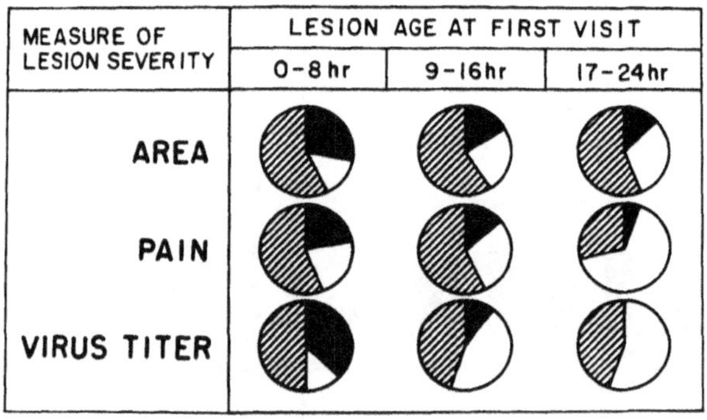

Fig. 2. Course of lesions of different age between the first and second clinic visits. ■, increase in severity; □, decrease; ▨, no change. (Adapted from ref. 15).

The data from our studies of lesion development permit one to draw several conclusions about the natural history of herpes labialis. First, the majority of lesions seen within 24 hours of onset will be the same or improved at the next visit, with or without some mode of therapy. Moreover, dissection of the first lesion day into three subgroups by time after onset showed the same pattern of early lesion maturity even among the youngest lesions, those less than eight hours of age. This latter finding supports a model of cutaneous pathogenesis in which HSV infection of the epidermis originates from multiple foci of inoculation (see below). Secondly, patients with established lesions can only expect to receive a small benefit from treatment. Even if seen within eight hours of lesion onset, only a quarter of the patients normally progress in the absence of therapy. While small improvements from antiviral treatment are likely achievable among patients whose disease is naturally waning, major clinical benefits in herpes labialis will require another approach.

A schematic representation of the multifocal model of herpes labialis pathogenesis is shown in Fig. 3. We have postulated that the area of a lesion represents a region of multiple inoculation sites corresponding to the distribution of the multiple nerve endings of an HSV-infected neuron. Growth of these sites of infection occurs during the prodrome, and coalescence of the foci results in a palpable lesion of established dimensions. Little further extension of the lesion occurs because of the early development of high titers of intralesional interferon and the rapid ingress of other host resistance factors (16). This model is further supported by a histologic study of prodromal-stage lesions in which the early cytopathic changes in the epidermis were found at multiple sites (17).

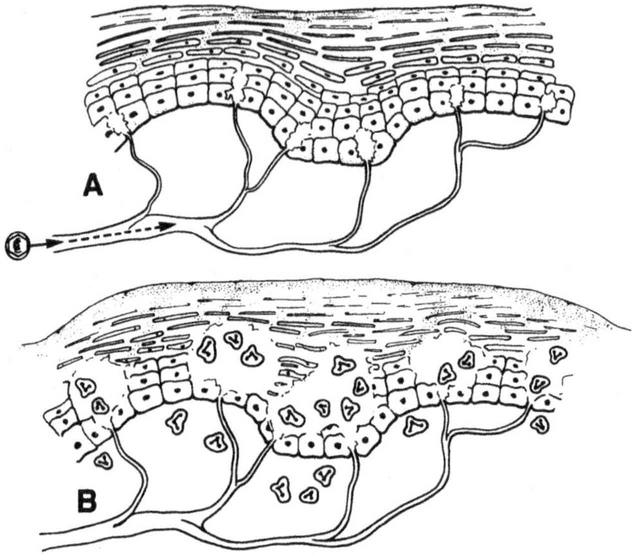

Fig. 3. Hypothetical representation of the sequence of events immediately following inoculation of HSV into the epidermis from an infected neuron. A, multifocal inoculation establishes numerous foci of infection in the basal cell layer. B, growth and coalescence of these foci determine the size of the clinically apparent lesion.

PHARMACEUTICAL ASPECTS OF TOPICAL ANTIVIRAL THERAPY

Doseage, drug delivery, formulation and stability of the test compound are pharmaceutical issues that are vital and frequently problematic in the development of a successful treatment for a microbial disease. Examples of important pharmaceutical issues in antibacterial chemotherapy include: the efficacy of intermittent doses of aminoglycosides used alone in the treatment of infections in neutropenic patients; inactivation of aminoglycosides by betalactam agents; and the use of intermittent dosing regimens in the treatment of tuberculosis. Antiviral chemotherapy is still in its infancy, and similar challenges await clinical investigators who seek to develop or improve the treatment of viral diseases.

In the development of a dosing schedule for the treatment of herpes labialis, one must consider the rapid evolution of the illness and the effectiveness of natural resistance factors; the cellular toxicity inherent in many agents which may potentially delay wound healing; the uniformly virus-inhibitory mode of action of the leading modes of chemotherapy; and the discrepancy between the continuous presence of the agent in standard viral sensitivity assays as compared with the intermittent dosing in the majority of clinical situations. The need for early initiation of treatment has been discussed above. The optimum frequency of doseage is poorly understood for cutaneous HSV disease. In clinical trials, different dosing frequencies have been compared in the prophylaxis of recurrent genital herpes, but not for its treatment (18). In animal models of mucocutaneous HSV infection, dosing frequency influences the outcome, and efficacy is significantly reduced when antivirals are administered fewer than 3 times per day (19). In the author's experience, dosing schedules for herpes labialis treatments have been selected somewhat arbitrarily based on human habits, convenience and cosmetic factors rather than from pharmacokinetic science. Of considerable theoretical concern is the interruption of therapy at night, particularly early in the course.

The duration of experimental treatments for herpes labialis has ranged from 4 to 14 days, often based on the conception of what constitutes a "course" of treatment, or a perception of the attitudes of regulatory agencies. In fact, it is hard to justify prolonging treatment of herpes labialis beyond the recognized duration of virus shedding, which is usually not more than 4 days (7). Furthermore, since healing of the skin requires proliferation of epidermal cells, and many antivirals possess antiproliferative activity, prolonged therapy would be potentially detrimental to lesion resolution in many instances.

Adequate delivery of an antimicrobial agent to the site of infection is a recognized principle of therapeutics. In the development of an oral antibacterial agent, the critical first studies are measurement of drug levels in serum in order to determine the feasibility of the oral route of administration. As the gastric mucosa may exclude drugs from the systemic circulation, so also the stratum corneum of the skin effectively blocks diffusion of many topically-applied chemotherapeutic agents to the living cells of the lower epidermis. While careful measurements of antiviral drug penetration through skin would therefore seem to be a logical step in the development of a topical treatment for recurrent herpes labialis, such studies have frequently not been performed. While the stratum corneum is eroded in the ulcerative stage of the disease, delay of treatment to this late time point insures an unsuccessful outcome.

In 1968, Tomlinson and MacCallum (20) reported that the efficacy of topical antiviral therapy for experimental cutaneous HSV infection was enhanced if dimethyl sulfoxide (DMSO), an agent which increases skin permeability, was used as the drug vehicle. We further developed this concept by quantitating the penetration of nucleoside antivirals in different vehicles through excised skin in vitro. For a given antiviral, the degree of penetration through skin has been predictive of the outcome of topical drug treatment in the Hubler model of dorsal cutaneous HSV infection (21,22). Because of the variable potency and different modes of action of antivirals, extension of these results to a general predictive model has been problematic. However, in 1986, Freeman et al described an index of in vitro drug characteristics, the $\log_{10}$ (drug flux through skin/ID_{50}), which was highly predictive of efficacy against experimental HSV infection for six different nucleoside antivirals in a variety of drug vehicles (23).

When human skin was studied in vitro, Freeman et al (24) again demonstrated that the nature of the drug vehicle has a major influence on the penetration of topically applied antivirals (Figure 4). Sheth et al have reported that assay of antiviral concentration in the skin by the adhesive tape procedure can be predictive of drug efficacy against experimental HSV infection (19). This is a simple, noninvasive test which can be used comfortably with human subjects. In summary, procedures for a thorough preclinical assessment of experimental topical antiviral formulations are available. Early identification of superior formulations should hasten the advent of effective products to the market.

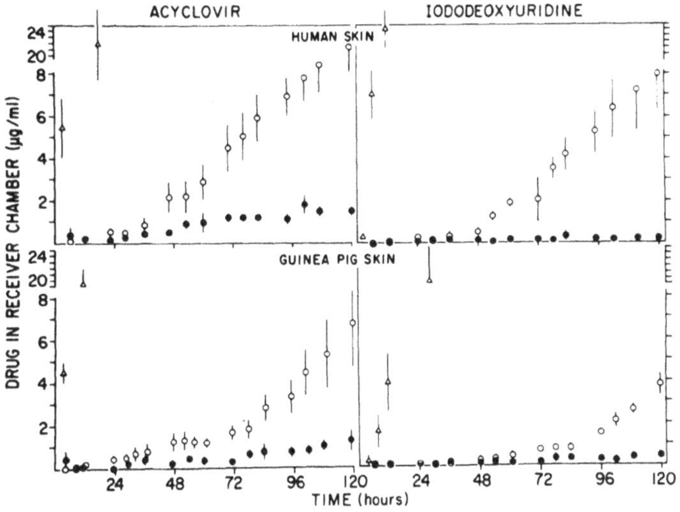

Fig. 4. Penetration of acyclovir and idoxuridine through human and guinea pig skin in different vehicles. △ , dimethyl sulfoxide; ○ , aqueous cream; ● , polyethylene glycol. (Reprinted with permission from ref. 24).

CHEMOTHERAPY OF HERPES SIMPLEX LABIALIS

The great majority of clinical trials examining treatments for herpes labialis have been negative. To understand this problem, one must first consider that the achievable degree of benefit from treatment is likely to be modest among patients with established or incipient lesions; and accordingly, only a well-designed study, with large numbers of subjects, employing an active antiviral compound, well-delivered to the site of infection, is likely to result in a positive outcome. In recent years, the utilization of potent, new antiviral compounds makes it likely, in the event of a negative study outcome, that the fault lies more with the drug delivery, numbers of patients, or some other aspect of study design, and not with the intrinsic properties of the test compound.

A proper study of herpes labialis must be carefully placebo-controlled and blinded because of the extensive placebo effect in this illness. In our experience at the University of Utah, 70% of patients who were retrospectively surveyed and who received placebo felt that the severity of the lesion course was reduced by treatment (25). Without an unmanipulated group for comparison, one cannot say if this response was subjective or real. Recently, Bolla et al found a striking placebo effect in their studies of Thymopentin (26). Treatment of patients with subcutaneous injections of drug vehicle reduced the frequency of herpes labialis by 40% ($p=0.01$).

The present review will examine selected clinical trials involving established, potent antiviral compounds and for which sufficient pharmacologic data exist to interpret the results of the trial. For a compilation of all treatment efforts against this disease, the reader is referred elsewhere (27).

Topical idoxuridine (IDU) in dimethyl sulfoxide (DMSO).

In 1966, MacCallum and Juel-Jensen (14) reported favorable results with the use of 5% IDU in DMSO among 16 otherwise healthy patients with herpes labialis. Subsequently, this formulation was made available commercially in Europe and has been widely used for the treatment of the disease. There is a compelling rationale for this treatment: the antiviral, IDU, has established efficacy in the treatment of herpetic keratitis; enhanced penetration of the drug through human skin has been demonstrated with DMSO as the vehicle (Figure 4); and IDU in DMSO has been highly effective in animal models of cutaneous HSV infection (20,23). Nevertheless, the clinical trial of MacCallum and Juel-Jensen is too small to support a firm conclusion about the efficacy of topical IDU in human subjects with herpes labialis. Moreover, the study is flawed by the inclusion of 5 retreated patients as individual cases in the data analysis. Further controlled, double-blind clinical trials examining the efficacy of IDU in DMSO for herpes labialis curiously have not been forthcoming, and experience with the formulation in recurrent genital herpes disease has been mixed (28,29). While this now "old" treatment has been passed over by clinical investigators, it retains its promise on the basis of preclinical data and limited clinical information.

Topical acyclovir (ACV) ointment.

The advent of ACV, an agent with selective activity in HSV-infected cells and a high toxic/therapeutic ratio, was a major advance in the development of antiviral chemotherapy (30). The initial topical formulation of this agent for commercial purposes consisted of 5% ACV suspended in a semisolid ointment vehicle of different molecular weight polyethylene glycol (PEG). PEG is a stable, non-toxic, cosmetically acceptable compound that is widely used as a base in dermatologic preparations. Preclinical evaluations of ACV in PEG in animal models of HSV infection were generally quite favorable; however, only marginal benefits could be demonstrated with the treatment in a dorsal cutaneous experimental infection which placed emphasis on penetration of the intact stratum corneum for efficacy (31). In vitro skin penetration studies of ACV in PEG have shown very low rates of transcutaneous drug diffusion from the PEG vehicle (24,31), providing an explanation for the animal model results. Poor skin penetration by other drugs in PEG (32), and markedly improved ACV penetration with the use of other vehicles (24) establish that the therapeutic shortcomings of ACV ointment are the property of the vehicle and not the antiviral compound.

ACV in PEG has undergone extensive clinical evaluation both in the United States and in Europe as a treatment for herpes labialis in otherwise healthy patients. In a clinic-initiated treatment study involving 208 subjects, a clear effect of the drug on virus shedding from lesions was demonstrated (11). A small benefit was noted in time to healing and time to loss of crust in selected subgroups, but these differences were not statistically significant. Subsequently, three trials were performed in which medication was prospectively dispensed and treatment was initiated by the patient, in order to begin therapy at an earlier stage of the disease. Among 69 patients in the United States, randomized to receive either ACV ointment or placebo, all of whom started treatment in either the prodromal or erythema stages (by history), no clinical benefit with drug therapy was observed (12). Of the ACV recipients, 91% progressed to the vesicle stage, compared with 71% in the placebo group. Two smaller studies in Great Britain reported a favorable experience with patient-initiated treatment protocols, but in one report of 29 patients (33), the results were not statistically significant, and in the second study involving only 13 patients (34), the data analysis was flawed because retreated patients were counted as individual cases. In summary, from a clinical experience totalling 319 subjects, no solid evidence has emerged that topical ACV in PEG is of clinical benefit to patients with recurrent herpes labialis.

Topical ACV cream.

A second topical formulation of ACV has been developed in which the vehicle is an aqueous cream containing a high concentration of propylene glycol (35). In experimental cutaneous HSV infections, the efficacy of 5% ACV cream has been superior to 5% ACV ointment, and this difference appears to be related to better skin penetration of ACV from the cream formulation (35, 35a).

Three therapeutic trials of ACV cream have been performed in Europe among otherwise healthy subjects with recurrent herpes labialis. These trials were all small (30-49 patients), randomized double-blind placebo-controlled, stressed early treatment by patient-initiation of therapy, and generally relied on the patient's history to establish the initial diagnosis. Interpretation of the data from these reports is hindered by retreatment of some of the original patients and analysis of the aggregate number of episodes as independent events; and failure to indicate if a one or two-tailed measure of statistical probability was employed. Among 49 patients treated once with ACV cream or placebo, Fiddian et al (36) found significantly more ACV treated subjects to have aborted lesions, and a 33% reduction in the median number of days to complete healing. Among 30 patients treated once, Van Vloten et al (37) found a similar reduction in the healing time among ACV recipients, and also reported that the duration of vesiculation was significantly less in the drug-treated patients (1.8 vs 2.7 days, p=0.016). More recently, the third trial by Shaw et al (38) observed no clinical benefit attributable to ACV cream.

The utility of ACV cream is supported by experimental data and selected clinical evidence to date has been encouraging. Further studies with this promising formulation are warranted.

Oral ACV

In a double-blind, placebo-controlled, patient-initiated trial of oral ACV for the treatment of herpes labialis, 173 otherwise healthy patients took 400 mg of ACV by mouth five times a day for five days beginning in the prodromal (70 patients), erythema (28 patients) or papular (75 patients) stage of the disease (13).Preliminary analysis of the data has revealed a trend toward reduction in lesion healing time among ACV patients (6.3 vs 6.9 days, p= 0.28), reduction in the duration of pain (2.7 vs 4.1 days, p= 0.04), and reduction in the duration of virus shedding from lesions (2.1 vs 3.6 days, p= 0.03).

Topical foscarnet.

Foscarnet (trisodium phosphonoformate) is an inhibitor of viral DNA polymerase which is under evaluation for the treatment of infections by HSV, cytomegalovirus, and human leukemia/lymphoma virus III. In a 3% aqueous cream formulation, it has been evaluated in two large clinical trials in Canada and Sweden for the treatment of recurrent herpes labialis in otherwise healthy subjects. In Sweden,

167 patients initiated treatment following a clinic visit, within 24 hours of lesion onset. Among 43 patients who began therapy in a prevesicular lesion stage, the number of patients with "active lesions" (vesicular or ulcer/crust stages) in the first four days was reduced by half in the foscarnet-treated group (39,40). In the Canadian study, in which treatment was patient-initiated, 144 subjects received 3% foscarnet cream or placebo (41). In a subgroup of 95 patients who began treatment in a prevesicular stage,there was a trend toward reduction in the duration of pain (3.1 vs 4.0 days, $p=0.08$), fewer patients treated with foscarnet progressed to vesicle formation ($p = 0.03$), and the time to end of virus shedding was reduced ($p=0.04$). In summary, a small clinical benefit has been shown for 3% foscarnet cream among subjects beginning treatment in the early lesion stages of herpes labialis.

Prophylactic application of chemotherapy.

Prophylaxis of mucocutaneous HSV disease has been pursued in herpes genitalis, to reduce disease transmission and patient anguish, and among immunosuppressed subjects, because of the attendant morbidity under these circumstances. Until recently, prophylaxis of recurrent herpes labialis in otherwise normal individuals has received scant attention. Recently, however, the efficacy of prophylactic treatment for oral herpes has been examined among selected subgroups of persons with unusually severe disease or at risk for complications.

In occasional individuals, disabling attacks of erythema multiforme occur 10-14 days after an episode of HSV infection. Two short reports now indicate that prophylactic oral ACV is effective in preventing recurrences of herpes labialis and the associated episodes of erythema multiforme (6,42). The necessary dose and duration of treatment is variable, and the consequences of prolonged administration of ACV, as in other clinical settings, remains to be fully elucidated.

Some individuals regularly trigger an episode of herpes labialis by exposure to ultraviolet light, dental manipulation, or upper respiratory infection. Under such circumstances, a short course of effective chemoprophylaxis might be considered worthwhile. To examine the feasibility of this hypothesis, we treated 146 skiers at high risk for sun-induced herpes labialis with oral ACV, 400 mgs twice daily for 7 days, or placebo (44). Placebo recipients developed herpes labialis in 18/73 subjects, compared with 5/73 episodes in those treated with ACV (p= 0.006). In addition, several of the episodes in ACV-treated patients occurred in the first two days after the start of prophylaxis. This successful study should encourage controlled trials of antiviral prophylaxis under other high-risk settings.

Thymopentin is a synthetic pentapeptide with the same biologic function on T-cells as its parent hormone, thymopoietin (45). Under the assumption that persons with recurrent HSV disease have a mild, underlying deficiency in some component of cell mediated immunity, Bolla et al (26) examined the effect of a six week course of subcutaneous thymopentin, 3 times a week, or placebo, on the subsequent frequency of herpes labialis among 36 patients suffering from extremely frequent recurrences (12 or more relapses per year). A 75% reduction in the frequency of recurrences was noted, a greater benefit than in the placebo-treated patients, who also responded. Further study of this compound seems justified.

Oral herpes in immunosuppressed patients.

Perioral infections in immunocompromised subjects may be treated with topical acyclovir ointment (5). If intraoral lesions are present, oral or intravenous acyclovir are effective (46,47).

SUMMARY AND CONCLUSIONS

Studies of the natural history of herpes labialis have assisted in the design and interpretation of clinical treatment studies. The author's synthesis of the histopathologic, immunologic and clinical course of herpes labialis is summarized schematically in Figure 5.

Histopathology begins well in advance of a palpable lesion, and is responsible for the "prodromal" symptoms of the disease. Clinical awareness of a lesion and maximal epidermal damage occur in close succession. The secondary immune response to recurrent virus infection begins early and results in curtailment of any lateral spread of the disease in the normal, non-immunocompromised host.

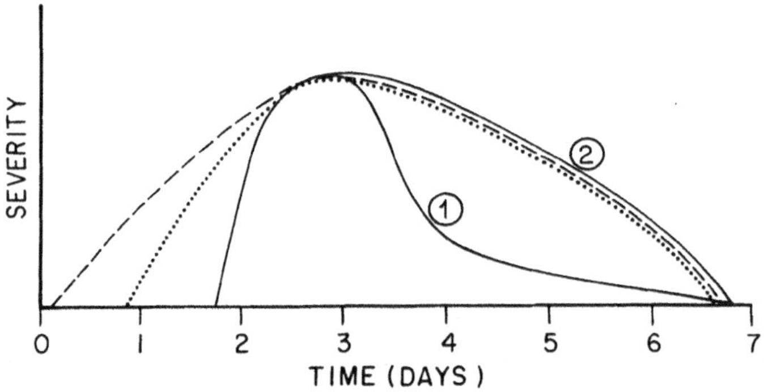

Fig. 5. Schematic representation of the possible course of events in the development of recurrent herpes simplex labialis. ----, histopathology; ...., immune response; ______, clinical disease, treated lesion (1) or untreated (2).

While the majority of investigative treatments for herpes labialis have been by the topical route of administration, cutaneous pharmacokinetic studies to select and evaluate topical formulations have been inconsistently applied. Clinical studies with ACV ointment for herpes labialis were disappointing, and retrospective investigations of the penetration of ACV through human and guinea pig skin have attributed the unfavorable results to the PEG vehicle. In contrast, penetration of ACV in cream formulation through human and guinea pig skin has been documented, and the results of clinical trials to date with this preparation show evidence of efficacy. Experimental formulations of nucleoside antivirals containing penetration-enhancing agents have been created which markedly enhance transcutaneous drug flux and which have correspondingly greater in vivo activity in an animal model of cutaneous HSV infection. Use of such formulations has not been widely accepted as yet by the pharmaceutical industry, in part because of skin irritation that may necessarily accompany modification of the stratum corneum, and also because of increased developmental risks and costs associated with formulations containing two active ingredients.

Examination of clinical trials of different antivirals for the treatment of herpes labialis appears to show that the disease is relatively resistant to therapy. While 10 years ago the failure of efforts at treatment could be ascribed to inadequate drug potency, drug delivery, or study design, the limited efficacy seen in clinical trials with oral ACV and topical cream preparations of ACV and foscarnet cannot be faulted on the same grounds. Evidence is emerging that the benefit of antiviral treatment may be a reduction of the vesicle and ulcer stages. While the shortening of lesion time effected may be small, the improvement could be clinically significant because vesicle and ulcer stages are the most painful and disfiguring periods of the eruption. Figure 5 shows the hypothetical course of a patient treated with an active antiviral preparation compared with untreated disease. While lesion duration is approximately the same in both instances, lesion severity declines more rapidly in the treated subject, such that the total lesion experience is more favorable.

As with genital herpes, prophylaxis of herpes labialis with antiviral agents provides a far superior response than does treatment of established disease. Selected clinical situations exist where this approach is justified, such as patients with herpes labialis-associated erythema multiforme reactions.

Because of the small benefits that occur with treatment and their restriction to patients seen in the earliest stages of a recurrent episode, the development and evaluation of new treatments is arduous and involves the performance of large-scale, patient-initiated clinical trials. Improvements in the clinical testing procedure would be of substantial value. We have had a favorable experience with the experimental induction of herpes labialis in volunteers using small doses of ultraviolet light (48). A pilot study with oral ACV is currently in progress to examine whether prophylaxis of experimentally induced lesions is a feasible method to screen the effectiveness of new antiviral products.

REFERENCES

1. Ship, I.I., Morris, A.L., Durocher, R.T., Burket, L.W.. Oral Surg. 13:1191-1202, 1960.
2. Embil, J.A., Stephens, R.G., Manuel, F.R. Can. Med. Assoc. J. 113:627-630, 1975.
3. Grout, P., Barber, V.E. J.Royal Col. Gen. Prac. 26:428-434, 1976.
4. Young, S.K., Rowe, N.H., Buchanan, R.A. Oral Surg. 41:498-507, 1976.
5. Whitley, R.J., Levin, M., Barton, N., Hershey, B.J., Davis, G., Keeney, R.E., Whelchel, J., Diethelm, A.G., Kartus, P., Soon, S-J. J. Infect. Dis. 150:323-329, 1984.
6. Lemak, M.A., Duvic, M., Bean, S.F. J. Am. Acad. Dermatol. 15:50-54, 1986.
7. Spruance, S.L., Overall, J.C., Kern, E.R., Krueger, G.G., Pliam, V., Miller, W. New Eng. J. Med. 297:69-75, 1977.
8. Bader, C., Crumpacker, C.S., Schnipper, L.E., Ransil, B., Clark, J.E., Arndt, K., Freedberg, I.M. J. Infect. Dis. 138:897-905., 1978.
9. Spruance, S.L. J. Clin. Microbiol. 19:675-679, 1984.
10. Spruance, S.L., Crumpacker, C.S., Haines, H., Bader, C., Mehr, K., MacCalman, J., Schnipper, L.E., Klauber, M.R., Overall, J.C. Jr. New Eng. J. Med. 300:1180-1184, 1979.
11. Spruance, S.L., Schnipper, L.E., Overall, J.C. Jr., Kern, E.R., Wester, B., Modlin, J., Wenerstrom, G., Burton, C., Arndt, K.A., Chiu, G.L., Crumpacker, C.S. J. Infect. Dis. 146:85-90, 1982.
12. Spruance, S.L., Crumpacker, C.S., Schnipper, L.E., Kern, E.R., Marlowe, S., Arndt, K.A., Overall, J.C. Jr. Antimicrob. Agents and Chemother. 25:553-555, 1984.
13. Spruance, S., Rowe, N., Stewart, J., Wenerstrom, G., Freeman, D. Davis, G. In: Abstracts of the 26th Interscience Conference on Antimicrobial Agents and Chemotherapy. p. 210, 1986.
14. MacCallum, F.O., Juel-Jensen, B.E. Brit. Med. J. 2:805-807, 1966.

15. Spruance, S.L., Wenerstrom, G. Oral Surg. Oral Med. Oral Path. 58:667-671, 1984.
16. Overall, J.C. Jr., Spruance, S.L., Green, J.A. J. Infect. Dis. 143:543-547, 1981.
17. Huff, J.C., Krueger, G.G., Overall, J.C. Jr., Copeland, J., Spruance, S.L. J. Am. Acad. Dermatol. 5:550-557, 1981.
18. Douglas, J.M., Critchlow, C., Benedetti, J., Mertz, G.J., Connor, J.D., Hintz, M.A., Fahnlander, A., Remington, M., Winter, C., Corey, L. New Engl. J. Med. 310:1551-1556, 1984.
19. Sheth, N.V., McKeough, M.B., Spruance, S.L. J. Invest. Dermatol. In press, 1986.
20. Tomlinson, A.H., MacCallum, F.O. Br. J. Exp. Pathol. 49:277-282, 1968.
21. Spruance, S.L., Freeman, D.J., Sheth, N.V. Antimicrob. Agts. Chemother. 28:103-106, 1985.
22. Spruance, S.L., McKeough, M.B., Sugibayashi, K., Robertson, F., Gaede, P., Clark, D.S. Antimicrob. Agts. Chemother. 26:819-823, 1984.
23. Freeman, D.J., Spruance, S.L. J. Infect. Dis. 153:64-70, 1986.
24. Freeman, D.J., Sheth, N.V., Spruance, S.L. Antimicrob. Agts. Chemother. 29:730-732, 1986.
25. Wenerstrom, G., Spruance, S.L. Unpublished data.
26. Bolla, K., Djawari, D., Kokoschka, E.M., Petres, J., Lidén, S., Gonseth, R., Amblard, P., Bernengo, M.G., Bonerandi, J.J., Claudy, A., Degreef, H., DeMaubeuge, J., Meynadier, J., Saurat, J.H., Schöpf, E., Höbel, W., Castaigne, J.P., Sundal, E. Surv. Immunol. Res. 4, Suppl. 1:37-47, 1985.
27. Overall, J.C. Jr. In: Antiviral Agents and Viral Diseases in Man, 2nd ed. (Eds. G.J. Galasso et al.), Raven Press, New York, 1984, pp. 247-312.
28. Parker, D.J. J. Antimicrob. Chemother. 3(Suppl.A):131-137, 1977.
29. Silvestri, D.L., Corey, L., Holmes, K.K. J. Am. Med. Assoc. 248:953-959, 1982.
30. Elion, G.B., Furman, P.A., Fyfe, J.A., DeMiranda, P., Beauchamp, L., Schaeffer, H.J. Proc. Natl. Acad. Sci. 74(21):5716-5720, 1977.
31. Spruance, S.L., McKeough, M.B., Cardinal, J.R. Antimicrob. Agts and Chemother. 25:10-15, 1984.
32. Sheth, N.V., Freeman, D.J., Higuchi, W.I., Spruance, S.L. Int. J. Pharmaceut. 28:201-209, 1986.
33. Yeo, J.M., Fiddian, A.P. J. Antimicrob. Chemother. 12(Suppl.B):95-103, 1983.
34. Fiddian, A.P., Ivanyi, L. Brit. J. Dermatol. 109:321-326, 1983.
35. Collins, P., Oliver, N. Am. J. Med. 73:96-99, 1982.
35a. Spruance, S.L., Freeman, D.J., Sheth, N.V. Antimicrob. Agts. Chemother. 30:196-198, 1986.
36. Fiddian, A.P., Yeo, J.M., Stubbings, R., Dean, D. Brit. Med. J. 286:1699-1701, 1983.
37. Van Vloten, W.A., Swart, R.N.J., Pot, F. J. Antimicrob. Chemother. 12(Suppl.B): 89-93, 1983.

38. Shaw, M., King, M., Best, J.M., Banatvala, J.E., Gibson, J.R., Klaber, M.R. Brit. Med. J. 291:7-9, 1985.
39. Wallin, J., Lernestedt, J-O., Lycke, E. In: Current Chemotherapy and Infectious Disease. Proc. 11th ICC and 19th ICAAC. American Soc. of Microbiology, Washington, 1980, pp. 1361-1362.
40. Wallin, J., Lernestedt, J-O. Personal communication.
41. Lawee, D., Rosenthal, D., Aoki, F.Y., Portnoy, J. Canad. Med. Assoc. J. In press, 1987.
42. Green, J.A., Spruance, S.L., Wenerstrom, G., Piepkorn, M.W. Annals Int. Med. 102:632-633, 1985.
43. Fawcett, H.A., Wansbrough-Jones, M.H. Brit. Med. J. 287:798-799, 1983.
44. Spruance, S.L., Hamill, M., Hoge, W., Davis, G., Mills, J. In: Abstracts of the 26th Interscience Conference on Antimicrobial Agents and Chemotherapy. p. 312, 1986.
45. DeMaubeuge, J., Haneke, E., Djawari, D., Wolff, K., Stingl, G., Molin, L., Schöpf, E., Stengel, R., Degreef, H., Panconesi, E., Wüthrich, B., Bolla, K. Surv. Immunol. Res. 4(Suppl. 1):30-36, 1985.
46. Shepp, D.H., Newton, B.A., Dandliker, P.S., Flournoy, N., Meyers, J.D. Annals Int. Med. 102:783-785, 1985.
47. Mitchell, C.D., Gentry, S.R., Boen, J.R., Bean, R., Groth, K.E., Balfour, H.H. Jr. Lancet 1:1389-1392, 1981.
48. Spruance, S.L. J. Clin. Microb. 22:366-368, 1985.

7

TREATMENT OF HERPES GENITALIS

S.L. SACKS

Division of Infectious Diseases, Department of Medicine,
Health Sciences Centre Hospital, The University of British Columbia,
Vancouver, Canada V6T 1W5

ABSTRACT

Effective treatment for genital herpes infections is now widely available in the form of systemically - administered acyclovir. Immunocompromised hosts may be treated, as well as selected individuals with normal immunity. In general, patients with severe infection will benefit most from treatment, especially those with first episodes or frequent recurrences. Demonstrating significant clinical benefits in recurrent herpes has been very difficult, partly because of the mild and relatively short clinical outcome of untreated disease, and partly because of several scientific pitfalls common to clinical trial design.

In this chapter, the results of clinical trials presented to date are reviewed, and the methods utilized in each trial are critically analyzed. Recommendations for future clinical explorations, based upon the information now available, suggest the need to limit comparisons to groups with similar clinical outcomes in the untreated state. The average patient with mild recurrences of genital herpes is still awaiting inexpensive, effective, and safe relief from the symptoms of infection, or, more importantly, from the worry of transmission of infection.

INTRODUCTION

In recent years, genital herpes simplex virus (HSV) infection has taken on major importance as a public health problem. While the onslaught of media attention is recent, this infection has probably been prevalent throughout history. Herpes (from the Greek, "to creep") was coined as a word during the era of Hippocrates (1). Possibly, the first reference to genital herpes in the medical literature was made by the French physician, John Astruc, in 1736 (2). The virus was first grown *in vitro* in 1925, although Lipshutz successfully inoculated vesicular fluid from genital lesions into volunteers in 1921, eliciting the expected clinical response (3,4). By the early 1960's, Schneweis (5) and Dowdle (6) established the existence of distinct antigenic types, and correlated their type with the site of virus recovery.

Unfortunately, the true prevalence of genital herpes is not accurately established. Data from metropolitan Toronto between 1978 and1980 suggests that 17.5% of the female population and 12.8% of the male population show evidence of seropositivity to herpes simplex virus type 2, and statistical correlations were made with young age at first intercourse and multiple sexual partners (7). The proportion of seropositive individuals with asymptomatic or symptomatic disease is unknown, however. Data collected by the U.S. National Disease and Therapeutic Index survey would suggest a rapidly increasing prevalence of infection (8). Between 1966 and 1981, the number of physician-patient consultations for genital herpes increased tenfold. This is corroborated by other circumstantial seroepidemiological evidence from Ann Arbor, Michigan (9), Auckland, New Zealand (10), Bankok, Thailand (11), and King County, Washington (12). Seropositivity may actually underestimate the incidence of genital herpes, since a sizeable proportion of first episodes are caused by herpes simplex virus type 1.

THE NATURAL HISTORY OF UNTREATED INFECTION

Definitions

For the purposes of this discussion, genital herpes is defined as virological proof of the presence of HSV, on at least one occasion, in an epithelial area provided with sensory innervation via the sacral plexus. Some investigators have also included patients with clinical disease consistent with genital herpes, who seroconvert during their first clinical episode. However, these patients are not as clearly defined. Acceptable virologic proof includes culture positivity or identification of viral antigens in clinical specimens or typical cytologic changes (13).

Clinical manifestations

Women commonly experience cervicitis or evidence of cervical virus shedding during their primary infection (14). External infection may be evident during primary episodes as well, and is the usual clinical location of recurrent disease. The most commonly affected genital sites during recurrent disease, in descending order of frequency, are listed in Table 1 (15):

Table 1. Lesion locations in recurrent genital herpes.

Women	Men
Labia minora (40%)	Penile shaft (67%)
Labia majora (36%)	Inner prepuce (10%)
Perineum (20%)	Outer prepuce (8%)
Mons (4%)	Glans penis (7%)
	Pubic region (7%)
	Perineum (1%)
	Scrotum (1%)

Other areas of importance in "genital" herpes include the buttock, thigh, and other lower extremity sites. These dry skin, nongenital lesions are likely to last longer and involve a larger area of skin than other sites (16), although precise data concerning the natural history of disease in different presenting sites are not available. Nongenital sites are included here as genital infection because they probably do not represent sites of autoinoculation, but rather result from recurrent episodes of infection whose latent sites in the sacral ganglia are indistinguishable from genital disease. Therapeutic trials may exclude nongenital areas from study (15), or specifically include them (14). However, the vast majority of clinical trials have not specifically reported presentation sites, beyond using the word "genital herpes" or "herpes of the genital tract".

Active infection with HSV may present to the physician in a variety of ways. Patients who deny a previous history of genital sores are classified as having initial episodes. When acute serologic determinations at the time of presentation fail to reveal the presence of HSV-specific antibody, the first episode is called primary infection (17). As will be discussed later, individuals with primary infection are more likely to have more numerous and more painful lesions, with longer duration, and increased susceptibility to complications of infection (18). If initial serology is positive, however, this is consistent with previous latent infection (17). Thus, their preexisting immune status often results in milder clinical disease than that observed for primary infection. Initially seropositive events are termed nonprimary, initial episodes. Persons with nonprimary, initial infection may have preexisting immunity to HSV-1 or HSV-2 (17,19). Seropositives to HSV-1 are usually presenting with their first genital infection, while seropositives to HSV-2 are presumed to be having recurrent disease which was previously asymptomatic. Primary genital infection is frequently caused by HSV-1 (10-40%, depending on the geographical area reporting).

For reasons that are not clear, HSV-1 genital primaries are less quickly and less frequently followed by recurrent infection than those caused by HSV-2 (20). This may stem from a reduced ability of HSV-1 to establish latency in the sacral ganglia, compared with HSV-2 (21). Accordingly, recurrences of genital infection are predominantly caused by HSV-2 (95+%) (18). Recurrent disease is usually mild and may be lesional or nonlesional (18,22). Generally, patients complain of prodromal symptoms, especially itching or, less commonly, tingling or burning at the site of lesion development. Remote prodromal symptoms such as sacral paresthesias or buttock or thigh pain will occur in a minority of patients. A general feeling of ill health or mental stress may be cited. Prodromal symptoms may be followed by lesion development (lesional episodes) or by their disappearance without lesion development (nonlesional episode). The latter are quite common and have been seen in as few as 5% (16) and as many as 40% of untreated episodes (22,23).

When lesions develop, they progress through a series of identifiable phases (16), some of which may be skipped. Skin redness may be accompanied by swelling in some individuals. As cellular swelling and lysis occurs, fluid-filled vesicles may be observed in areas of keratinized epithelium. Vesicles are poorly formed on mucous membranes such as the labia minora or cervix where lesions may first appear as ulcers. Wet ulcers also follow the vesicular stage when the lesion is unroofed. These ulcers may become covered by a wet serous crust which then dries as underlying skin reepithelializes. The dry crust may fall off quickly or linger for several days, leaving healed skin. Reepithelialized (healed) skin may remain discolored (residual erythema, residual hypopigmentation) for a short or prolonged period, unrelated to viral shedding.

First-episode infection

Genital herpes infections may be complicated by either acute or long-term sequelae. Physical manifestations of a severe nature, however, are usually limited to the primary episode. Corey has reported (12) that nearly 40% of men and 70% of women report fever, headache and myalgias. Mindel and his colleagues (24) report systemic signs in 75% of these patients. Vaginal or urethral discharge, along with local lesions are common. Virtually all lesions are painful and symptoms such as itching and dysuria (Seattle data: 44% men; 83% women) are the rule. Corey describes new lesion formation in 68% (HSV-1) to 75% (HSV-2) of primary episodes (12) overall, and in up to 82% of patients in one study (25). New lesions continue to form for about 10 days (12), but one study reports a mean duration of new lesion formation of 10.6 days in men and 10.1 days in women, with one woman forming new lesions 23 days into her episode (14). Mertz et al (26) described new lesion formation continuing longer than 48 hours after presentation, in 58% and Bryson et al (27) in 44% of placebo-treated patients. Tender inguinal lymphadenopathy may begin after initial presentation, and may be the last symptom to resolve (12). It is present in 80% of patients (18) and lymph nodes may be tender for a mean of 3.9 days (27) or longer (18). Durations of primary genital lesions have varied widely during studies by different investigators. Corey suggests a mean duration of 16.5 days in men and 19.7 days in women (18). Mertz et al (26) describe a median time to healing of 16 days when treated with placebo, while Bryson's placebo-treated women had a median time to healing of 16.2 days compared with 21 days in men (27). Mindel (28) describes a median of 14 days in all patients and 12.5 days in women, although seropositives (non-primaries) are included in this analysis. A topical treatment study of primary disease demonstrated healing times of 14.3 days in the placebo group (25). Extragenital infection is not uncommon during the primary episode, the pharynx (10%)

and the fingers (6%) being the most common sites. Aseptic meningitis may occur in as many as 33% of cases, although usually symptoms are mild and short-lived (12).

Long-term complications of genital herpes are less common. Prolonged urinary hesitancy has been reported even after complete resolution of lesions (29). Transverse myelitis has been rarely reported (29,30). Transmission of virus infection antepartum has been documented (31,32), but the usual route of neonatal infection is through direct contact with infected lesions during vaginal delivery (33). The incidence of neonatal herpes infection is rising in both the U.S. (34) and Canada (35), in parallel with the increased incidence of genital herpes infection. Carcinoma of the cervix (36) and vulva (37) have also been associated with genital herpes infection, although cause and effect relationships have not been proven.

Recurrent infection

The most common and clinically bothersome sequel to primary genital infection is recurrent genital herpes infection - both symptomatic and asymptomatic. Symptomatic recurrent infection occurs in nearly 100% of patients who develop initial or primary HSV-2 infection. While the median number of yearly recurrences is 4, nearly 40% of patients will experience 6 or more episodes per year (18). Recurrent disease is mild. It follows the same stage progression as for primary infection but rarely reproduces systemic symptoms or a severe local response.

Lesions may be painful or pruritic. Dysesthesias may also occur (12). If near the urethral meatus, external dysuria is common. Tender lymphadenopathy may recur, but with less severity and frequency than for primary disease. Patients are usually concerned more with avoiding transmission, however, than they are with their symptoms (38). New lesions will commonly form during recurrent disease in 16% (39) to 24% (40) of episodes. In one study of recurrent disease, up to 65% of patients developed new lesions (25). It is not determined whether these new lesions represent the natural history of one recurrent episode, or the coincidental occurrence of a second recurrent episode.

Generally, clinical disease outlasts the period of virus-shedding from lesions, and in as many as 50% of episodes, may develop, progress, and heal, in the absence of a positive lesion viral culture (41). Culture-negative lesions are poorly understood. They heal significantly sooner than their culture-positive counterparts (15). The precise duration of recurrent episodes is not established, since each study is unique in its design, inclusion criteria, and objective standards for followup. Table 2 displays the disease durations reported in several recent studies, along with important features of study design which may have influenced the outcome. In general, earlier studies were associated with longer recurrent disease durations.

Table 2. Times in Untreated or Placebo-treated Recurrent Genital Herpes

Treatment Modality	No. Subjects	Allowable disease duration prior to enrollment (hours)	Times to Healing (days) Men	Women	Unique Study Features	First Author	Year (Ref. no)
None	9	24	8.8	8.8		Adams	1976(14)
None	13	48	7.7	9.4		Adams	1976(14)
Topical Placebo	7	24	6.8	7.0	Petrolatum-Base predisposed to new lesion formation	Adams	1976(14)
	13	48	7.7	8.0		Adams	1976(14)
Miconazole Aqueous Cream	7	96	men not studied	12.0	100% CP[2]	Blough	1979(64)
None, placebo or inactive	362	48	10.6	9.3	100% CP Total of several studies.	Corey	1976 to 1983(18)
Polyethylene Glycol	56	48	8.4	7.6	100% CP	Corey	1982(25)
Oral placebo	43	48	6.0	not reported	CI[3] 84% CP	Nilsen	1982(53)
Saline	35	72	10.7[4]		100% CP	Silvestri	1982(62)
Polyethylene Glycol	58	48	6.3	4.3	100% CP	Reichman	1983(39)
Aqueous Cream in Propylene Glycol	41	24 from prodrome	6.0 6.0[5]	3.0 5.0[5]	73% CP	Fiddian	1983(58)
None	32 340 episodes	nil	6.9	6.3	Previous CP	Sacks	1984(22)
Polyethylene Glycol	162	24 from prodrome	6.0 6.6[5]	3.9 4.3[5]	Previous CP Incl. nonlesional episodes. 45% CP	Luby	1984(41)
Oral placebo	106	48	7.0[4] 7.4[5]		CI 86% CP	Reichman	1984(76)
Oral placebo	83	24	6.5[4] 7.2[5]		PI[6] 77% CP	Reichman	1984(76)
Aqueous cream	45	24	5.3	4.6	Previous CP	Wallin	1985(68)
Oral placebo	51	24	8.0	4.0	CI Previous CP	Ruhnek-Forsbeck	1985(78)
Oral placebo	43	24	7.0	5.0	PI	Ruhnek-Forsbeck	1985(78)
Aqueous cream	74	24	7.8	7.0	76% CP	Douglas	1986(40)
Aqueous cream	20	6	6.04	4.85	100% CP	Sacks	1987(15)
Aqueous cream	103	6	2.96	2.77	0% CP	Sacks	1987(15)

1 Reference Number (this chapter)

2 Culture-positive patients studied

3 Clinic-initiated trial

4 Figures include both sexes -- sex not specified in report.

5 Results for all lesions--includes new lesions developing during study. Other values represent lesions only or details were not specified in report.

6 Patient-initiated trial

Other features which may help to explain the decrease in observed disease duration over the last few years include:

- The tendency for patients with milder infection to report to the physician because of "herpes", where earlier patients may have sought medical help only for especially bothersome symptomatology.
- The inclusion of specific patient instructions, when participating in studies, which require patients to report with their next episode, regardless of severity.

CLINICAL TRIALS WITH ANTIVIRALS

Cure

To date, attempts to modify or eradicate latent infection, once established, have met with failure. This problem may result from the fact that potent antivirals generally require virus-specific enzyme production in order to exert their effects. HSV-specified thymidine kinase activity, however, may be intermittently expressed in latently-infected ganglionic cells (42). Furthermore, the possibility exists that latent infection is repeatedly reestablished with each passing day of a primary episode or each recurrence. Accordingly, one might presume that effective treatment of primary disease would modify subsequent recurrence rates. Accordingly, a number of treatment studies with acyclovir have looked at times to subsequent recurrences as a function of drug effect. Bryson followed her first episode patients treated with oral acyclovir (27) for two years after the study's completion (43). Although the numbers are small, and the patients self-reported their episodes, she found a significant diminution of episodes in the second year of disease in patients treated with acyclovir for primary infection. No benefits were observed in the first year after disease onset in this group. The nonprimary, first episode patients treated with acyclovir did not experience any different outcome than placebo-treated patients in this long-term study. Long-term benefits of systemic acyclovir treatment were also sought by Corey et al (44). Ten months' followup in 61 patients treated with intravenous acyclovir or placebo for first episode infection showed no differences in long-term outcomes between the two groups in this analysis. Recurrence suppression has not beneficially altered outbreak frequency after drug withdrawal (23,45-49). In fact, increased lesion severity (45) and more rapid lesion development (47) have occasionally been reported in acyclovir recipients during the immediate post-suppression period.

Since latent infection is established soon after infection and before disease onset (50), early treatment of established disease may not be expected to alter more than the treated episode (with the possible exception of Dr. Bryson's study:43). In certain situations, however, it may be theoretically possible to prevent epithelial infection. This would include the use of barrier contraception or possibly, topical antivirals during intercourse. After inoculation, latent infection might be prevented by the use of antivirals, eg. in the setting of post-exposure treatment of uninfected sexual partners, health care workers with needlestick exposures, and neonates delivered vaginally or after prolonged membrane rupture, in the presence of a genital maternal lesion. These approaches have not been clinically established as efficacious, although post-exposure treatment of certain animal models will prevent disease and latency in a proportion of treated animals (51).

Non-cure

Treatment of first episodes. Treatment of active disease is quite a different matter. Recent evidence suggests that the clinical course of primary genital herpes can be modified by treatment after lesions are established. Specifically, Mindel et al (28) found that intravenous acyclovir shortened the duration of lesions during first episodes from 14 days to 7 days. Data analyses in this report did not separate true primaries from non-primary first episodes, although this unreported analysis was shown to be in favor of acyclovir for primary patients. Interestingly, they found major differences in healing times of internal lesions, but "no significant differences in the duration of external lesions between the two (treatment) groups". In a followup report the next year, however, these authors clarified their data (24). When primary patients were reported separately, healing times were again reduced (15 days for placebo vs. 9.0 days for acyclovir), but the *P* value was less striking ($P<0.05$). The duration of new lesion formation was reduced to 0.0 days for acyclovir recipients ($P<0.01$), but total symptomatology was not benefitted statistically. In 1983, Corey et al (52) reported on their results with 31 patients treated for 5 days with intravenous acyclovir or placebo. They added a unique exclusion criterion in this study, to decrease the number of non-primary episodes, i.e., a history of oral-labial as well as genital herpes infection. Of the 31 patients, 27 had primary disease, serologically-verified. In this group, virus shedding from lesions was reduced from 13 days to 2 days and from the cervix was reduced from 9 days to 1 day ($P<0.001$ for either). Virus shedding from the throat and urethra was also reduced. Both complete crusting (13 days vs. 6.0 days; $P<0.001$) and complete healing (21 days vs. 9 days; $P=0.007$) were benefitted by treatment, as were pain (7 days vs. 3 days; $P=0.03$), itching (8 days vs. 2 days; $P=0.01$), vaginal discharge (11 days vs. 4 days; $P=0.002$), dysuria (7 days vs. 4 days; $P=0.04$), and sore throat (10 days vs. 2 days; $P=0.03$). Constitutional symptoms were not affected statistically (7 days vs. 4 days; $P=0.17$). Because clinical relapse was noted after therapy withdrawal, this study suggested a 7-10 days' therapy course for primary disease might be more appropriate.

Oral acyclovir treatment of primary disease has also been quite effective. Bryson et al (27) studied 48 subjects of whom 23 had primary infection. Women on oral acyclovir (200 mg 5 times daily for 10 days) healed sooner (16.2 days vs. 10 days;

P=0.015), crusted sooner (15 days vs. 8.8 days; P=0.01), and shed virus from external lesions for less time (14.7 days vs. 4.9 days; P=0.001). Parallel findings for primary infections in men were observed with somewhat less impressive statistical benefits, possibly because of the lower number of subjects. New lesion formation after 48 hours was seen in one male acyclovir recipient compared with 44% of placebo recipients (P=0.004). Patients with dysuria, adenopathy and moderate malaise at presentation had reduced durations of these symptoms, although pain and malaise, in general, were not shown to improve with treatment. Mertz et al (26) reported their findings on 119 primary and 31 nonprimary first episodes in 1984. The treatment, dosage, and duration were identical with Bryson's (27). In the true primary group, virus shedding from genital lesions (9 days vs. 2 days; P<0.01), and from the cervix (8 days vs. 0 days; P<0.01) was reduced, as was time to crusting (10 days vs. 7 days; P<0.01), time to healing (16 days vs. 12 days; P<0.01), and pain duration (7 days vs. 5 days; P<0.05). In contrast to Bryson's findings, however, dysuria and constitutional symptoms were not benefitted by treatment in this study. Once again, however, new lesion formation after 48 hours was dramatically reduced (62% vs. 18%; P<0.001). In their nonprimary group, virus shedding was reduced from all genital lesions (6 days vs. 0 days; P<0.01) but no other parameters of efficacy, including new lesion formation, approached statistical significance. Another report on initial disease which did not attempt to separate the primary and nonprimary episodes for analyses (53) showed a reduction in viral shedding (13 days vs. 1 day; P<0.001), crusting time (6 days vs.4 days; P<0.05), healing times assessed by patient interview (11 days vs. 6 days; P<0.01), itching duration (6 days vs. 1.5 days; P<0.01), and pain duration (7 days vs. 3 days; P<0.01). New lesion formation was reduced from 55% to 0% (P<0.01).

Acyclovir was also tested as a 5% topical preparation in polyethylene glycol. This offered minor benefits in improved times to crusting and reduction of viral shedding during primary infection (25), when compared with placebo. Corey et al studied 77 first episodes with this formulation. For the 49 true primaries, virus shedding of presenting lesions was reduced (6.7 days vs. 2.8 days; P<0.01) as was time to crusting (10.5 days vs. 7.1 days; P<0.01). Time to healing was not statistically benefitted by treatment (12.3 days vs. 10.6 days; P=0.11). Topical acyclovir did not influence the frequency of new lesion formation or alter the duration of pain (8.8 days vs. 6.2 days; P=N.S.), although pain severity may have been reduced in one center. Treatment of nonprimary first episodes did result in a reduction of viral shedding from presenting genital lesions from 1.0 day for placebo to 0.2 days for acyclovir (P<0.05), although no statistical benefit was seen when all external lesions were considered. Acyclovir-treated patients actually had a slightly longer time to healing and crusting of initial lesions than did those treated with placebo, and pain was not significantly altered. By repeated-measure analysis, the investigators were able to show a significant reduction in viral titers between acyclovir and placebo-treated patients for both the primary and the nonprimary groups.

Thin and his colleagues separately reported their results on topical acyclovir (54). In this study, 18 patients (14 culture-positive) received acyclovir ointment and 22 (20 culture-positive) received placebo. Viral shedding was reduced from all lesions from 9 days to 2 days ($P<0.05$) and time to crusting from 7 days to 5 days ($P<0.05$). Healing times were 13.5 days for all lesions and 12 days for external lesions in the placebo group, compared with 9 days and 7 days in the acyclovir group ($P<0.05$ and $P<0.01$, respectively). In the U.K., topical acyclovir was then reformulated into a 5% aqueous cream vehicle containing 40% propylene glycol, which shows better absorptive characteristics in the guinea pig (Hubler) model (55,56). Kinghorn et al (57) treated 53 patients with first episodes of whom 25 were true primaries. Once again, efficacy data were grouped, preventing separate analyses of primary vs. nonprimary first episodes. Time to healing was reduced from 11 days to 8 days ($P<0.01$) for original lesions and from 14 days to 8 days for all lesions ($P<0.001$). New lesion formation was reduced in acyclovir recipients ($P<0.001$), as was duration of pain (8 days vs. 4 days; $P<0.05$) and duration of virus shedding (7 days vs. 3 days for original lesions; $P<0.01$ and 11 days vs. 4 days for all lesions; $P=0.001$). Itching, dysuria, discharge and crusting were unaffected. Two patients in their placebo group were culture-negative as was one patient in their acyclovir group, and their acyclovir group had a slightly longer duration of symptoms (0.5 day) prior to entry. These factors could affect outcome only slightly, however. Yet, the mixture of two distinct groups (primaries and non-primaries) in the analysis makes it difficult to extract clear information about drug effects. An apparent expansion of these data to a multicenter analysis of 101 patients suggested even more of a beneficial effect (56). With the additional patients, however, serologic testing was not performed. In this analysis, acyclovir was associated with reduced duration of viral shedding from presenting external lesions, as well as all lesions. Reduced times to first crusting, healing of lesions, and loss of symptoms were also observed in the acyclovir group.

These striking differences between studies are difficult to explain on the surface. Just examining topical acyclovir in polyethylene glycol, alone, one study showed minor differences in the primary group and virtually no differences in the nonprimary group (25). By contrast, a much smaller study demonstrated highly significant results using the same formulation. One can only assume that the "blending" of primaries and nonprimaries in the analyses has confused the data. Even a few extra nonprimary patients included with the acyclovir group would have markedly reduced the duration of disease parameters for that group. However, careful analysis of serological status may not completely obviate the problem, since antibody responses may be partially established by five days (the usual enrollment time limit). This could mean that some true primary disease will be included in the nonprimary group. Furthermore, lack of seropositivity in some individuals with proven genital herpes simplex infection is a known phenomenon (20) and may cause a small percentage of nonprimaries to be counted in the primary group for analysis.

Nevertheless, since at the time of the patient's clinical presentation, serologic status is unknown, both Canada and the U.S.A. gave licence to topical acyclovir ointment for the treatment of "first episodes", knowing that a significant proportion of such patients (the nonprimaries) might not receive benefit from its application. Because oral acyclovir may prevent systemic complications (27) of primary infection and/or reduce subsequent recurrence rates (43), it should be used in all patients with first episode disease. Combining oral acyclovir with topical acyclovir does not enhance the efficacy of the oral preparation (59).

Treatment of active recurrent episodes. The first reported success with antiviral therapy against HSV infections after study in a randomized, double-blind trial, appeared in 1966. MacCallum (60) reported on the therapeutic outcome in 21 patients with oral-labial infection utilizing idoxuridine (IDU) in dimethyl sulfoxide (DMSO). Subsequently, Davidson-Parker examined the course of genital herpes in a double-blind trial using this combination (61). Mixtures of 20% IDU or 5% IDU or 0% IDU in a DMSO base were compared. Healing times were 3.9 days (20% IDU), 5.7 days (5% IDU) and 5.7 days (DMSO alone) such that $P<0.05$. Viral shedding was also reduced from 4.1 days (DMSO alone) to 2.9 days (5% IDU) and 2.1 days (20% IDU). Accordingly, Silvestri et al (62) examined a 30% IDU in DMSO mixture compared to DMSO alone and saline placebos, after application of drug four times daily for 7 days. The duration of viral shedding was again reduced (1.4 days for DMSO; 2.0 days for saline; 0.6 days IDU-DMSO). No effects were observed on symptom duration, new lesion formation, healing times, or risk of subsequent recurrence. Toxicity (burning, contact dermatitis, vulvar carcinoma *in situ*) was observed, however. While toxicity would suggest that this approach to therapy would be ill-advised, certain problems exist in the trial design. Specifically, patients with recurrent genital herpes were included up to 3 days after recurrence onset, although subsequent stratification to patients enrolled within 24 hours of lesion onset did not alter the outcome. Symptoms and times to healing were unaffected, although the reported healing times are unusually long (11.0 days for IDU-DMSO; 9.8 days DMSO; 11.3 days saline). On the other hand, viral shedding was reduced in patients receiving 30% IDU-DMSO to 2.3 days compared with 3.3 days for

DMSO alone or 3.8 days for saline ($P<0.05$). All patients analyzed for efficacy were culture positive on day 1 (24 of 124 excluded because of negative culture).

Adenine arabinoside (vidarabine; ara-A; Vira-A®) was studied in a 3% formulation in petrolatum by Adams et al (14). Untreated patients were included, along with a placebo ointment group in this genital study. Patients with thigh and buttock lesions were included in the analysis of 7 days' treatment, utilizing four applications daily. In the middle of this study an aqueous gel base became available and women were switched to this mode of treatment. Followup visits were 3 and 8 days after enrollment, and weekly thereafter. All patients were culture-positive at some point during the treatment course. Treatment with ara-A ointment had no beneficial effect, and the data suggested that occlusive ointment predisposed to new lesion formation. Untreated healing times from disease onset were 9.7 days in women and 10.0 days in men. The more soluble 5'-monophosphate derivative of ara-A (ara-AMP) was subsequently tested in a 10% gel formulation in 134 patients. In this double-blind, placebo-controlled trial of treatment four times daily for 7 days, only 5% of drug recipients failed to heal by day 6, compared with 18% of placebo recipients (P=0.01). Other efficacy parameters were not reported (63).

A controversial publication examined the effects of topical 2-deoxy-D-glucose (64). These investigators studied a 0.19% formulation in miconazole cream compared with placebo D-glucose in the same cream base. Lesion duration was reduced from 12 days to 6.8 days in recurrent infection ($P<0.001$) and virus shedding, which outlasted lesions in the placebo group (15.3 days), was reduced to 4.4 days in the treatment group ($P<.001$). These reported durations were added to an average lesion duration of 4.0 days prior to study entry. The authors went on to claim a marked reduction in subsequent lesion frequency in the treatment group. Interestingly, they limited the number of placebo recipients to 7 because of "ethical and moral considerations". This study describes a small group of patients who received an extraordinary drug benefit. However, the times to healing and viral shedding in their limited placebo group were 12 days and 15.3 days respectively. On top of an average of 4 days before enrollment, this suggests 1.5 to 3 times the durations seen in other studies of recurrent disease. Furthermore, placebo-treated entrants were severely limited in number, suggesting that the study was not properly randomized. The prolonged duration of viral shedding 3 days beyond duration of genital lesions was unexplained. These controversial matters were addressed in a subsequent letter to the editor (64). It is not possible to draw a final conclusion. Follow-up studies in animal models (65) have failed to show any beneficial effect from 2-deoxy-D-glucose treatment. Properly controlled, randomized clinical trials with this agent remain to be performed.

Topical treatment studies with acyclovir have also led occasionally to conflicting results. The first report from Corey et al (25) employed a study of 111 culture-positive recurrent episodes, comparing 5% acyclovir in polyethylene glycol to the ointment base alone, after 4 applications for 5 days, allowing for enrollment up to 48 hours after the onset of symptoms. Viral shedding was reduced in acyclovir-treated men from 2.3 days (all lesions) to 1.2 days ($P<0.05$) and from 1.5 days (presenting lesions) to 0.8 days ($P<0.05$). This effect was not observed for women. An improvement trend in healing times was also observed for men, but this was not statistically significant (8.4 vs. 6.9 days). Two followup studies using these preparations applied 6 times daily were then undertaken. Reichman et al (39) looked at 88 episodes in a clinic-initiated format (less than 48 hours enrollment). These investigators stated that 88 patients were entered and subsequently confirmed as culture-positive. One infers from this that culture-negative episodes were not described. Acyclovir-treated men demonstrated a modest reduction in viral shedding (3.2 days for placebo versus 2.0 days for acyclovir; $P<0.05$) while acyclovir-treated women shed virus slightly longer than their placebo-treated counterparts (not significant). Times to healing were 6.6 days (acyclovir-men), 6.3 days (placebo-men), 5.2 days (acyclovir-women) and 4.3 days (placebo-women). In this study, males shed virus for a longer period of time than females ($P=0.003$) and took longer to heal. No clinical benefits were observed with acyclovir, therefore, in this clinic-initiated trial. Thus, a prodrome-initiated trial was conducted and reported by Luby et al (41). Of 547 enrollees, 309 patients subsequently presented, during treatment, for evaluation. Patients were instructed to apply drug as soon as possible after the appearance of the prodrome and to visit the clinic within 24 hours of the onset of treatment which was continued for 5 days (6 times daily). Of 173 cultured men, 80 were culture-positive, while 59/130 cultured women were culture-positive. Because this was a patient-initiated study, culture-positive and culture-negative patients were analyzed together. Healing times were 6.0 days for presenting lesions in men receiving either acyclovir or placebo, while placebo-treated women healed in 3.9 days compared with 4.5 days for acyclovir. The duration of virus excretion was 1.8 days for both treatment groups in men and 2.0 days versus 1.1 days ($P<0.05$) for women treated with placebo and acyclovir, respectively. No drug effects were demonstrated on lesions at the time of presentation, suggesting that acyclovir had no influence on the rate of nonlesional prodromes. Thus, despite earlier treatment, no clear benefits were demonstrated with topical acyclovir.

Acyclovir displays better cutaneous absorption when applied in an aqueous cream (56). Accordingly, acyclovir was retested in an aqueous cream formulation in several studies from the U.K. Kinghorn et al (57) described 41 patients with prior culture-proven disease, using 5% acyclovir in aqueous cream compared with cream placebo, in a patient-initiated trial. In this trial, 100% of placebo recipients were culture-positive on day 1 compared with only 67% of acyclovir recipients, all seen within 24 hours of starting treatment. While this is a *possible* drug effect, it is difficult to draw the conclusion that such an observation is entirely drug-induced, since this would mean that 1/3 of patients ceased to shed virus after less than 24 hours of topical treatment. Data obtained on day 0 (presentation) may be difficult to interpret since virus shedding may have been influenced by drug therapy or alternatively, two distinct populations (culture-negative, culture-positive) may have been studied. An antiviral effect of this magnitude has not been observed with any other therapeutic modality, including systemic acyclovir (26). Several studies of recurrent disease have now demonstrated up to a 50% negative virus isolation rate during recurrent episodes (41). This has been seen most commonly in patient-initiated trials, not unlike that of Dr. Kinghorn's in design (57), where both placebo and treatment groups present with a high proportion of negative cultures. Furthermore, if this is not a drug-induced effect, but rather selection bias, it would call to question any other positive observations of drug effects, since culture-negative recurrent genital infection is milder and shorter than culture-positive disease in the absence of treatment (15).

Potential solutions to the design difficulties inherent in patient-initiated trials might include:

- ◊ Use of larger, multicenter trials where chance randomization bias would be less likely to occur.
- ◊ Complementation of patient-initiated trials with clinic-initiated trials separating culture-positive and culture-negative patients for analysis.
- ◊ Use of home cultures prior to self-initiating therapy in a patient-initiated design. (Requires delaying treatment past prodrome stage.)

Arildone is an aryl diketone compound which displays antiherpesviral activity *in vitro*. Douglas et al (40) recently reported the effects of this compound in recurrent genital herpes in a multicenter, randomized, placebo-controlled, clinic-initiated trial. Allowable time from symptom onset to enrollment was 24 hours. Of 145 evaluable patients, 91 had positive cultures on their first visit. The duration of viral shedding was reduced in women receiving arildone from 4.0 days to 3.2 days ($P<0.05$). Viral shedding was unaltered in men. Symptoms were reduced slightly in duration for men, as was the time to crusting. Shorter healing times were observed in arildone-treated men (6.8 days vs. 7.8 days), but these were not statistically significant. Women were not benefitted by treatment for any efficacy parameter.

One of the most effective topical agents in the guinea pig model has been foscarnet (67). This DNA polymerase inhibitor was compared to its placebo cream base using a 0.3% formulation in a study performed by Wallin and his colleagues (68). They studied

129 episodes in 86 patients, such that each patient was randomized separately for each of 1 or 2 episodes. The first episode for study was clinic-initiated for patients presenting within 24 hours of lesion onset, and patient-initiated for the next episode in the others. While all patients had had past proven culture-positive disease, virus cultures were not followed during the treatment episode. The bulk of the efficacy data was provided by a home record, and, indeed, of 86 episodes studied at the first treatment course, 34 had no clinic followup beyond the home record. The remainder were seen from one to three times. Healing times were recorded as shortened for men from 5.0 days (placebo) to 4.0 days (foscarnet) which was highly significant ($P<0.002$). Placebo-treated women healed in 4.6 days compared with 4.4 days for foscarnet (not significant). Patients who participated in two study episodes and received foscarnet in one and placebo in the other, were analyzed according to the difference between results from one episode to the other. In this way, improved clearance of redness, swelling, blisters, ulcers, and time to total healing were significantly benefitted by treatment. This study's strong dependence on patient-recorded data, and it's lack of culture information during the treatment episode makes analysis difficult. Accordingly, several multicenter trials were established in 1984 to study the effects of foscarnet treatment utilizing objective criteria. A Canadian collaborative study group of 7 centers studied 309 patient episodes of recurrent herpes in a clinic-initiated trial (15). Patients were enrolled only if their lesions had been present for under 6 hours. The 0.3% foscarnet cream was tested in men while a 1% cream was tested in women. Patients were followed daily. A clear antiviral effect was seen in foscarnet-treated men, but no other significant clinical benefits from treatment were observed. Nevertheless, several important observations were made in this study. The longer duration of disease in men which was also observed by Reichman (39) and Luby (41), was statistically confirmed. Men presented with larger lesions which took longer to heal (6.04 days vs. 4.85 days), but complained of fewer symptoms (lower pain and itching scores). Culture-confirmed, placebo-treated women healed in 4.85 ± 0.60 days vs. 2.77 ± 0.55 days for women who were negative by culture (P=0.0001). Culture-confirmed, placebo-treated men healed in 6.04 ± 0.68 days compared with culture-negative episodes, which healed in 2.96 ± 0.57 days (P=0.00001). Barton et al (69) have also reported no clinical efficacy for this agent in the treatment of recurrent genital herpes.

Since, with rare exception, hospitalization for recurrent herpes infections is not indicated in the normal host, intravenous treatment has not been utilized in that setting. Several orally administered agents have been tested, however. Ribavirin was administered to 48 patients at a dose of 800 mg daily for 10 days and patients were required to report within 24 hours of onset (69). Significance levels and methods for evaluation were not reported and virological assessments were not performed. The authors state that new vesicle formation and time to crust formation and healing were reduced by the therapy. Interestingly, the placebo patients were often not healed after 10 days of treatment, suggesting an unusually long duration of recurrent disease in this group. Furthermore, benefits seemed to diminish with each subsequent recurrence. Side-effects of elevated liver enzymes and anemia were observed. L-lysine has been a favored treatment for many years by a number of patients with recurrent disease (71). Nevertheless, clinical trials have failed to demonstrate either a therapeutic or a suppressive effect (72). Immune modulators have also been tried. Levamisole had an apparent suppressive effect in recurrent mucocutaneous infection in one study (73), but this has not been confirmed (74). Transfer factor offered no clinical benefit in reducing recurrence rates of genital herpes (75).

Oral acyclovir has been tested by a number of investigators in the treatment of recurrent disease. Nilsen et al studied 85 patients treated with 200 mg five times daily for 5 days (53). Patients with clinically diagnosed genital herpes were studied if they presented within 48 hours. Lesion cultures were positive in 31/42 (74%) of acyclovir recipients and 36/43 (84%) of placebo recipients. Virus shedding was reduced from 2.0 days to 1.0 days ($P<0.001$). Healing times were assessed by discussion with the patient and were reduced from 6 days to 5 days ($P<0.001$) in the tested group. New lesion formation was also reduced by acyclovir treatment. Itching was slightly shorter and pain slightly longer in the acyclovir group. Neither difference achieved statistical significance. This study was followed by a large, multicenter study in the U.S. and Canada, involving 250 patients (76). For the initial phase of the study, an identical "clinic-initiated" treatment program was employed. However, after the first phase, patients were given the same treatment for a second "patient-initiated" phase, where they self-administered oral acyclovir or placebo at the first sign or symptom of a recurrence, and reported within 24 hours for evaluation. A positive culture from either episode was considered a valid inclusion criterion for analysis of both episodes. A total of 212 persons were available for analysis for the clinic-initiated phase and 165 persons were available for analysis of the patient-initiated phase. Initial lesions healed in 6.3 days in the acyclovir-treated group and 7.0 days in the placebo-treated group in the clinic-initiated phase ($P<0.05$), but in 5.5 days and 6.5 days, respectively, in the patient-initiated ($P<0.001$). New lesion formation was reduced by acyclovir and, thus, "all" lesions healed sooner in the acyclovir group in both phases (6.3 vs. 7.4 days for clinic-initiated; $P<0.01$, and 5.7 vs. 7.2 days for patient-initiated; $P<0.001$). While the mean durations of itching and pain were less in the acyclovir group in both phases, neither parameter was shown statistically as a treatment benefit. Interestingly, 83% of acyclovir, and 85% of placebo-treated patients were culture-positive in the clinic-initiated phase, while only 59% of acyclovir vs. 77% of placebo-treated patients were culture-positive in the patient-initiated phase. This difference in the patient-initiated phase may have resulted from the drug therapy, or alternatively, may have favored less serious (culture-negative) disease in the drug-treatment group (15). Other studies with oral acyclovir by Salo (77) and Ruhnek-Forsbeck (78) failed to assess the virologic status of outbreaks during the study, although previous culture-confirmation was required. The Swedish study (77) demonstrated enhancement of healing times in the clinic-initiated phase from 7.5 days to 4.0 days overall ($P<0.001$). However, this reduction was entirely limited to men (8.0 days to 5.0 days). This was a crossover study where patients switched from acyclovir to placebo (or vice versa) for the second phase. The trial design concomitantly changed from clinic-initiated to patient-initiated. In the patient-initiated phase, acyclovir reached the 0.05 level of significance for lesion healing in women, but not in men. Symptom duration was also reduced in women (5 days vs. 2 days; $P<0.001$), but not in men. Nonlesional (aborted) episodes were more common in the acyclovir group (16% vs. 32%; $P=0.017$). Since these episodes were included in the overall analyses, they may account for much of the positive drug effects, since patients who were "healed" at presentation were used for calculating duration of lesions and symptoms. This was the only oral acyclovir study of treatment during the prodromal phase to show an effect on nonlesional prodromes. These investigators also asked patients to assess their treatment overall, and 89% of patients receiving acyclovir vs. 41% of patients receiving placebo felt they were receiving active drug ($P<0.001$).

Suppression of frequently recurrent infection. Once recurrent disease has established a pattern in any individual, it becomes appropriate to think in terms of emotional adjustment with no drug treatment, episodic treatment of recurrences, or suppression. Approximately 50% of patients will eventually notice a decrease in recurrence frequency with time, but this is by no means a predictable pattern (38). Treatment of widely-spaced episodes which would significantly alter symptoms or times to healing, or better yet, induce episodes to abort during the prodromal phase, would be a major addition to the physician's therapeutic armamentarium. As stated above, however, such a situation was reported in only one previous trial (77). Anecdotally, it seems that oral acyclovir, as treatment for established disease, is useful mainly for highly selected patients with infrequent recurrences who complain of consistently long recurrent outbreaks (>7days), preceded by extended prodromal periods (>12-24hours). In these situations, it may be possible to establish an effective circulating blood level of acyclovir at the point it still works best, i.e. before new lesion development. This specific subpopulation has never been studied in controlled trials, however.

In patients whose outbreak frequencies are increasing, the severity of emotional sequelae may increase with time (38). With each recurrence, many patients will suffer a relapse of anger, depression, and sleep disruption. In addition, high recurrence rates may significantly disrupt romantic interludes and one's sexual functioning. Therefore, the degree of impact on any individual's life is partly a function of recurrence frequency. A person with 2 or 3 outbreaks per month may be physically affected from 50-100% of the time, compared with a person with 2-3 outbreaks yearly who has active problems from herpes approximately 5% of the time. In order to offer assistance to patients with high frequency rates, several studies have been performed looking at chronic disease suppression using oral acyclovir. Among the first to be published were two 4 month studies from the U.S. Straus and his colleagues (45) evaluated oral acyclovir 200 mg capsules in a 3 times daily regimen in 32 evaluable patients with recurrence rates of 1 episode per month. Four of 16 (25%) recurred during 125 days of oral acyclovir treatment while 16 of 16 recurred during placebo treatment. A simultaneous report from Douglas and his colleagues (47) evaluated patients who gave a history of at least 6 episodes per 12 months, although the mean reported frequency was 13 episodes in that period. Of 153 patients enrolled, 143 completed 4 months' therapy including 47 placebo recipients, 45 recipients of five 200 mg capsules daily (acyclovir-5) and 51 recipients of two capsules daily (acyclovir-2). Ignoring the first week of therapy, placebo recipients experienced only a 2% "prevention" rate while acyclovir-5 recipients experienced a 76% prevention rate and acyclovir-2 recipients a 66% prevention rate during the 4 months of this clinical trial. A 75% reduction in recurrences was noted by 87% of acyclovir-5 recipients and 71% of acyclovir-2 recipients, compared with 13% of placebo recipients. Mindel and his associates (46) studied 4 doses daily with similar results. The 56 patients in this study had at least 4 recurrences yearly and received drug for 12 weeks. The mean recurrence rate per month was 1.4 in placebo-treated patients compared with 0.05 in the acyclovir group. This group was observed after treatment withdrawal where recurrence frequency was 1.09 per month for previous acyclovir recipients, compared with 1.25 per month for previous placebo recipients. A crossover study in patients with 8 recurrences yearly was subsequently reported by Thin and colleagues (48), again using 4 tablets daily for 84 days, followed by crossover. In this study 88% of patients on placebo developed lesions compared with 13% of acyclovir recipients. Another 42% developed symptoms and/or erythema only, while 45% of patients remained entirely symptom free on acyclovir. Halsos et al (49) used a 4 times daily schedule for 12 weeks and prevented lesions entirely in 77% of recipients compared with 10% on placebo. Sacks et al (23) studied 47 patients for 6 months of treatment, followed by a 6 month period of no prophylaxis. Intermittent therapy of active lesions was utilized rather than placebo alone. A patient home record was also added in order to motivate lesion reporting. Either because of the longer study duration, or because of the option for self-reporting of lesions, 71% of acyclovir recipients reported one or more lesion episodes during the study period. Nevertheless, the lesional recurrence rate was markedly reduced with acyclovir (0.29 per person per month vs. 1.16 per person per month for intermittent therapy). Reduction of acyclovir dosing from 200 mg 3 times daily to 400 mg 3 times daily on weekends only resulted in an increase in breakthrough failures (79), suggesting that daily dosing is necessary.

A variety of other approaches have been taken in an attempt to modify the course of established disease, including BCG vaccine (80) and smallpox vaccine (81). These agents are ineffective and potentially dangerous. Polio vaccine (82) and influenza vaccine have also been tried (83). Isoprinosine (84), systemic (85-87) and topical (88,89) interferons and their inducers (90) have also been subjected to clinical trials without major success. Human leukocyte interferon, however, has been shown to modify the course of oral-labial infection when given prophylactically after trigeminal nerve surgery (91). A recent report by Eron and his colleagues (92) outlined the effects of thrice-weekly subcutaneous injections of 3 x 10^6 IU of $alpha_{2b}$ interferon versus placebo in 37 patients with recurrent genital herpes, treated for 12 weeks. All patients had frequent recurrences (8 or more per year). Only 3 of 18 interferon patients remained lesion-free during the study period and no differences in lesion frequency or duration were observed in the interferon group compared with placebo. By contrast, Kuhls et al (93) found a moderate reduction of recurrence frequency, lesion duration, viral shedding, and lesion symptoms, utilizing the same preparation (identical lot numbers), dose, routes, and treatment period as Eron. Kuhls et al, however, compared $3x10^6$ I.U. to $1x10^6$ I.U. and to placebo. The lower dose was no more effective than placebo.

Friedman-Kien et al (94) recently reported on a new formulation using a combination of agents with demonstrated *in vitro* synergy. Their preparation contained either 10^5 or 10^7 IU of human leukocyte interferon per gram of cream base with 1% nonoxynol 9, a nonionic surfactant. A third group received placebo in this double-blind, randomized trial on 128 patients. Analyzed by log-rank test, no significant advantages were observed in cessation of viral shedding or lesion healing. Indeed, high dose interferon delayed healing significantly, presumably as a result of its antiproliferative effects. Scabbing of lesions was also seen less frequently in those treated with low dose interferon. Low dose interferon (but not high dose) was associated with a reduction in new lesion formation (P=0.006). The end of new lesion formation, as well as scabbing and lesion healing were all statistically superior in patients treated with low dose compared with high dose interferon. Further work with dosage adjustments and treatment periods, along with more frequent patient evaluations may help to elucidate the role for topical interferons and/or combinations in the future.

LESSONS FROM PREVIOUS STUDIES

The problems and successes with drug trials that have preceded, teach us that in designing a chemotherapeutic trial for a new compound, major categories of disease with proven (or suspected) differences in disease natural history must be separately studied. This will not only help us to avoid the pitfalls of completing biased studies, but will also help us to design each trial with objectives appropriate to the therapeutic expectations. Major categories of disease requiring separate analysis are outlined in Table 3.

Table 3. Categories of genital herpes requiring separate analysis

First-episode disease	Recurrent Disease
- Primary infection as defined by seronegativity at onset.	- Culture negative vs. culture-positive.
- Nonprimary, initial infection.	- Men vs. women.
- Men vs. women.	- Patient-initiated vs. clinic-initiated and/or prodrome-initiated vs. lesion-initiated.
- HSV-1 vs. HSV-2?	- Genital vs. nongenital (buttock, thigh).
	- Lesion size at presentation.

When studying recurrence suppression, patients need to be categorized on the basis of both their frequency and duration of infection before the study. Several parameters of efficacy have been utilized in clinical trials. These are outlined in Table 4.

The most crucial efficacy parameters in any study are times to healing and loss of viral shedding. Either parameter satisfied alone is insufficient without the other, since virus shedding must be reduced as an objective demonstration of efficacy, but an antiviral effect alone offers the patient no measurable clinical benefit. Ideally, an agent will also modify the duration of symptoms, in order to be of symptomatic benefit. New lesion formation may be relevant, in that it may be a more sensitive marker of prophylactic antiviral effect. However, in a trial of a topical agent, it should only be applied to new lesions in the treated area. Since the minority of recurrent episodes are associated with new lesion formation, this parameter might be more appropriately categorized as a new episode, although for primary infection, new lesion formation would seem to be a natural part of the disease evolution process in most patients.

Another potentially important efficacy parameter might be the analysis of nonlesional episodes. In a patient-initiated, prodrome-initiated trial, this could be a potential outcome of successful therapy. Unfortunately, this parameter has been under-reported. Its incidence may range from 5% to 40% with no treatment. Therefore, an efficacy trial should account for this random distribution in its projections for patient numbers. One would assume that a beneficial drug effect would consistently raise the

Table 4. Parameters of efficacy utilized in clinical trials.

Presenting lesion characteristics	Prodromes
Time to healing	Incidence at the genital site.
Time to crusting.	Incidence at remote site.
Duration of vesicle/ulcer.	Culture status during prodrome.
Time to near-healing (loose crust).	Duration of prodrome.
Duration of virus shedding (or change in viral titre).	Incidence of evolution into lesions.
Duration of pain/tingling.	
Duration of itching.	
New lesion characteristics*	**Related nonlesional disease* - incidence, severity**
Incidence of formation in treated area.	Lymphadenopathy incidence.
Incidence of formation in untreated areas.	Headache.
All parameters as for presenting lesion.	Dysuria.
	Sacral paresthesia.
	Fever.
	Cervicitis.
	Pharyngitis.
	Malaise

* Probably important to follow only in primary infection.

number of nonlesional episodes above the reported "normal range", or alternatively, establish a more proper normal range through careful, large-scale analyses of new trials. What does one do about the nonlesional prodrome in the analyses for other efficacy parameters? If culture-negative and excluded, this might bias strongly against a therapy directly proportional to its degree of effectiveness. If included in the analysis, should the data be recorded as zero time to healing?; or one day?; or varied based on the duration of prodromal symptoms? If included in the overall analyses, healing times may be reduced to unrealistically short intervals (41). Another option might be to ask patients to obtain cultures themselves, at home, prior to starting therapy, although the number of culture-positive prodromes may be low. Furthermore, what should be done about remote prodromes? In other words, where should topical treatment be applied, e.g., during sacral paresthesias? If the usual site where drug is applied turns out to be different from the actual lesion site, it would be a difficult task to separate them for analysis since, on the one hand they might be counted as new lesions in an untreated area, or, on the other hand, as original lesions. Another trial design alternative would be to delay treatment in a patient-initiated trial to the onset of lesions. This, combined with home cultures would obviate the selection bias problem, but works against the need to direct drug to the site of viral replication at the earliest possible moment.

It may be necessary to perform at least two trials for analysis of drug efficacy. The first, a lesion-initiated trial of established lesions will provide good data on duration of viral shedding and time to healing in groups separated by their culture status before therapy. If early events are missed, a prodrome-initiated trial will be used to assess for nonlesional episodes. Indeed, these trials can be conveniently done on sequential episodes in one patient. If the first episode is clinic-initiated, there is no need to confirm culture status before enrollment, since only the culture-positives will be analyzed for efficacy in that episode, and rerandomized for the patient-initiated phase.

Men and women should be separately analyzed and, therefore, separately randomized. Several studies (15,39) have now demonstrated that men take longer to heal than women. It is no more acceptable to lump them together in analyses than to lump any other mixed population, such as culture-positives and culture-negatives or primaries and nonprimaries. This may present a problem in studies of first episodes where it may be unrealistic to expect enough male patients to present for analysis. I would propose that demonstration of efficacy in male first episodes not be required, if efficacy is demonstrated in females. There should be little problem in finding sufficient numbers of both men and women with recurrent disease.

Patients should report for treatment in less than 24 hours and, ideally, in under 6 to 12 hours for recurrent disease. It is probably impossible to change the course of a 5 or 6 day outbreak after 24-48 hours have passed before treatment. This adds inconvenience to a study, but more closely reflects the "real-life" situation of treatment since most patients will be able to apply drug at home quite quickly after symptom onset. First episode patients have been generally enrolled up to 5 days after symptom onset. From a practical point of view, this 5 day allowance is usually a necessity. However, nonprimary disease cannot be adequately assessed under these conditons, and this may be reflected in the generally poor showing of agents in the treatment of this condition. One solution would be provided by the availability of rapid tests for serologic status at the time of enrollment. It may be necessary, however, to reduce allowable enrollment to 48 hours or less in all first episodes. From the ethical point of view, first episode trials should no longer be placebo controlled, since oral acyclovir has been shown to reduce complications (27) and, possibly, long-term recurrence rates (43).

Followup regimens for recurrent disease should likely be held on a daily basis, until healing, in order to avoid false delays in defining times to healing and other efficacy parameters, solely based on gaps between study visit days. If an effective drug will improve times by only 1 day, it makes little sense to skip days during the trial. Ideally, hours of the day will be recorded, as well, in order to define times as clearly as possible. This means recording the time of prodrome onset, lesion onset, and clinic followup visits.

Home records are of demonstrated value (22,23), but must be interpreted only in conjunction with objective data. It is reasonable to accept an educated patient's self-recorded healing time, if and only if, the record is reviewed daily and the investigator agrees with the assessment the next day.

It remains to be determined if lesion size determines outbreak duration or severity. However, for now, it makes sense to separately analyze outcome on the basis of lesion size at presentation. If no differences occur among groups, then collective analyses would be logical. This would also apply to areas where larger lesions are typical, e.g., the buttocks and thigh areas, and other "nongenital" sites.

SUMMARY

Unfortunately, eradication of latent infection remains impossible once disease is established. Nevertheless, major advances in chemotherapy have allowed us to modify the natural history of genital infection. At this time, treatment is appropriate for selected patients:

Immunocompromised hosts

Treatment of active disease is indicated during periods of peak compromise using intravenous or oral acyclovir. Topical acyclovir has only limited benefit, and is of little practical consideration. Prophylaxis of seropositives in high risk groups with oral acyclovir, or alternatively, intravenous acyclovir where mucositis prevents oral treatment.

First episode infection in the normal host

Early systemic treatment benefits primaries who may be indistinguishable from nonprimaries on initial presentation. Oral acyclovir is especially useful for prevention of new lesion formation. It may shorten the course of painful lymphadenopathy and/or dysuria, and some investigators feel that it is associated with a long-term reduction of lesion frequency in certain patients.

Topical therapy with polyethylene-glycol-based acyclovir is only of minor benefit in primary infection. Systemic treatment is superior. The aqueous cream base looks more promising, but further studies are needed to verify the results. Adding topical therapy to systemic therapy is of no benefit.

Active treatment of recurrent disease

Intravenous acyclovir is not of practical use. Oral acyclovir shortens the duration of lesions and the period of viral shedding by about 1.0 to 1.5 days. Symptoms are not consistently altered by treatment. Its major effects are on prevention of new lesion formation. Distributions of culture-positive and culture-negative episodes have frequently biased studies in favor of drug recipients. One study showed an increase in the percentage of episodes which are nonlesional when treatment is prodrome-initiated. Topical acyclovir in polyethylene glycol is of no benefit in recurrent infection. The aqueous cream formulation offers theoretical benefits, but only one study has been reported, which is inconclusive because of its small size and unusual distribution of culture-positives and culture-negatives.

Suppression of frequent recurrences

Beneficial and highly reproducible reductions in frequency rates of lesional episodes are seen during the therapy period. The long-term effects are unknown. Effective dosing ranges from 200 mg twice daily to five times daily, or 400 mg twice daily. Breakthrough lesions and asymptomatic virus shedding may occur infrequently on therapy. Nonlesional prodromes may actually increase in some patients.

CONCLUSIONS

The state of the art suggests that the most severely affected individuals will receive the greatest benefits from intervention. Following selected guidelines in the conduct of clinical trials will allow us to test new agents with care, so that the best possible and most realistic effects are recorded.

The next major step forward must make chemotherapy available to the average patient with genital herpes. Without cure, or, alternatively, absolutely safe and totally complete control, patients with herpes continue to seek reduction in symptoms. For most, however, prevention of transmission to partner or newborn is the first priority and, to date, no therapeutic agent offers more protection than barrier contraception and avoidance of symptomatic lesion contact. The average recurrent herpes patient is not symptomatically affected to a degree which calls for the use of an expensive, systemic nucleoside analogue. If oral acyclovir in recurrent herpes could be shown to consistently reduce lesion development or symptom duration with prodromal use, the situation might be different. Eventually, acyclovir's long-term safety record, or modification of its pharmacokinetic problems through use of its prodrug (95) may make its use more appropriate for the mild to moderately affected person with herpes. For now, the vast majority of patients will benefit most from proper diagnosis, education, counselling, and control of treatable genital infections. Therapy in these cases should be limited to prevention of complications, i.e., autoinoculation, neonatal herpes, transmission to partners, and/or complications of inappropriate therapy such as steroids, antibiotics, and occlusive ointments, as well as the elimination of fruitless ingestion of "alternative substances".

In order to provide symptomatic relief without attendant risk, many compounds are under continued study in topical formulations. So far, none are of proven benefit in recurrent disease. A successful agent will provide a cosmetically acceptable formulation which is well-absorbed into skin, but poorly absorbed into the general circulation. Ideally, it should have no significant local or systemic adverse effects, while providing acceptable antiviral activity to the site of infection. A reduction in virus shedding duration and time to healing should be demonstrated. Symptom relief should be more rapid with drug use and, ideally, the agent will reduce the number of lesional episodes when used early in a prodromal period. Such a practical tool has remained elusive to date - partly because we await new invention, and partly because we are still learning how to demonstrate the effects of new agents in clinical trials.

REFERENCES

1. Wildy, P. In: The Herpesviruses (Ed. A.S. Kaplan), Academic Press, New York, 1973, pp. 1-25.
2. Astruc, J. In: De Morbis Venereis: Libri Sex (III:VIII), Apud Guillelmum Cavelier, Paris, 1736, p. 254 (III).
3. Lipschutz, B. Arch. Dermatol. Syph. (Berl.) 136:428, 1929.
4. Parker, F., Jr., and Nye, R.N. Am. J. Pathol. 1:337-340, 1925.
5. Schneweis, K.E. Z. Immunitatsforsch. 124:24-48, 1962.
6. Dowdle, W.R., Nahmias, A.J., Harwell, R.W., and Pauls, F.P. J. Immunol. 99:974-980, 1967.
7. Stavraky, K.M., Rawls, W.E., Chiavetta, J., Donner, A.P., and Wanklin, J.M. Amer. J. Epidemiol. 118:109-121, 1983.
8. Becker, T.M., Blount, J.H., and Guinan, M.E. JAMA. 253:1601-1603, 1985.
9. Britto, E., Dikshit, S.S., and Naylor, B. Univ. Mich. Med. Ctr. J. 42:152-154, 1976.
10. MacDougall, M.L. N.Z. Med. J. 82:333-335, 1975.
11. Srivannaboon, S. J. Med. Assoc. Thai. 62:201- , 1979.
12. Corey, L. In: Sexually Transmitted Disease (Eds. K.K. Holmes, P.A. Mardh, P.F. Sparling, and P.J. Wiesner), McGraw-Hill, Inc., New York, 1984, pp. 449-474.
13. Sacks, S.L. Med. N. Amer. 6:526-532, 1983.
14. Adams, H.G., Benson, E.A., Alexander, E.R., Vontver, L.A., Remington, M.A., and Holmes, K.K. J. Infect. Dis. 133 (Suppl.): A151-A159, 1976.
15. Sacks, S.L., Portnoy, J., Lawee, D., Schlech, W., Aoki, F., Tyrrell, D.L., and Poisson, M. J. Infect. Dis. 1987, In press.
16. Overall, J.C., Jr. In: Antiviral Agents and Viral Diseases of Man (Eds. G.J. Galasso, T.C. Merigan, Jr., R.A. Buchanan), Raven Press, New York, 1979, pp. 305-384.
17. Corey, L., Reeves, W.C., and Holmes, K.K. N. Engl. J. Med. 299:986-991, 1978.
18. Corey, L., Adams, H.G., Brown, Z.A., and Holmes, K.K. Ann. Intern. Med. 98:958-972, 1983.
19. Bernstein, D.I., Lovett, M.A., and Bryson, Y.J. Amer. J. Med. 77:1055-1060, 1984.
20. Reeves, W.C., Corey, L., Adams, H.G., Vontver, L.A., and Holmes, K.K. N. Engl. J. Med. 305:315-319, 1981.

21. Richards, J.T., Kern, E.R., Overall, J.C., Jr., and Glasgow, L.A. J. Infect. Dis. 144:464-471, 1981.
22. Sacks, S.L. J. Infect. Dis. 150:873-877, 1984.
23. Sacks, S.L., Fox, R., Levendusky, P., and Stiver, H.G. In: Recent Advances in Chemotherapy (Ed. J. Ishigami), University of Tokyo Press, Tokyo, 1985, pp. 1961-1962.
24. Mindel, A., and Sutherland, S. J. Antimicrob. Chemother. 12 (Suppl. B): 51-59, 1983.
25. Corey, L., Nahmias, A.J., Guinan, M.E., Benedetti, J.K. Critchlow, C.W., and Holmes, K.K. N. Engl. J. Med. 306:1313-1319, 1982.
26. Mertz, G.J., Critchlow, C.W., Benedetti, J., Reichman, R.C., Dolin, R., Connor, J., Redfield, D.C., Savoia, M.C., Richman, D.D., Tyrrell, D.L., Miedzinski, L., Portnoy, J., Keeney, R.E., and Corey, L. JAMA. 252:1147-1151, 1984.
27. Bryson, Y.J., Dillon, M., Lovett, M., Acuna, G., Taylor, S., Cherry, J.D., Lamar Johnson, B., Wiesmeier, E., Growdon, W., Creagh-Kirk, T., and Keeney, R. N. Engl. J. Med. 308:916-921, 1983.
28. Mindel, A., Adler, M.W., Sutherland, S., and Fiddian, A.P. Lancet I:697-700, 1982.
29. Caplan, L.R., Kleeman, F.J., and Berg, S. N. Engl. J. Med. 297:920-921, 1977.
30. Craig, C., and Nahmias, A. J. Infect. Dis. 127:365-372, 1973.
31. South, M.A., Tompkins, W.A.F., Morris, C.R., and Rawls, W.E. J. Pediatr. 75: 13-18, 1969.
32. Hanshaw, J.B. Am. J. Dis. Child. 126:546-555, 1973.
33. Nahmias, A.J., Josey, W.E., Naib Z.M. Freeman, M.G., Fernandez, R.J., and Wheeler, J.H. Am. J. Obstet. Gynecol. 110:835-837, 1971.
34. Sullivan-Bolyai, J. JAMA. 250:3059-3062, 1983.
35. Todd, M.J., Hockin, J.C., and Jessamine, A.G. Can. Dis. Weekly Rep. 11:161-163, 1985.
36. Nahmias, A.J., and Sawanabori, S. Prog. Exp. Tumor Res. 21:117-139, 1978.
37. Kaufman, R.H., Dreesman, G.R., Burek, J., Korhonen, M.O., Matson, D.O., Melnick, J.L., Powell, K.L., Purifoy, D.J.M., Courtney, R.J., and Adam, E. N. Engl. J. Med. 305:483-488, 1981.
38. Sacks, S.L.and Koss, M. In: Programs and Abstracts of the 26th Interscience Conference on Antimicrobial Agents and Chemotherapy, American Society for Microbiology, Washington, 1986, Abstr. 533.
39. Reichman, R.C., Badger, G.J., Guinan, M.E., Nahmias, A.J., Keeney, R.E., Davis, L.G., Ashikaga, T., and Dolin, R. J. Infect. Dis. 147:336-340, 1983.
40. Douglas, J.M., Judson, F.N., Levin, M.J., Bosso, J.A., Spruance, S.L., Johnston, J.M., Corey, L., McMillan, J.A., Weiner, L.B., and Frank, J.A., Jr. Antimicrob. Agents Chemother. 29:464-467, 1986.
41. Luby, J.P., Gnann, J.W., Jr., Alexander, W.J., Hatcher, V.A., Friedman-Kien, A.E., Klein, R.J., Keyserling, H., Nahmias, A., Mills, J., Schachter, J., Douglas, J.M., Corey, L., and Sacks, S.L. J. Infect. Dis. 150:1-6, 1984.
42. Yamamoto, H., Walz, M.A., and Notkins, A.L. Virology 76:866-869, 1977.
43. Bryson, Y.J. Scand. J. Infect. Dis. 47:70-75, 1985.
44. Corey, L., Mindel, A., Fife, K.H., Sutherland, S., Benedetti, J., Adler, M.W. Sex. Transm. Dis. 12:215-218, 1985.
45. Straus, S.E., Takiff, H.E., Seidlin, M., Bachrach, S., Lininger, L., DiGiovanna, J.J., Western, K.A., Smith, H.A., Nusinoff Lehrman, S., Creagh-Kirk, T., and Alling, D.W. N. Engl. J. Med. 310:1545-1550, 1984.
46. Mindel, A., Faherty, A., Hindley, D., Weller, I.V.D., Sutherland, S., Fiddian, A.P., and Adler, M.W. Lancet II: 57-59, 1984.

47. Douglas, J.M., Critchlow, C., Benedetti, J., Mertz, G.J., Connor, J.D., Hintz, M.A., Fahnlander, A., Remington, M., Winter, C., and Corey, L. N. Engl. J. Med. 310:1551-1556, 1984.
48. Thin, R.N., Jeffries, D.J., Taylor, P.K., Ayra, O.P., Rodin, P., Yeo, J., and Fiddian, A.P. J. Antimicrob. Chemother. 16:219-226, 1985.
49. Halsos, A.M., Salo, O.P., Lassus, A., Tjøtta, E.A.L., Hovi, T., Gabrielsen, B.O., and Fiddian, A.P. Acta. Derm. Venereol. (Stockh.) 65:59-63, 1985.
50. Cook, M.L., and Stevens, J.G. Infec. Immun. 7:272-288, 1973.
51. Park, N.-H., Pavan-Langston, D., and McLean, S.L. J. Infect. Dis. 140:802-806, 1979.
52. Corey, L., Fife, K.H., Benedetti, J.K., Winter, C.A., Fahnlander, A., Connor, J.D., Hintz, M.A., and Holmes, K.K. Ann. Intern. Med. 98:914-921, 1985.
53. Nilsen, A.E., Aasen, T., Halsos, A.M., Kinge, B.R., Tjøtta, E.A.L., Wikstrom, K., and Fiddian, A.P. Lancet II:571-573, 1982.
54. Thin, R.N., Nabarro, J.M., Parker, J.D., and Fiddian, A.P. Br. J. Vener. Dis. 59:116-119, 1983.
55. Hubler, W.R., Jr., Felber, T.D., Troll, D., and Jarratt, M. J. Invest. Dermatol. 62:92-95, 1974.
56. Freeman, D.J., and Spruance, S.L. J. Infect. Dis. 153:64-70, 1986.
57. Kinghorn, G.R., Turner, E.B., Barton, I.G., Potter, C.W., Burke, C.A., and Fiddian, A.P. Antiviral Res. 3:291-301, 1983.
58. Fiddian, A.P., Kinghorn, G.R., Goldmeir, D., Rees, E., Rodin, P., Thin, R.N.T., and de Konig, G.A.J. J. Antimicrob. Chemother. 12 (Suppl. B):67-77, 1983.
59. Kinghorn, G.R., Abeywickreme, I., Jeavons, M., Barton, I., Potter, C.W., Jones, D., and Hickmoft, E. Genitourin Med. 62:186-188, 1986.
60. MacCallum, F.O., and Juel-Jensen, B.E. Br. Med. J. 2:805-807, 1966.
61. Davidson-Parker, J. J. Antimicrob. Chemother. 3 (Suppl. A):131-137, 1977.
62. Silvestri, D.L., Corey, L., and Holmes, K.K. JAMA. 248:953-959, 1982.
63. Eron, L.J., Harvey, L., Gleason, J., Toy, C., Marcus, E. In: Programs and Abstracts of the 26th Interscience Conference on Antimicrobial Agents and Chemotherapy, American Society for Microbiology, Washington, 1986, Abstr. 624.
64. Blough, H.A., and Giuntoli, R.L. JAMA 241:2798-2801, 1979.
65. Corey, L., and Holmes, K.K. JAMA. 243:29, 1980.
66. Kern, E.R., Glasgow, L.A., Klein, R.J., and Friedman-Kien, A.E. J. Infect. Dis. 146:159-165, 1982.
67. Alenius, S., Berg, M., Broberg, F., Eklind, K., Lindborg, B., and Oberg, B. J. Infect. Dis. 145:569-573, 1982.
68. Wallin, J., Lernestedt, J.-O., Ogenstad, S., and Lycke, E. Scand. J. Infect. Dis. 17:165-172, 1985.
69. Barton, S.E., Munday, P.E., Kinghorn, G.R., van der Meijden, W.I., Stolz, E., Notowicz, A., Rashid, S., Schuller, J.L., Essex-Cater, A.J., Kuijpers, M.H.M., and Chanas, A.C. Genitourin Med. 62:267-269, 1986.
70. Bierman, S.M., Kirkpatrick, W., and Fernandez, H. Chemother. 27:139-145, 1981.
71. Walsh, D.E., Griffith, R.S., and Behforooz, A. J. Antimicrob. Chemother. 12:489-496, 1983.
72. DiGiovanna, J.J., and Blank, H. Arch. Dermatol. 120:48-51, 1984.
73. Jose, D.G., and Minty, C.C.J. Med. J. Aust. 2:390-394, 1980.
74. Bierman, S.M. Cutis 21:352-354, 1978.
75. Starr, S.E. Cutis 20:596-598, 1977.
76. Reichman, R.C., Badger, G.J., Mertz, G.J., Corey, L., Richman, D.D., Connor, J.D., Redfield, D., Savoia, M.C., Oxman, M.N., Bryson, Y., Tyrrell, D.L., Portnoy, J., Creigh-Kirk, T., Keeney, R.E., Ashikaga, T., and Dolin, R. JAMA. 251:2103-2107, 1984.

77. Salo, O.P., Lassus, A., Hovi, T., and Fiddian, A.P. Eur. J. Sex. Transm. Dis. 1:95-98, 1983.
78. Ruhnek-Forsbeck, M., Sandstrom, E., Andersson, B., Eriksson, G., Hersle, K., Lovhagen, G.-B., Mobacken, H., Hillstrom, L., and Svensson, L. J. Antimicrob. Chemother. 16:621-628, 1985.
79. Straus, S.E., Seidlin, M., Takiff, H.E., Rooney, J.F., Lehrman, S.N., Bachrach, S., Felser, J.M., DiGiovanna, J.J., Grimes, G.J., Jr., Krakauer, H., Hallahan, C., and Alling, D. Antiviral Res. 6:151-159, 1986.
80. Douglas, J.M., Vontver, L.A., Stamm, W.E., Reeves, W.C., Critchlow, C., Remington, M.L., Holmes, K.K., and Corey, L. Antimicrob. Agents Chemother. 27:203-206, 1985.
81. Kern, A.B., and Schiff, B.L. J. Invest. Dermatol. 33:99-102, 1959.
82. Tager, A. Dermatologica 149:253-255, 1974.
83. Miller, J.B. Ann. Allergy 43:295-305, 1979.
84. Kalimo, K.O.K., Joronen, I.A., and Havu, V.K. Arch. Dermatol. 119:463-467, 1983.
85. Kuhls, T.L., Sacher, J., Wiesmeier, E., Santomauro, D., Pineda, E., and Bryson, Y. In: Program and Abstracts of the 25th Interscience Conference on Antimicrobial Agents and Chemotherapy, American Society for Microbiology, Washington, 1985, Abstr. 94.
86. Radolf, J., Wofsy, C., Niland, N., Lovett, M., Dillon, M., Fall, H., Hauer, L., Bigley, J., Rosenbaum, D., Gordin, F., Nadler, P., Mills, J., and Bryson, Y. In: Program and Abstracts of the 25th Interscience Conference on Antimicrobial Agents and Chemotherapy, American Society for Microbiology, Washington, 1985, Abstr. 91.
87. Eron, L.J., Toy, C., Harvey, L, and Santomauro, D. In: Program and Abstracts of the 25th Interscience Conference on Antimicrobial Agents and Chemotherapy, American Society for Microbiology, Washington, 1985, Abstr. 92.
88. Friedman-Kien, A.E., Lafleur, F., Glaser, R., Rosenberg, J., and Zang, E. In: Program and Abstracts of the 25th Interscience Conference on Antimicrobial Agents and Chemotherapy, American Society for Microbiology, Washington, 1985, Abstr. 93.
89. Eron, L., Toy, C., Wood, D., and Nadler, P. In: Recent Advances in Chemotherapy (Ed. J. Ishigami), University of Tokyo Press, Tokyo, 1985, pp. 1967-1968.
90. Crane, L.R., Levy, H.B., and Lerner, A.M. Antimicrob. Agents Chemother. 21:481-485, 1982.
91. Pazin, G.J., Armstrong, J.A., Lam, M.T., Tarr, G.C., Jannetta, P.J. and Ho, M. N. Engl. J. Med. 301:225-230, 1979.
92. Eron, L.J., Harvey, L., Toy, C., and Santomauro, D. Antimicrob. Agents Chemother. 30:608-610, 1986.
93. Kuhls, T.L., Sacher, J., Pineda, E., Santomauro, D., Wiesmeier, E., Growdon, W.A., and Bryson, Y.J. J. Infect. Dis. 154:437-442, 1986.
94. Friedman-Kien, A.E., Klein, R.J., Glaser, R.D., Czelusniak, B.A. J. Amer. Acad. Dermatol. 15:989-994, 1986.
95. Krenitsky, T.A., Hall, W.W., De Miranda, P., Beauchamp, L.M., Schaeffer, H.J., and Whiteman, P.J. Proc. Natl. Acad. Sci. (USA) 81:3209-3213, 1984.

8

TREATMENT OF HERPES SIMPLEX VIRUS INFECTIONS IN IMMUNOSUPPRESSED PATIENTS

R. SARAL and P. LIETMAN

Department of Oncology and the Division of Clinical Pharmacology of the Department of Medicine and of the Department of Pharmacology and Molecular Sciences, The Johns Hopkins University School of Medicine, Baltimore Maryland, USA

ABSTRACT

Herpes simplex virus (HSV) infection is a major cause of morbidity in immunosuppressed patients. In rare instances HSV infection also may be a primary cause of death. The natural history of these infections has been well described in selected patient populations. The temporal occurrence of HSV infections is predictable in many immunosuppressed patients. Advances in antiviral therapy have led to the development of compounds which inhibit HSV replication. Carefully conducted, controlled, randomized, double-blind clinical trials have demonstrated that acyclovir is currently the most effective therapy for the treatment of established HSV infections in immunosuppressed patients.

Viral infections are major causes of morbidity and mortality in immunosuppressed patients (1). The herpes viruses have been among the most carefully studied groups of viruses causing infections in this heterogenous patient population. This manuscript will discuss therapeutic approaches to herpes simplex virus (HSV) infections in immunosuppressed patients.

The natural history of HSV infections in immunosuppressed patients stands is marked contrast to that seen in normal individuals. While primary infections with HSV may be severe in normal individuals, reactivation disease is generally self-limited. Studies done by Spruance (2) and Guinan (3) have defined the natural history of reactivated oral and genital HSV in normal individuals and demonstrated that viral shedding in lesions is short lived, pain is not severe and healing occurs 8 to 12 days after onset. In immunosuppressed patients primary infections with HSV are rarely encountered. Reactivated HSV infections, however may cause significant morbidity and rarely mortality (1). The natural history of these infections has also been well studied, particularly as a part of carefully designed, randomized, double blind, placebo controlled clinical trials evaluating therapeutic approaches to HSV infections in immunosuppressed patients (4,5). The placebo arms of these studies show that viral shedding in lesions may persist for 2 to 3 weeks. Pain may be very severe and healing may not occur for 3 to 4 weeks. The tissue destruction caused by reactivation of HSV in the immunosuppressed patient results not only in pain but in disruption of natural barriers in the oral and genital area increasing the risk for both bacterial and fungal infections. Since many of these patients may have altered granulocyte function or absence of granulocytes, related to their disease or its treatment, disruption of these natural barriers may create sites for local infection and may also increase the risk of bacteremia and fungemia. HSV esophagitis is another clinical problem seen in immunosuppressed patients with reactivated infection. In addition to the risks of local and disseminated bacterial and fungal infections, the pain created by reactivated HSV infections of the oral cavity and esophagitis reduces oral intake and may create the need to increase parenteral fluid supplementation and to provide or increase nutritional supplementation. In rare instances HSV pneumonia may occur, leading to respiratory failure (6). Viremia also may occur, leading to disseminated HSV infection. In the pre-antiviral era, pneumonia and viremia were universally fatal complications.

An important feature of reactivated HSV infections in immunosuppressed patients is their predictability of occurrence (17). While this generalization may not be applicable to all patient populations, several different groups of immunosuppressed patients who have been carefully studied demonstrate remarkably similar patterns of reactivation of HSV. In all groups studied the occurrence of HSV infection appears to be reactivation of HSV rather than primary infection. Patients at risk are those who have evidence of latent virus determined by preexisting antibody. One of the first group of patients studied were renal transplantation recipients (8). Reactivated HSV infections were generally seen in the first month following transplantation. Similar observations have been made in cardiac transplant recipients (9). We and others have used bone marrow transplant recipients as a model for studying herpes virus infections. In this group of patients, HSV infections occur in 70 to 80% of all seropositive individuals. The median day of onset is 18 days after treatment with chemotherapy or 8 days following bone marrow transplantation. We extended our observations to patients with acute leukemia receiving intensive chemotherapy and found that seropositive patients with pretreatment antibody titers of 1:16 or greater had a 65% incidence of HSV infection occurring a median of 17 days following initiation of chemotherapy (1). This temporal occurrence is remarkably similar to observations made in bone marrow transplant recipients and to those made in organ transplant recipients. There are other groups of patients who would appear to be at risk for reactivated HSV infections who have not yet been carefully studied. Patients with non-Hodgkin's lymphoma frequently receive repetitive intensive chemotherapy and, although we have not prospectively studied HSV infections in this patient population, we have seen patients who have reactivated HSV with clinically significant infections following chemotherapy in a time frame similar to that observed in leukemics and bone marrow transplant recipients. The exact mechanisms that trigger reactivations of latent HSV in immunosuppressed patients are complicated and incompletely understood. However, the temporal occurrence is of importance for two reasons. Firstly, it allows the clinician to evaluate patients carefully during the "period of risk" to define such lesions so that early antiviral therapy might be initiated. Secondly, we believe that the occurrence of predictable disease leads to an alternative approach to the treatment of established HSV infection, namely suppression of the virus with antiviral therapy during the "period of risk".

If this chapter had been written ten years ago, the primary purpose of a chapter on the therapy of HSV infection in immunosuppressed patients could be dispensed within a few sentences. Fortunately, that is not the case at present. The early experience with antiviral therapy for herpes simplex virus infections is of importance mainly because the initial and subsequently unsubstantiated reports of success were based on anecdotal case studies. Cytosine arabinoside and iodoxuridine were felt to be effective antiherpes agents when given systemically to patients with established infection (10,11). Yet, when controlled clinical trials were conducted each of these agents was found to be highly toxic and of no clinical value of the treatment of herpes virus infections (12,13). In fact our leukemia patients currently receiving cytarabine as therapy for their leukemia and seropositive patients (antibody titer of 1:16 or greater) have a 65% rate of reactivation of HSV attesting to its lack of efficacy as an anti-HSV drug. The true test of efectiveness of anti-HSV therapy can best be determined by examining the results of prospective, controlled, randomized, double-blind clinical trials. In fact there are such trials and they demonstrate the efficacy of antiviral agents in the treatment of HSV infections in immunosuppressed patients.

Vidarabine was the first drug approved by the Food and Drug Administration for the treatment of serious HSV infections in man. Although the clinical trials of vidarabine as treatment of mucocutaneous HSV infection in the immunosuppressed patient were published after trials evaluating acyclovir, in this review we will examine vidarabine prior to acyclovir since it was available for clinical trials much earlier. Eighty-five immunosuppressed patients with documented mucocutaneous HSV infection were randomized to receive vidarabine or placebo (14). Patients who received vidarabine had more pain relief than placebo recipients ($P = 0.0099$) and a shorter duration of fever ($P = 0.03$). However, no statistical significance was noted in several other important variables although trends favoring the vidarabine group emerged in this trial. These variables included cessation of viral shedding, time of cessation of new lesion formation ($P = 0.12$) and earlier time to crusting ($P = 0.2$). The results of this trial, when compared to trials evaluating acyclovir in the treatment of HSV infections, are clearly less impressive. However, to truly place the role of vidarabine in perspective with acyclovir as a treatment for HSV infections in the immunosuppressed patient, a direct comparison of these two compounds in clinical trials would be optimal. When one considers the other advantages of acyclovir, however, it is unlikely that such a clinical trial will be performed.

The synthesis of acyclovir and early *in vitro* studies evaluating its mechanisms of action suggested that this compound would be a major development in HSV therapy (15). Clinical trials have confirmed the early speculations and acyclovir is currently the treatment of choice for HSV infections in immunosuppressed patients. The intravenous formulation of acyclovir was initially evaluated in a multicenter clinical trial which was prospective, controlled, randomized and double-blinded. Ninety-seven patients with established lesions that were HSV culture positive were randomized to receive acyclovir or placebo (5). The most dramatic effect of acyclovir was on virus shedding. Patients receiving acyclovir were free of virus in lesions after a median of 2.8 days compared to a median of 16.8 days for patients receiving placebo ($p<0.0002$). There was also a statistically significant effect of acyclovir treatment on pain ($p<0.01$), scabbing ($p<0.004$) and time to healing ($p<0.004$). In this clinical trial acyclovir was administered in a dose of 250 mg/M^2 every eight hours.

Topical acyclovir has also been evaluated in the treatment of HSV infection of the lips and skin in 63 immunosuppressed patients (4). In this placebo controlled clinical trial patients who received acyclovir had more rapid cessation of viral shedding ($p<0.001$), pain loss ($p<0.004$) and healing ($p<0.038$). The results of this trial are very similar to those reported by investigators evaluating intravenous acyclovir.

Oral acyclovir has been extensively studied in the treatment and prevention of HSV infection in normal individuals. Only one study evaluating oral acyclovir in immunosuppressed patients exists in the literature but the results are very similar to those achieved with the intravenous or topical formulations (16). Twenty-one bone marrow transplant recipients were randomized to receive either 800 mg of oral acyclovir five times a day or placebo. Acyclovir significantly shortened the duration of viral shedding ($p<0.008$), new lesion formation ($p<0.003$), time to first decrease in pain ($p<0.04$) and healing ($p<0.01$).

In summarizing the clinical trials of acyclovir, it is clear that all formulations of the drug (intravenous, topical and oral) are potent inhibitors of HSV infections.

For immunosuppressed patients with documented HSV infections the clinical presentation should probably determine the mode of therapy prescribed. In hospitalized patients we prefer the use of intravenous acyclovir using a dose of 250 mg/M^2 every 8 hours, acknowledging that the optimal dose and schedule may not yet have been defined. If lesions are only external, then topical acyclovir as a 5% ointment may be used six times daily. The trial evaluating oral acyclovir for the treatment HSV infections in immunosuppressed patients used a dose of 800 mg five times daily. In other patient populations the dose of oral acyclovir used to treat established HSV infections was 200 mg five times a day. Therefore in immunosuppressed outpatients who develop HSV infection the optimal oral dose of acyclovir to treat dose established HSV infection has not yet been defined in clinical trials.

In immunosuppressed patients who have predictable patterns of HSV reactivation and disease, we recommend the use of acyclovir prophylactically as outlined in previous studies (17,18). This approach is discussed in greater detail in another chapter in this monograph.

Side effects attributable to acyclovir have been infrequently reported. The major side effects have been noted with the intravenous formulation of acyclovir. Rarely reversible renal dysfunction secondary to intravenous drug may occur but may be prevented with adequate hydration and infusion of the drug over an hour. Central nervous system toxicity also has been associated with the intravenous formulation and is characterized by aqitation, tremor, altered mental studies and obtundation. Most of the patients experiencing this side effect have been extremely complicated patients receiving multiple medications with multiple medical problems. The oral formulation produces very few side effects, but nausea, vomiting, headache, and diarrhea have been reported and these same symptoms have been noted with intravenous acyclovir. The use of topical acyclovir may be associated with local irritation. Only with greater use of the drug and careful monitoring will rare or identified and placed in perspective long term side effects.

A potential problem with the use of acyclovir as treatment for established HSV infection is the development of drug resistance. The definition of resistance has varied from laboratory to laboratory and is quite dependent on the host cell used for viral growth, the medium in which the cells are grown and the multiplicity of viral infection. Although no precise definition has emerged, an arbitrary definition of resistance using <u>in vitro</u> susceptibility (ID50, the concentration of drug required to inhibit viral growth by 50%) would classify HSV strains as resistant if the ID50 is greater than 2 ug/ml. In one clinical study using acyclovir to treat established HSV infections in bone marrow transplant recipients, 3 of 52 patients were found to have relatively resistant viral isolates present after initiation of therapy (19).

Perhaps more important to the potential pathogeneicity of resistant strains is the biochemical basis for resistance. There are three mechanisms whereby resistance may occur. The resistant virus may lead to the productional a reduced amount of a normal thymidine kinase, an altered thymidine kinase or an altered DNA polymerase. The majority of resistant isolates cultured from humans have been thymidine kinase deficient strains which are relatively non-virulent when studied in animal models. Although HSV strains with an altered thymidine kinase have been isolated in man and neurovirulence in animal models has been demonstrated, the pathogeneicity of these strains in humans has not been determined (20). To date no HSV strains with altered DNA polymerase have been recovered in man.

Although "resistant" strains have been isolated from patients during, therapy they do not seem to persist.

Patients who have these strains generally heal their lesions and when they reactivate the HSV that is cultured is sensitive to acyclovir suggesting that in man the resistant strains isolated to date cannot establish latency.

One theoretical way to minimize the emergence of acyclovir resistance is to use the drug prophylactically rather than as therapy for established HSV infection in the immunosuppressed patient. The use of acyclovir when viral burden is low or not present is less likely to result in the emergency of drug resistance, than if the drug is used to treat active HSV infection when viral burden is high (21). In fact we have given acyclovir prophylactically to over 200 patients and have observed only one patient who had one isolated culture positive for HSV without lesions while on the drug. In addition, we have never isolated a resistant strain of virus from patients who received the drug prophylactically.

The fact that, to date, resistant strains have not emerged as a significant clinical problem in immunosuppressed patients should not stop ongoing research to determine whether such isolates will emerge with the increasing use of acyclovir.

With the availability of acyclovir the need for additional antiviral agents directed at HSV might at first glance be questioned. However, studies done on resistant HSV strains have shown that they may remain sensitive to other nucleoside analogues or to pyrophosphate analogues (20). Therefore it is important to search for additional anti-HSV agents which may be used in future situations where resistant strains are responsible for significant infections in immunosuppressed patients.

At present there are other drugs, apart from acyclovir and vidaribine, that are available as experimental agents and that inhibit HSV replications <u>in vitro</u>. These include BVDU, phophonoformic acid, the halogenated pyrimidine nucleosides (FIAC, FIAU, and FMAU) and interferon. However, not all these drugs have been evaluated in controlled trials; hence the question arises whether they are efficacious in the treatment of HSV infections. Furthermore, there are suggestions that they are potentially more toxic than acyclovir.

One potential supplement to acyclovir in the treatment of HSV infections might be vaccination as an adjunct to treatment and prophylaxis. Studies in bone marrow transplant recipients demonstrate that immunization with "recall" antigens like tetanus on the day of transplantation stimulates both a humoral and cellular response to the specific antigen. If any appropriate HSV vaccine were available one approach to preventing recurrent HSV disease might be to immunize the patient while concurrently treating or suppressing the patient. Presumably the immunization might boost the immune response while the drug (acyclovir) exerted its antiviral effect. Theoretically this approach may reduce recurrences of HSV infection. This is currently speculation but might be one of the areas of clinical research to pursue in the future.

Clearly the past few years have seen major advances in antiviral therapy. These advances have been translated into effective and safe therapy of HSV infections in immunosuppressed patients. With the availability of such therapy the morbidity and mortality associated with HSV infections have been markedly reduced.

REFERENCES

1. Saral, R., Burns, W.H., Prentice, H.G., Clin. in Hemat. 13:3, 645-660, 1984.
2. Spruance, S.L., Overall, J.C., Jr., Kern, E.R., Krugger, G.G., Pliam, V., Miller, W. N. Engl. J. Med. 297: 69-75, 1977.
3. Guinan, M.E., MacCalman, J., Kern, E.R., Overall, J.C., Jr., Spruance, S.L. N. Engl. J. Med. 304: 759-63, 1981.
4. Whitley, R.J., Levin, M., Barton, N., et al. J. Infect. Dis. 150: 323-329, 1984.

5. Meyers, J.D., Wade, J.C., Mitchell, C.D., et al. Amer. J. Med. 73(Suppl.): 229-235, 1982.
6. Ramsey, P.G., Fife, K.H., Hackman, R.C. et al. Ann. Int. Med. 97: 813-820, 1982.
7. Elfenbein, G.J., Saral, R. In Infection and the compromised Host (ed.). Allen, J.C. Baltimore, Williams & Wilkens. 157-196, 1981.
8. Pass, R.F., Whitley, R.J., Whelchel, J.D. J. Infect. Dis. 140: 487-492, 1979.
9. Rand, K.H., Rasmussen, L.E., Pollard, R.B., et al. N. Engl. J. Med. 296: 1372-1377, 1977.
10. Breeden, C.J., Hall, T.C., Tyler, H.R., Ann. Int. Med. 65: 1050-1056, 1966.
11. Chow, A.W., Ronald, A., Fiala, M., et al. Antimicrob. Agents Chemother. 3: 412-417, 1973.
12. Stevens, D.A. Jordan, G.W., Waddell, T.F., Merigan, T.C., N. Engl. J. Med. 289: 873-878, 1973.
13. Boston Interhospital Virus Study Group and the NIAID-Sponsored Cooperative Antiviral Clinical Study. N. Engl. J. Med. 292: 599-603, 1975.
14. Whitley, R.J., Spruance, S., Hayden, F.G. J. Infect. Dis. 149: 1-8, 1984.
15. Elion, G.B., Furman, P.A., Fyfe, J., et al. PNAS, USA, 74: 5716-20, 1977.
16. Shepp, D.H., Newton, B.A., Dandliker, P.S., Flournoy, N., Meyers, J.D. Ann. Int. Med. 102: 783-785, 1985.
17. Saral, R., Burns, W.H., Laskin, O.L., et al. N. Engl. J. Med. 305: 63-67, 1981.
18. Saral, R., Ambinder, R.F., Burns, W.H., et al. Ann. Int. Med. 99: 773-776, 1983.
19. Wade, J.C. McClaren, C., Meyers, J.D. J. Infect. Dis. 148: 1077-1082, 1983.
20. McLaren, C., Chen, M.S., Ghazzouli, I., Saral, R., Burns, W.H. Antimicrob. Agents Chemother. 28: 740-744, 1985.
21. Ambinder, R.F., Burns, W.H., Lietman, P.S., Saral, R. Lancet 1: 1154-1155, 1984.

9

ANTIVIRAL THERAPY OF VARICELLA-ZOSTER VIRUS INFECTIONS

J. ENGLUND and H.H. BALFOUR, JR.

Department of Pediatrics, Laboratory Medicine and Pathology, University of Minnesota Medical School, Minneapolis, Minnesota 55455, U.S.A.

ABSTRACT

Treatment of varicella-zoster virus (VZV) infections must be based on multiple factors, including the patient's immune status, age, and clinical presentation. Intravenous acyclovir therapy is the preferred treatment for VZV infections in immunocompromised hosts and those normal individuals with severe disease. Immunocompetent hosts with herpes zoster may be managed as outpatients with high doses of oral acyclovir. At present, topical therapy has no role in the treatment of VZV infections. Further studies with acyclovir and its derivatives, as well as several unrelated antiviral agents, may enable oral therapy of VZV infections in the future.

INTRODUCTION

The importance of varicella-zoster virus (VZV) infections has become increasingly appreciated over the past decade due to the sizable morbidity and mortality associated with infection in high-risk patients. Although the incidence of other childhood diseases has been dramatically reduced in developed countries, varicella remains a common clinical disease today. Complications in normal children are not uncommon. Reactivation of VZV, or herpes zoster, in both the normal and immunocompromised host is frequently seen by many primary care providers. The growing number of high-risk patients, the lack of a licensed vaccine in most countries, and the relative ineffectiveness of immunoglobulin prophylaxis have contributed to the demand for effective antiviral therapy.

MECHANISMS OF ACTION

A number of antiviral compounds are currently known to be active in vitro against VZV and other herpesviruses. The lack of a good animal model for VZV has limited the development and testing of antiviral compounds directed against this virus. The antiviral effect of these agents can be compared in various cell culture systems, but many factors need to be considered before drugs can be introduced into clinical studies. In vitro methods have been devised to compare a drug's ability to terminate viral replication with the toxic effect of the drug on host cellular metabolism. One such measure, a "selectivity index", contrasts a drug's ability to inhibit viral replication with its effect on human fibroblast DNA synthesis (Table 1) (1).

Determination of the mechanism and site of action of the antiviral agent is important in evaluating and predicting possible drug efficacy. A major class of antiviral compounds active against the herpes family of viruses inhibits replication of viral DNA directly. These include vidarabine (adenine arabinoside, ara-A), acyclovir [9-(2-hydroxyethoxymethyl)guanine], and phosphonoacetic acid (PAA). Both acyclovir and PAA inhibit the viral DNA polymerase, whereas vidarabine may affect several sites of viral DNA replication. Acyclovir, a purine analogue, acts as a substrate for viral thymidine kinase. Acyclovir is phosphorylated by thymidine kinase to acyclovir-monophosphate and further phosphorylated to acyclovir-triphosphate by cellular host enzymes. Acyclovir-triphosphate is the active inhibitor of viral DNA polymerase (2). Phosphonoformate (PFA) is a close analogue of PAA and has similar mechanisms of action (3). BVDU [E-5-(2-bromovinyl)-2'-deoxyuridine] is another antiviral

Table 1. In vitro drug efficacy against VZV[a]

Compound	ID_{50} (μg/ml) VZV	ID_{50} (μg/ml) HEF DNA synthesis	Selectivity index
IVDU	0.0015	100	66,667
BVDU	0.0024	100	41,666
BVDC	0.023	100	4,348
Acyclovir	4.64[b]	180	38.8
Phosphonoacetic acid	8.74	130	14.9
Vidarabine	1.62	5	3.1
Idoxuridine	1.38	0.8	0.58

[a]Adapted from Shigeta et al. (1).
[b]Median sensitivity for patient isolates in our laboratory and from literature is 0.93 μg/ml.
HEF = human embryonic fibroblast cells.
IVDU = E-5-(2-iovinyl)-2'-deoxyuridine.
BVDU = E-5-(2-bromovinyl)-2'-deoxyuridine.
BVDC = E-5-(2-bromovinyl)-2'-deoxycytidine.

agent which is phosphorylated by viral thymidine kinase to become an active inhibitor of the viral DNA polymerase (4). Other antiviral agents, such as idoxuridine (5-iodo-2'-deoxyuridine), act via competitive inhibition with thymidine at phosphorylation steps. Idoxuridine eventually becomes IdTTP and is incorporated into DNA, which may result in the loss of protein synthesis (3).

Naturally occurring cellular derivatives have also been investigated for use as antiviral agents. Interferon, an antiviral protein made in human leukocytes and other cells infected by viruses, has been shown to inhibit viral replication through the induction of an antiviral protein which interferes with viral transcription. Adenine monophosphate (AMP), a purine nucleotide, is an intermediate in cellular metabolism and nucleic acid synthesis. This compound is dephosphorylated by the cell plasma membranes in vitro. The exact mechanisms for this agent remain unclear, but it is believed that the antiviral activity is caused by the production of a viral subunit protein that alters viral transcription (5,6).

Knowledge of both the mechanism of action and the in vitro sensitivity are often insufficient due to the complexity of the interactions among the virus, host, and drug. For example, vidarabine has a reasonably low ID_{50} (a 50 % reduction in the yield of infectious virus) for VZV in vitro, and the selectivity ratio as determined by Shigeta et al. is about 10-fold lower than that of acyclovir. However, this compound is rapidly deaminated in vivo to hypoxanthine arabinoside, a metabolite with minimal antiviral effect against VZV (1).

TREATMENT OF PRIMARY VARICELLA INFECTIONS

Immunocompromised host

The advent of effective chemotherapeutic regimens for the treatment of childhood diseases and malignancies has dramatically changed the clinical impact of VZV infections. The morbidity and mortality in immunocompromised children who are primarily infected with VZV are now well appreciated. A study of 60 children undergoing chemotherapy at the time of primary varicella infection revealed a visceral dissemination rate of 32 % and a mortality rate of 7 % (10). Fatality rates of 21 % to 50 % have been reported in patients with visceral varicella who are not treated with antiviral drugs (7-10). Life-threatening complications, including pneumonitis, hepatitis, coagulopathy, small bowel obstruction, encephalitis, and secondary bacterial complications contribute to morbidity and mortality. Acute leukemia and lymphopenia, with an absolute lymphocyte count of less than $500/mm^3$, have been associated with pneumonitis and a higher fatality rate (10).

Passive immunization with varicella zoster immune globulin (VZIG) has been found to decrease the incidence of pneumonitis and other forms of visceral dissemination, but has not significantly changed the attack rate of VZV in immunocompromised children (10). High titered zoster immunoglobulin has been shown to be an ineffective treatment for varicella (11). Although there are anecdotal reports concerning the use of polyvalent intravenous immunoglobulin in the treatment of varicella (12), no controlled trial has been published. However, antiviral therapy of primary varicella in immunocompromised hosts has been actively investigated. Many antiviral compounds are effective in vitro against VZV, but well organized, placebo-controlled, double-blind studies have been carried out with only three drugs: vidarabine, acyclovir, and alpha-interferon.

Vidarabine. Vidarabine (ara-A, adenine arabinoside) was the first antiviral compound with an acceptable therapeutic index to be effective against varicella in controlled clinical trials. The initial double-blind crossover trial of vidarabine therapy for chickenpox revealed that viral replication was reduced, cutaneous healing was accelerated, and toxicity of the drug was negligible (13). A subsequent placebo-controlled trial of 34 patients, who began treatment within 72 hours of onset of disease, clearly demonstrated the superiority of vidarabine treatment over placebo (8). A significant reduction in the apperance of new lesions and duration of fever was noted. Most importantly, the incidence of varicella-related complications was reduced from 62 % to 5 %. Vidarabine toxicity includes bone marrow suppression, and hepatic and CSN dysfunction. These findings are relatively infrequent in adults, and even less common in children. Disadvantages of vidarabine therapy include the relatively large volume of fluid required for administation of the drug, and the necessity of intravenous administation.

A recent prospective trial comparing acyclovir with vidarabine for the treatment of varicella in immunocompromised children had to be terminated prematurely after severe and irreversible neurotoxicity developed in a child given vidarabine (14). When all 34 cases of varicella treated with vidarabine were evaluated, six documented cases (16 %) of vidarabine-associated neurotoxicity were found. Severe sequelae, including intellectual and motor dysfunction, and irreversible encephalopathy, were noted in some of these children. One case of fatal vidarabine-associated neurotoxicity in an adult with normal renal function has also been reported (15). Vidarabine appears to be clinically efficacious against VZV, but the potential for neurotoxicity makes it a less desirable first-line agent in the treatment of primary VZV infections.

Acyclovir. Acyclovir has been shown to be an effective treatment for varicella. A double blind, placebo-controlled trial in immunocompromised children demonstrated that intravenous acyclovir given at a dose of 500 mg/m^2 every eight hours for seven days prevented the development of pneumonitis or visceral dissemination in all seven acyclovir treated patients as compared with five of 11 controls (16). No differences in viral shedding healing, or defervescence were noted in this small clinical trial. Subsequent uncontrolled studies of intravenous acyclovir in the treatment of varicella in immunocompromised children have been reported (17,18). A retrospective review of antiviral therapy in immunocompromised children with varicella revealed that both acyclovir and vidarabine halted the progression of skin lesions but that pneumonitis developing after two days of therapy was significantly more common in patients receiving vidarabine than acyclovir (29 % vs. 0 %) (10). Most remarkably, acyclovir therapy has been shown to be beneficial even in disseminated disease when therapy was initiated after the second day of illness (17). However, fatalities due in part to viral dissemination were noted in two of four patients who began acyclovir therapy more than five days after onset of disease. Recurrence of VZV in severely immunocompromised patients after acyclovir therapy has also been reported (19). Because VZV strains resistant to acyclovir have not been documented during or after acyclovir therapy (20), viral resistance is not the major factor responsible for recurrent disease in these individuals.

The transient elevation of serum creatinine due to cyrstallization of the drug in the renal tubules has been demonstrated with high dose intravenous therapy (21). This finding appears to be aggrevated by dehydration or impaired renal function (22). Reversible neurological symptoms associated with acyclovir administration have been reported in bone marrow transplant patients (23) and an adult on chemotherapy (24), although this appears to be a rare occurrence. Patients who developed neurotoxicity were characteristically on multiple therapeutic agents, and the dosage of acyclovir varied considerably. The use of acyclovir in other studies of high risk children and adults has not substantiated this finding (10,14,17).

The value of oral acyclovir for the treatment of varicella has not yet been determined. The ID-50 of VZV in vitro is approximately 4 μM, nearly 100 times higher than that of herpes simplex virus. This must be compared with peak serum levels of nearly 7 μM obtained with the oral form of acyclovir at a dose of 800 mg five times daily (25). Anecdotal accounts of oral acyclovir in high risk children with varicella have been reported (26), and controlled trials are currently underway. Until results are available, we cannot recommend oral acyclovir for treatment of primary VZV in immunocompromised patients. In the future, new formulations of acyclovir derivatives with increased bioavailability may have greater clinical utility.

Alpha-interferon. Alpha-interferon has been studied in a randomized trial of primary VZV infection in 44 children undergoing chemotherapy (9). Treatment with either 4.2×10^4 or 2.55×10^5 U/kg every 12 hours was initiated within 72 hours of the appearance of exanthema. This trial demonstrated a significant decrease in the mean number of days to cessation of new vesicle formation, but the survival rates in both groups were approximately 90 %. However, the incidence of life-threatening complications in the survivors was reduced significantly in the treated group. One fatal case of recurrent viremia was noted at the end of the treatment period in a patient treated with high dose interferon. Adverse effects previously seen with interferon therapy, including leukopenia or thrombocytopenia, were not noted in this study or in anecdotal case reports of interferon therapy in children (27).

Other agents. Limited data from well controlled tirals of other antivirals are available. The need for easily administered, effective antiviral therapy is apparent due to the increasing number of immunocompromised children at risk for varicella, and the development of disease after unknown exposures or failure of VZIG prophylaxis. The antiviral BVDU is known to be an effective agent *in vitro* against VZV, and has been used orally at a dosage of 15 mg/kg/day for 5 days in an uncontrolled study of varicella and zoster in immunocompromised children (29). All 21 treated children recovered promptly and without sequelae, and no toxic side effects of the drug were observed. Further studies comparing available antiviral agents, and particularly oral formulations of these antivirals, must be performed to determine optimal therapy of varicella in these extremely high risk individuals.

Conclusions. Clinical efficacy of vidarabine and acyclovir in the treatment of varicella in immunocompromised children appears to be similar. Alpha-interferon has not been extensively studied. A comparison of vidarabine and acyclovir in the treatment of herpes zoster in immunocompromised patients has demonstrated that acyclovir is superior (28). No large scale study for the treatment of primary varicella infection has been completed, due in part to the neurotoxicity of vidarabine in immunocompromised children. Based on available data, intravenous acyclovir must be considered the treatment of choice for VZV infections in the immunocompromised patient. However, it is clear that any antiviral chemotherapy must be instituted as rapidly as possible, preferably within three days after onset of illness. Therapy should ideally be initiated prior to visceral dissemination to prevent unnecessary mortality and morbidity.

TREATMENT OF PRIMARY VARICELLA INFECTIONS

Normal host

Treatment of primary varicella in the normal host has not been extensively evaluated. In children, the disease is often benign, and infections in adults, although often more severe, are uncommon in developed countries where the prevalence of antibody against VZV by accurate assays may approach 100 % (30). Varicella in the high-risk age groups, including individuals less than one month of age or older than 19 years, accounts for less than 3 % of all reported cases of chickenpox in the United States (31). However, complications do occur in normal people. According to the Centers for Disease Control, an estimated 4500 to 6000 children (170/100,000 cases) are hospitalized in the United States each year because of varicella. Significant complications, including Reye syndrome, encephalitis, or death, occur at a rate of 2-3 per 100,000 cases. According to death certificates, an associated malignancy has been reported in only 16 % of the deaths related to varicella (31).

Complications of varicella in the normal person may vary in significance and severity. Well known complications include secondary bacterial infection, pneumonia, encephalitis, hepatitis, otitis media, cerebellar ataxia, thrombocytopenia, and Reye syndrome. Acute cerebellar ataxia is the most common neurologic complication of varicella and occurs in approximately one in 4000 cases among children less than 15 years (32). In adults, the development of varicella pneumonia is the most common complication and results in hospitalization about once in every 400 cases.

Varicella infections in pregnant women are associated with significant morbidity. In a recent study, morbidity was noted in nine of 43 pregnant women with varicella, and four cases of varicella pneumonia and one fatality were reported (33). Serious congenital malformations have been reported in offspring born to mothers who acquire primary varicella during pregnancy (33-35). Although acyclovir is not teratogenic in mice, rats, or rabbits (36), its use in pregnancy to decrease the risk of congenital varicella cannot be recommended at this time. The use of antiviral therapy as treatment for life-threatening disease in pregnant women may be necessary, and should be considered in cases of visceral dissemination.

Neonates born to women who develop varicella within five days prior to delivery or two days after delivery are known to be at risk for serious or fatal disease (37). Immunoglobulin prophylaxis with VZIG, while recommended, does not always prevent fatalities (38). Antiviral therapy should be seriously considered in all infected newborns in this extremely high risk situation.

Antiviral treatment of normal individuals is not currently standard practice. Clearly, not all children with varicella merit antiviral therapy. The use of antiviral therapy in adults with VZV infections has not yet been determined. Concern about the type and adequacy of the immune response after antiviral therapy has been raised as an argument against the use of antiviral therapy in otherwise healthy individuals. The only published placebo-controlled trial of intravenous acyclovir in normal adults with chickenpox demonstrated no significant effect on the duration of symptoms or evolution of disease (39). Complications related to varicella or therapy were not seen in either group, and the overall therapeutic benefit was minimal. Case reports describing antiviral therapy in normal adults with severe disease have been published (40), but the use of antiviral therapy in normal individuals, even in age groups at high risk, remains controversial. Additional controlled trials of oral or intravenous acyclovir for the treatment of chickenpox in immunocompetent children and adults are needed. Factors that should be evaluated include the type and duration of the immune response and complication rates in drug-treated patients compared with placebo recipients.

TREATMENT OF HERPES ZOSTER

Immunocompromised host

Reactivation of VZV, or herpes zoster, can be associated with substantial morbidity and mortality, particularly in immunocompromised, elderly, or debilitated patients. The incidence of zoster has been shown to increase steadily with age, with an incidence of greater than 400 per 100,000 person-yeras for those individuals over the age of 75 (41). An increased incidence of herpes zoster has been found in normal children who are infected with VZV during infancy (42,43). Complications associated with zoster infections in immunocompromised patients include cutaneous and visceral dissemination, pneumonia, CNS involvement producing encephalitis, cranial nerve palsies, and motor neuropathy, and postherpetic neuralgia (43). Postherpetic neuralgia, which may persist for years following acute herpes zoster, is responsible for much of the morbidity of zoster (44). Mortality rates of 14 % are reported in bone marrow transplant patients who develop disseminated herpes zoster during the first nine months post transplant (45). Pneumonia is the most common and serious clinical manifestation in these patients. The often unpredictable course of zoster in immunocompromised and normal hosts has led to many undocumented or controversial claims about efficacy of various therapeutic agents. Because the natural history of the disease is spontaneous recovery, studies evaluating therapy for herpes zoster must be reproducible, well controlled, and based on large numbers of patients who receive frequent and close follow-up by the investigators.

A number of clinical trials have examined antiviral agents for the treatment of herpes zoster in the immunocompromised host. The first controlled study demonstrating efficacy of an antiviral agent in the treatment of herpes zoster was the 1976 study of intravenous vidarabine that utilized a placebo-controlled crossover design (46). A subsequent placebo-controlled study in which intravenous therapy was initiated within 72 hours of onset of rash confirmed that vidarabine accelerated cutaneous healing and reduced pain associated with acute neuritis (47). A decrease in cutaneous and visceral dissemination, an important development, was also demonstrated. The incidence of postherpetic neuralgia was not affected, although the duration appeared to be decreased. Adverse reactions, although limited, were reported in 18 vidarabine recipients compared with eight placebo recipients of 121 patients followed. Because of possible drug toxicity and lack of a clear decrease in postherpetic neuralgia, the investigators concluded that the use of vidarabine should be reserved for the treatment of acute herpes zoster in immunocompromised patients.

Human alpha-interferon also has been used in controlled clinical trials for the treatment of herpes zoster in patients with malignancies. The highest of several intramuscular dosing schedules reduced the rate of cutaneous and visceral dissemination in 90 patients who were treated within 72 hours after onset of disease (48). There was also a reduction in severity of postherpetic neuralgia. However, a subsequent study of the same high dose of interferon given every 12 hours for four doses demonstrated no effect on acute pain or progression of disease and only a modest effect on distal spread and postherpetic neuralgia (49).

The efficacy of intravenous acyclovir in the treatment of herpes zoster in the immunocompromised host has now been firmly established. A double-blind, placebo-controlled study of 94 patients with either localized or disseminated cutaneous zoster evaluated clinical efficacy (50). Therapy was initiated within three days of onset of rash, unless patients were still forming new lesions. This study demonstrated clear decreases in the frequency of cutaneous dissemination (17 % vs. 2 %) and visceral dissemination (9.5 % vs. 0 %). Progression of disease was halted even in patients whose therapy was initiated three days after the onset of rash. Subsequent studies have confirmed these results in immunocompromised children and adults (51,52).

A randomized prospective comparison of acyclovir with vidarabine therapy in 22 severely immunocompromised zoster patients concluded that acyclovir therapy was superior (28). Acyclovir was shown to shorten the duration of viral shedding from skin lesions, and to decrease the time to pain diminution, crusting, and healing of all lesions. In addition, the duration of fever was significantly shortened in the patients receiving acyclovir treatment. The lack of toxicity in hydrated patients and ease of administration make acyclovir an excellent antiviral agent. Therefore, acyclovir is currently regarded as the drug of choice in the treatment of herpes zoster in immunocompromised patients.

Other promising antiviral agents that have undergone at least initial clinical evaluation for the therapy of zoster include BVDU, FIAC, and adenosine monophosphate (AMP). BVDU, a highly potent agent *in vitro* against members of the herpes family of viruses, is administered orally at a dose of 7.5 to 15 mg/kg/day for 5 days (29,53). With this regimen, plasma levels of greater than 1 µg/ml are obtained, which are in excess of the 0.01 µg/ml required for inhibition of VZV *in vitro* (53). In a small uncontrolled clinical trial, a prompt response to therapy with the lower dose was noted in 11 immunocompromised patients with herpes zoster or disseminated zoster, and evidence of cessation of infection was noted within 24 hours of therapy (53). There was no evidence of drug toxicity in this study. However, many patients were neutropenic due to their primary disease and therefore bone marrow suppression by the drug could have been masked. This preliminary study serves as an impetus for further comparative studies to evaluate oral antiviral therapy.

Another promising new antiviral, FIAC, is a pyrimidine nucleoside analogue with excellent *in vitro* activity against VZV. A double-blind clinical trial comparing intravenously administered FIAC with ara-A was conducted in 34 immunocompromised patients (54). In this well-controlled study, a significant reduction in pain, accelerated crusting, and reduced time to appearance of last new lesion was associated with FIAC therapy. Only minimal toxicity of FIAC was noted, including a transient rise in liver function tests and mild nausea.

TREATMENT OF HERPES ZOSTER

Normal hosts

Systemic antiviral therapy of herpes zoster in the normal host has recently been extensively evaluated. The only published trial of alpha-interferon for the treatment of zoster in normal adults was an uncontrolled investigation (55). More rapid healing of lesions in the primary dermatome was reported in subjects receiving intramuscular interferon compared with controls, but these results have not been confirmed. Intravenous acyclovir has been demonstrated to accelerate cutaneous healing and decrease the severity and duration of acute pain in herpes zoster in the normal host (56-58). Although high-dose intravenous therapy (10 mg/kg/dose administered TID) may be considered more likely to be efficacious due to the higher serum levels obtained, there has been no real differences in several controlled clinical trials which utilized lower dose (5 mg/kg administered TID) (56) or higher dose intravenous acyclovir (57,58). These studies have demonstrated an improved rate of healing, shortened period of acute pain, and in one study, a decreased period of viral shedding in acyclovir recipients as compared with placebo controls (58). Acyclovir therapy was associated with a transient rise in serum creatinine levels in that trial, perhaps related to dehydration in these patients. Treatment with the drug demonstrated no effect on the development of postherpetic neuralgia in these studies.

An improved rate of healing has been reported in some studies of high dose (800 mg five times daily) and lower doses (400 mg five times daily) oral acyclovir, although findings have not been consistent (25,59-62). A recent double-blind, placebo-controlled study of high dose oral acyclovir in over 200 elderly immunocompetent patients demonstrated significantly accelerated healing when treatment began within 48 hours of appearance of rash (59). Acyclovir therapy was associated with significant reduction in pain during treatment, and no important adverse effects were noted. Development of postherpetic neuralgia is still being evaluated. Another recent double-blind placebo-controlled study, which evaluated oral acyclovir at a dose of 600 mg five times daily, reported a decreased incidence and severity of herpes zoster ophthalmicus (61). This was more apparent in patients treated within 72 hours of onset of lesions. Again, no effects on the development of postherpetic neuralgia were noted. Preliminary results of a multi-center, placebo-controlled trial which utilized 800 mg of acyclovir orally five times daily for 10 days, beginning within 72 hours of onset of rash, show a significant difference in time to clinical improvement (62).

The prevention of postherpetic neuralgia remains a prime consideration in the therapy of herpes zoster in the immunocompetent patient, although beneficial effects in healing and pain reduction are important. The mechanism by which VZV induces postherpetic neuralgia is not known. The development of postherpetic neuralgia has been studied in several trials of both intravenous and oral acyclovir. No significant effect on the incidence or duration of postherpetic neuralgia has been shown to date.

A placebo-controlled study of intravenous and oral acyclovir for the treatment of postherpetic neuralgia was begun by our group at the University of Minnesota in 1983 for patients who had postherpetic neuralgia for at least six months prior to entry. Although acyclovir was well tolerated by all nine patients, there was no dramatic improvement in pain scores of individual patients. A major goal in therapy of acute herpes zoster must be the prevention of postherpetic neuralgia because it is very difficult to manage that condition once it is established.

Several other agents have been studied in clinical settings for the treatment of herpes zoster in immunocompetent patients. Adenine monophosphate (AMP) was utilized in a randomized placebo-controlled trial in 32 patients whom were given intramuscular injections of AMP prepared in an aqueous gelatin base three times weekly for up to four weeks (6). In this trial, AMP was shown to significantly reduce viral shedding and pain after four weeks. No side effects or toxicity of this compound were noted. In vitro inhibitory studies on this natural cellular metabolism have not yet been performed. Because preliminary evidence indicates that the incidence of postherpetic neuralgia may be reduced, further studies with this agent should be undertaken.

The use of steroids for the treatment of acute zoster and the prevention of postherpetic neuralgia is controversial. A trial of topical acyclovir compared with steroids for the treatment of herpes zoster keratitis-uveitis demonstrated significant differences in time of resolution of corneal disease and in the number of patients experiencing a recurrence of infection, favoring acyclovir therapy (63). The NIH Antiviral Study Group, headed by R.J. Whitley of the University of Alabama, is currently conducting a randomized blinded study to evaluate the role fo steroids in the treatment of herpes zoster.

Topical therapy of herpes zoster with several antiviral agents has resulted in varying clinical results. The effect of idoxuridine in dimethylsulfoxide (DMSO) has been studied for the treatment of uncomplicated herpes zoster, but results remain unclear. A study comparing 40 % idoxuridine in DMSO to 40 % idoxuridine in saline or garlic placebo demonstrated a significant reduction in pain (64). In contrast, other investigators found no effect of 40 % idoxuridine in DMSO on pain duration in thoracic zoster, but noted some decrease in trigeminal zoster (65). A review of antiviral therapy by Nicholson concluded that "irrespective of efficacy, the cost, side-effects, and inconvenience of topical treatment with idoxuridine limits its usefulness" (66). Topical treatment of herpes zoster with acyclovir has been shown to have some effect on healing, but no significant decrease in lesion virus titers or resolution of pain (67). Topical therapy should not be used at the present time as an alternative to systemic therapy.

FUTURE DEVELOPMENTS

The need for effective antivirals that are easily administered and have good therapeutic indices is obvious. The increasing number of immunocompromised patients, including those being treated for malignancies, those receiving immunosuppression, or those with AIDS, will magnify the problems of zoster management. Although vaccination of all young children to prevent chickenpox is not yet possible in the United States, it may soon be a reasonable goal. However, antiviral therapy will still be necessary because of rare cases of disease associated with vaccination and the frequency of reactivation of VZV in immunocompromised and normal hosts.

The development of acyclovir derivatives with increased bioavailability, such as desiclovir (BWA515U) (68), as well as other new compounds, may advance the therapeutic potential of oral drugs. This would, in turn, allow for earlier treatment of patients at particularly high risk. In many cases, early treatment could mean that patients would not require hospitalization for VZV or its complications. Efforts are underway to determine which groups of normal patients might benefit from antiviral therapy. Postherpetic neuralgia remains a serious problem. New therapeutic strategies, such as vaccination with attenuated virus or subunit vaccines or new antivirals, may hopefully lesser morbidity from this complication in the future.

SUMMARY

Treatment of varicella-zoster virus infections must take into account the individual's immune competency, age, past medical history, and the clinical course of the infection. Acyclovir therapy is presently the recommended treatment for VZV infections in immunocompromised hosts and in otherwise normal patients with severe VZV disease (Table 2). All immunocompromised patients with either varicella or zoster should be treated to prevent serious complications, especially visceral dissemination. Patients at risk for disseminated disease include transplant recipients or those undergoing chemotherapy for malignancy, as well as patients with an underlying disease associated with an immunocompromised state, such as systemic lupus erythematosus. At this time, patients receiving systemic steroids for conditions such as asthma or eczema appear to be at minimal risk for disseminated VZV infections. However, such patients should be followed closely and intravenous acyclovir therapy initiated if any signs of toxicity or dissemination develop. Therapy of zoster infections in

Table 2. Recommendations for treatment of varicella-zoster infections

VZV infection	Type of patient	Therapy
Uncomplicated varicella	Immunocompromised	Intravenous acyclovir[a]
	Neonate	Intravenous acyclovir[a]
	Normal children, adults	Under investigation
Varicella with visceral dissemination	Immunocompromised or normal	Intravenous acyclovir[a]
Localized zoster	Immunocompromised	Intravenous acyclovir[a]
	Normal	Oral acyclovir[b]
Disseminated zoster	Immunocompromised or normal	Intravenous acyclovir
Postherpetic neuralgia	Immunocompromised or normal	None available

[a]Patients with normal renal function should receive 7.5 to 10 mg acyclovir/kg of ideal body weight every 8 hours. The dose should be infused over 1 hour at a concentration no greater than 5 mg/ml. Total duration of therapy is 5 to 10 days, depending upon rate of clinical improvement. For dosage modification in renal failure, obtain acyclovir serum levels, or use the table in the package insert, substituting 7.5 mg/kg/dose instead of 5.0 mg/kg/dose for creatinine clearances from 10 to > 50 ml/min/1.73 m^2.

[b]Adults should receive 800 mg 4 to 5 times daily for 5-10 days. The oral dose for children has not been established.

normal individuals with high-dose oral acyclovir therapy is now possible, but the relatively decreased bioavailabity of the drug must be noted. Topical therapy plays no role in the treatment of varicella or zoster infections. The advent of new therapeutic agents or different formulations of available drugs will hopefully facilitate oral therapy of VZV infections in the future.

REFERENCES

1. Shigeta, S., Yokota, T., Iwabuchi, T., Baba, M., Konno, K., Ogata, M. and De Clercq, E. J. Infect. Dis. 147 : 576-584, 1983.
2. Elion, G.B., Furman, P.A., Fyfe, J.A., De Miranda, P., Beauchamp, L. and Schaeffer, H.J. Proc. Natl. Acad. Sci. USA 74 : 5716-5720, 1977.
3. Elion, G., Pagano, J., Whitley, R.J. and Crumpacker, C. In: Human Herpesviruses (Eds. A.K. Nahmias, W.R. Dowdle and R.F. Schinazi), Elsevier Press, New York, 1981, pp 583-585.
4. Shigeta, S., Yokota, T. and De Clercq, E. Antiviral Res., suppl. 1 : 35-44, 1985.
5. Sherlock, C.H. and Corey, L. J. Am. Med. Assoc. 253 : 1444-1445, 1985.
6. Sklar, S.H., Blue, W.T., Alexander, E.J. and Bodian, C.A. J. Am. Med. Assoc. 253, 1427-1430, 1985.
7. Feldman, S., Hughes, W.T. and Daniel, C.B. Pediatrics 56 : 388-397, 1975.
8. Whitley, R., Hilty, M., Haynes, R., Bryson, Y., Connor, J.D., Soong, S.-J., Alford, C.A. and the National Institute of Allergy and Infectious Diseases Collaborative Antiviral Study Group. J. Pediatr. 101 : 125-131, 1982.
9. Arvin, A.M., Kushner, J.H., Feldman, S., Baehner, R.L., Hammond, D. and Merigan, T.C. New Eng. J. Med. 306 : 761-765, 1982.
10. Feldman, S. and Lott, L. J. Ped. : in press, 1987.
11. Groth, K.E., McCullough, J., Marker, S.C., Howard, R.J., Simmons, R.L., Najarian, J.S. and Balfour, H.H. Jr. J. Am. Med. Assoc. 239 : 1877-1879, 1978.
12. Sulliger, J.M., Imbach, P., Barandun, S., Gugler, E., Hirt, A., Lüthy, A., Rossi, E., Tönz, O. and Wagner, H.P. Helv. Paediat. Acta 39 : 63-70, 1984.
13. Whitley, R.J. and Alford, C.A. In: Human Herpesviruses (Eds. A.K. Nahmias, W.R. Dowdle and R.F. Schinazi), Elsevier Press, New York, 1981, pp 478-490.
14. Feldman, S., Robertson, P.K., Lott, L. and Thornton, D. J. Infect. Dis. 154 : 889-893, 1986.
15. Van Etta, L., Brown, J., Mastri, A. and Wilson, T. J. Am. Med. Assoc. 246 : 1703-1705, 1981.
16. Prober, C.G., Kirk, L.E. and Keeney, R.E. J. Pediatr. 101 : 622-625, 1982.
17. Balfour, H.H. Jr. J. Pediatr. 104 : 134-136, 1984.
18. Shulman, S.T. Am. J. Dis. Children 139 : 137-140, 1985.
19. Oblon, D.J., Elfenbein, G.J., Rand, K. and Weiner, R.S. South. Med. J. 79 : 256-257, 1986.
20. Cole, N.L. and Balfour, H.H. Jr. J. Infect. Dis. 153 : 605-608, 1986.
21. Brigden, D., Rosling, A.E. and Woods, N.C. Am. J. Med. 73(1A) : 182-185, 1982.
22. Bean, B. and Aeppli, D. J. Infect. Dis. 151 : 362-365, 1985.
23. Wade, J.C. and Meyers, J.D. Ann. Intern. Med. 98 : 921-925, 1983.
24. Krigel, R.L. J. Infect. Dis. 154 : 189, 1986.

25. McKendrick, M.W., McGill, J.I., Bell, A.M., Hickmott, E. and Burke, C. Lancet ii : 925, 1984.
26. Novelli, V.M., Marshall, W.C., Yeo, J. and McKendrick, G.D. J. Infect. Dis. 151 : 372, 1985.
27. Kim, B.S. Acta Med. Okayama 38 : 71-78, 1984.
28. Shepp, D.H., Dandliker, P.S. and Meyers, J.D. New Engl. J. Med. 314 : 208-212, 1986.
29. Benoit, Y., Laureys, G., Delbeke, M.-J. and De Clercq, E. Eur. J. Pediatr. 143 : 198-202, 1985.
30. Muench, R., Nassim, C., Niku, S. and Sullivan-Bolyai, J.Z. J. Infect. Dis. 153 : 153-155, 1986.
31. Preblud, S.R., Orenstein, W.A. and Bark, K.J. Pediatr. Infect. Dis. 3 : 505-509, 1984.
32. Guess, H.A., Broughton, D.D., Melton III, L.J. and Kurland, L.T. Pediatrics 78 (suppl.) : 723-727, 1986.
33. Paryani, S.G. and Arvin, A.M. New Engl. J. Med. 314 : 1542-1546, 1986.
34. Preblud, S.R., Cochi, S.L. and Orenstein, W.A. New Engl. J. Med. 315 : 1416-1417, 1986.
35. Lagrew, D.C. Jr., Furlow, T.G., Hager, W.D. and Yarrish, R.L. J. Am. Med. Assoc. 252 : 2058-2059, 1984.
36. Moore, H.L., Szczech, G.M., Rodwell, D.E., Kopp, R.W., DeMiranda, P. and Tucker, W.E. Fundamental Appl. Toxicol. 3 : 560-568, 1983.
37. Meyers, J.D. J. Infect. Dis. 129 : 215-217, 19874.
38. King, S.M., Gorensek, M., Ford-Jones, E.L. and Read, S.E. Pediatr. Infect. Dis. 5 : 588-589, 1986.
39. Al-Nakib, W., Al-Kandari, S., El-Khalik, D.M.A. and El-Shirbiny, A.M. J. Infect. 6 (suppl. 1) : 49-56, 1983.
40. van der Meer, J.W.M., Thompson, J., Tan, W.D. and Versteeg, J. Lancet ii : 473-474, 1980.
41. Ragozzino, M.W., Melton III, L.J., Kurland, L.T., Chu, C.P. and Perry, H.O. Medicine 61 : 310-316, 1982.
42. Guess, H.A., Broughton, D.D., Melton III, L.J. and Kurland, L.T. Pediatrics 76 : 512-517, 1985.
43. Baba, K., Yabuuchi, H., Takahashi, M. and Ogra, P.L. J. Pediatri. 108 : 372-377, 1986.
44. Mazur, M.H. and Dolin, R. Am. J. Med. 65 : 738-744, 1978.
45. Atkinson, K., Meyers, J.D., Storb, R., Prentice, R.L. and Thomas, E.D. Transplantation 29 : 47-50, 1980.
46. Whitley, R.J., Ch'ien, L.T., Dolin, R., Galasso, G.J., Alford, C.A. Jr. and the Collaborative Study Group. New Engl. J. Med. 294 : 1193-1199, 1976.
47. Whitley, R.J., Soong, S.-J., Dolin, R., Betts, R., Linnemann, C. Jr., Alford, C.A. Jr. and the NIAID Collaborative Antiviral Study Group. New Engl. J. Med. 307 : 971-975, 1982.
48. Merigan, T.C., Rand, K.H., Pollard, R.B., Abdallah, P.S., Jordan, G.W. and Fried, R.P. New Engl. J. Med. 298 : 981-987, 1978.
49. Merigan, T.C., Gallagher, J.G., Pollard, R.B. and Arvin, A.M. Antimicrob. Agents Chemother. 19 : 193-195, 1981.
50. Balfour, H.H. Jr., Bean, B., Laskin, O.L., Ambinder, R.F., Meyers, J.D., Wade, J.C., Zaia, J.A., Aeppli, D., Kirk, L.E., Segreti, A.C., Keeney, R.E. and the Burroughs Wellcome Collaborative Acyclovir Study Group. New Engl. J. Med. 308 : 1448-1453, 1983.
51. Meyers, J.D., Wade, J.C., Shepp, D.H. and Newton, B. Transplantation 37 : 571-574, 1984.

52. Serota, F.T., Starr, S.E., Bryan, C.K., Koch, P.A., Plotkin, S.A. and August, C.S. J. Am. Med. Assoc. 247 : 2132-2135, 1982.
53. Tricot, G., De Clercq, E., Boogaerts, M.A. and Verwilghen, R.L. J. Med. Virol. 18 : 11-20, 1986.
54. Leyland-Jones, B., Donnelly, H., Goshen, S., Myskowski, P., Donner, A.L., Fanucchi, M., Fox, J. and the Memorial Sloan-Kettering Antiviral Working Group. J. Infect. Dis. 154 : 430-436.
55. Emödi, G., Rufli, T., Just, M. and Hernandez, R. Scand. J. Infect. Dis. 7 : 1-5, 1975.
56. Peterslund, N.A., Seyer-Hansen, K., Ipsen, J., Esmann, V., Schonheyder, H. and Juhl, H. Lancet ii : 827-830, 1981.
57. Juel-Jensen, B.E., Khan, J.A. and Pasvol, G. J. Infect. Dis. 6 (suppl.) : 31-36, 1983.
58. Bean, B., Braun, C. and Balfour, H.H. Jr. Lancet ii : 118-121, 1982.
59. McKendrick, M.W., McGill, J.I., White, J.E. and Wood, M.J. Brit. Med. J. 293 : 1529-1532, 1986.
60. McKendrick, M.W., Care, C., Burke, C., Hickmott, E. and McKendrick, G.D.W. J. Antimicrob. Chemother. 14 : 661-665, 1984.
61. Cobo, L.M., Foulks, G.N., Liesegang, T., Lass, J., Sutphin, J.E., Wilhelmus, K., Jons, D.B., Chapman, S., Segreti, A.C. and King, D.H. Ophthalmology 93 : 763-770, 1986.
62. Bean, B., Aeppli, D., Huff, J.C., Laskin, O.L., Connor, J.D., Corey, L., Bryson, Y. and Balfour, H.H. Clin. Res. 34 (abstracts) : 512A, 1986.
63. McGill, J. and Chapman, C. Brit. J. Ophthalmol. 67 : 746-750, 1983.
64. Juel-Jensen, B.E., McCallum, F.O. and MacKenzie, A.M. Brit. Med. J. ii : 776, 1970.
65. Wildenhoff, K.E., Esmann, V., Ipsen, J., Harving, H., Peterslund, N.A. and Schønheyder, H. Scand. J. Infect. Dis. 13 : 257-262, 1981.
66. Nicholson, K.G. Lancet ii : 677-682, 1984.
67. Levin, M.J., Zaia, J.A., Hershey, B.J., Davis, L.G., Robinson, G.V. and Segreti, A.C. J. Amer. Med. Dermatol. 13 : 590-596, 1985.
68. Selby, P., Powles, R.L., Blake, S., Stolle, K., Mbidde, E.K., McElwain, T.J., Hickmott, E., Whiteman, P.D. and Fiddian, A.P. Lancet ii : 1428-1430, 1984.

10

TREATMENT (BROMOVINYLDEOXYURIDINE) OF VARICELLA-ZOSTER VIRUS INFECTIONS

S. SHIGETA[1] and E. DE CLERCQ[2]

[1]Department of Bacteriology, Fukushima Medical College, Fukushima 960, Japan and [2]Rega Institute for Medical Research, Katholieke Universiteit Leuven, B-3000 Leuven, Belgium

ABSTRACT

Bromovinyldeoxyuridine (BVDU) is a highly potent and selective inhibitor of varicella-zoster virus (VZV) *in vitro*. It is specifically phosphorylated by the virus-induced thymidine kinase and appears to be targeted at the viral DNA polymerase and viral DNA replication (upon incorporation into the viral DNA). As demonstrated by a number of open clinical studies, oral BVDU holds great promise for the treatment of various VZV infections, including varicella (chickenpox) and zoster (shingles) in immunocompromised patients. The blood drug levels achieved by BVDU upon oral administration at 3 x 2.5 mg/kg/day are far in excess of those required for inhibition of VZV replication *in vitro*. The efficacy of oral BVDU in the treatment of VZV infections remains to be corroborated by double-blind controlled clinical studies, now in progress.

INTRODUCTION

Varicella-zoster virus (VZV) usually causes a mild infection which is epidemic in infancy and childhood. However, VZV infections may follow an aggravated course in immunocompromised patients, especially in those undergoing immunosuppressive therapy upon allogeneic bone-marrow transplantation (1,2) (Table 1). Human interferons, transfer factor and human anti-VZV immunoglobulins have been used for the therapy of VZV infections. Whether they are effective is questionable, however (3-5). An attenuated live VZV vaccine has been developed by Takahashi *et al*. and proven successful in the prophylaxis of chickenpox in leukemic children (6,7). However, vaccination would obviously be of no avail after the infection.

We have investigated several anti-herpes compounds for their effects on VZV replication *in vitro*, and from these studies (*E*)-5-(2-bromovinyl)-2'-deoxyuridine (BVDU) emerged as the most potent and most selective inhibitor of VZV replication (8,9). BVDU is activated by the VZV-encoded thymidine (dThd)

Table 1. Severe VZV infections

<u>Chickenpox</u> Congenital varicella (upon infection during late pregnancy) Varicella in immunocompromised patients - receiving immunosuppressive therapy - suffering from an inborn immunodeficiency
<u>Herpes zoster</u> Zoster in immunocompromised patients - after receiving bone-marrow transplantation Zoster in cancer patients - receiving immunosuppressive therapy

kinase (TK). It is inhibitory to TK-positive VZV strains but inactive against TK-negative VZV variants (9,10). BVDU triphosphate is inhibitory to VZV-DNA polymerases from both TK-positive and TK-negative VZV strains, but not inhibitory or much less inhibitory to cellular DNA polymerases (11). Thus the selective inhibitory effect of BVDU on VZV replication depends on two virus-specific enzymatic actions : (*i*) phosphorylation of BVDU by the VZV-induced TK and (*ii*) inhibition of VZV-DNA polymerase by BVDU triphosphate. As has been monitored with $[^{125}I]$-labelled (*E*)-5-(2-iodovinyl)-2'-deoxyuridine (IVDU), the twin compound of BVDU and also a potent inhibitor of VZV replication, IVDU is incorporated into DNA of cells infected with TK-positive VZV but not into DNA of TK-negative VZV-infected or mock-infected cells (12). Within the TK-positive VZV-infected cells IVDU is incorporated to a much greater extent into VZV-DNA than cellular DNA. These data thus explain the selective inhibitory effects of IVDU (and BVDU) on VZV replication *in vitro* (9). The extent of incorporation of IVDU into VZV-DNA, as could be deduced from the buoyant density of the viral DNA, increases proportionally with the concentration of IVDU added to the cell culture medium. Substitution of 0.1 to 1.0 % of the thymidine residues by IVDU in VZV-DNA suffices for IVDU to achieve an inhibitory effect on VZV replication in human embryo fibroblasts (our unpublished data).

There are no appropriate animal models for VZV infection *in vivo*. Although some investigators reported experimental infection of guinea pigs by VZV, the animals did not die or show any sign of clinical illness (13-15). The clinical use of BVDU in VZV infections has been prompted by its excellent activity against VZV *in vitro* as well as its efficacy in experimental HSV-1 (herpes simplex virus type 1) infections.

CLINICAL STUDIES

BVDU therapy of eye infections due to HSV-1 or VZV

Topical use of 0.1 % BVDU drops for herpetic eye infection has been reported by Maudgal *et al.* (16). BVDU was administered 9 times a day at 1 hr-intervals during the day to patients with dendritic or geographic corneal ulcers or stromal keratitis. Among 76 patients with dendritic keratitis, 44 patients who had not been cured following treatment with either iododeoxyuridine (IDU), trifluorothymidine (TFT), adenine arabinoside (ara-A) or acyclovir (ACV), healed within an average time of 8.6 days after BVDU treatment was installed. Twenty-six patients with geographic corneal ulcers, who were clinically resistant to IDU, TFT, ara-A or ACV, healed within an average time of 11.7 days after BVDU treatment. Thirty-two patients with stromal keratitis, who were clinically resistant to IDU, TFT or ACV, healed within an average time of 30 days upon BVDU therapy. Before BVDU treatment there were recurrences in 44.7 %, 71.4 % and 76 % of the patients with dendritic or geographic corneal ulcers or stromal disease, respectively, whereas 27.6 %, 45.7 % and 42.8 % of these patients had recurrences following BVDU treatment. Thus, topical application of 0.1 % BVDU drops was efficacious in the treatment of herpetic keratitis, even if the latter had become clinically resistant to other antivirals, i.e. IDU, TFT, ara-A or ACV (16). An additional 15 patients with ophthalmic herpes zoster were treated with oral BVDU capsules at 375 mg/kg for 5 days, combined with topical 0.1 % BVDU eyedrops. Both the skin eruption and eye diseases responded promptly to this treatment (16). Toxic side effects were not observed.

BVDU therapy of chickenpox (varicella) in children

Benoit *et al.* (17) performed an open clinical trial with oral BVDU in children with cancer suffering from an intercurrent chickenpox infection. The underlying infection was either acute lymphoblastic leukemia, Hodgkin's disease or another malignant hematologic disease, although all patients were in complete remission when they developed varicella. They were treated with BVDU capsules orally at 15 mg/kg/day for 5 days. All patients responded promptly to BVDU treatment and recovered completely from varicella without complications. Fever resolved within 3 to 5 days and new lesions ceased to develop from the 2nd to the 5th day onwards. In the 2 to 33 month follow-up period after treatment, none of the patients showed a recurrence of the VZV infection. As determined by a bioassay, based upon the inhibition of HSV-1 cytopathogenicity in primary rabbit kidney cell culture (18), the highest BVDU concentrations attained in serum and urine were 0.3-2.0 μg/ml and 2-20 μg/ml, respectively.

These BVDU levels were 125 to 833 times higher than the 50 % inhibitory dose (ID_{50}) of BVDU for VZV *in vitro* (9).

BVDU therapy of zoster in cancer patients

Benoit *et al*. (17) also reported the results for oral BVDU treatment (15 mg/kg/day for 5 days) of severe herpes zoster in 11 children with cancer (8 with leukemia and 3 with another malignant tumor). All patients responded rapidly to BVDU, and new lesions ceased to develop within 1 to 5 days after treatment was started. With the exception of one girl, none of the patients having received BVDU had a recurrence of herpes zoster during the follow-up period (up to 33 weeks).

Wildiers and De Clercq (19) reported on the use of oral BVDU treatment of severe herpes zoster in adult patients with cancer. BVDU was administered to 20 patients with either localized or disseminated herpes zoster at a dosage of 7.5 mg/kg/day for 5 days. Upon BVDU treatment a rapid cessation of the acute herpes zoster episode was noted in all but one patient. According to their response to BVDU treatment the patients could be divided into 4 categories (excellent, good, moderate or poor). The corresponding numbers of patients were 9, 3, 5 and 1, respectively (for two patients no definitive evaluation could be made).

Tricot *et al*. (20) reported similarly favorable results with oral BVDU (7.5 mg/kg/day for 5 days) in the therapy of VZV infections in severely immunocompromised patients. In most patients, new lesions ceased to form within 1 day after BVDU treatment was started (Fig. 1).

Our recent experience with oral BVDU treatment of severe herpes zoster is summarized in Table 2. It is best illustrated by patient S.S. (no 9 in Table 2). This patient had lung cancer as the underlying disorder. He developed a severe herpes zoster at dermatomes Th 4-5, and 4 days after onset of the disease treatment with oral BVDU (7.5 mg/kg/day for 5 days) was started. At that time he had 82 vesicles at dermatomes Th 4-5. One day after BVDU therapy was initiated only 1 new eruption had appeared and from the second day onwards no new lesions appeared. The existing vesicles evolved to crusts within 5 days, and after 10 days all the lesions had healed. The patient did not develop any postherpetic neuralgia. The physicians judged the response of the patient as "excellent" (rapid cessation of new vesicle formation, prompt disappearance of pain, prompt crust formation, prompt healing).

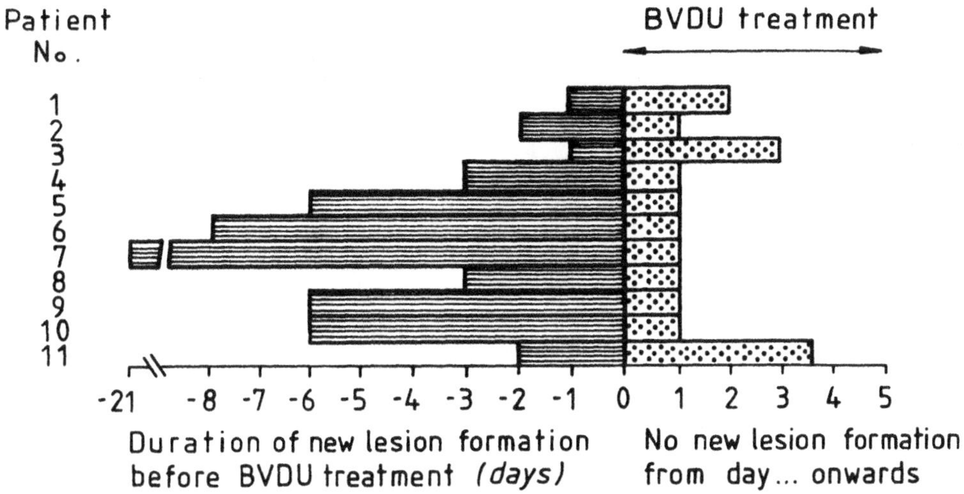

Fig. 1. Diagrammatic presentation of the duration of formation of new VZV lesions before and after BVDU treatment was initiated. According to Tricot et al. (20).

Therapy of herpes zoster in patients without cancer

We have administered BVDU (7.5 mg/kg/day for 5 days) by the oral route to two patients with herpes zoster at the forehead and back of the head, respectively. Neither of these 2 patients had an underlying malignant disorder. The 84-year old patient (no 8 in Table 2) who suffered from herpes zoster in the occipital area, recovered completely from the disease within 6 days after BVDU treatment was begun. The response was rated as "excellent". The other patient, aged 64, who had herpes zoster at the right forehead (no 6 in Table 2), did not form new eruptions after BVDU treatment was started, but this patient experienced post-herpetic neuralgia 3 weeks after onset of the disease. Laboratory examinations (erythrocyte, leukocyte and thrombocyte counts, hemoglobin, hematocrit, transaminase, phosphatase, urea, and creatine determinations) did not reveal any abnormalities which could be attributed to BVDU.

Table 2. Oral BVDU therapy in herpes zoster patients

Patient* No.	Age	Sex	Affected dermatome	Number of days to complete healing	Number of days to cessation of new lesion formation	Postherpetic neuralgia	Judgment
1	19	M	Th 7	7	-	-	excellent
2	29	M	Th 4	17	-	-	excellent
3	31	M	Th 5-6	10	-	-	excellent
4	59	M	facialis	20	-	-	good
5	61	F	C 3-4	11	-	±	good
6	64	M	facialis	9	1	++	good
7	72	M	C 8, Th 1-3	7	1	±	good
8	84	M	C 1-2	6	1	-	excellent
9	62	M	Th 4-5	10	2	-	excellent

* Patients no 7 and 9 had an underlying malignant disease, whereas the other patients had not.
Excellent : Prompt cessation of new lesion formation and pain; prompt crust formation and healing.
Good : Prompt cessation of new lesion formation as well as prompt crust formation; but either longer duration of healing process or residual neuralgia.

At Fukushima Medical College Igari *et al.* treated six herpes zoster patients with oral BVDU (7.5 mg/kg/day for 5 days). Their herpes zoster episode subsided completely within 7 to 20 days after BVDU treatment was started. The response of three patients was judged as being "excellent", the three others as "good" (no 1-5 and 7 in Table 2).

BVDU therapy of Ramsey Hunt disease

At Shirakawa Kosei hospital (Fukushima, Japan) Omata *et al.* used oral BVDU (7.5 mg/kg/day for 5 days) in the treatment of patients with Ramsey Hunt disease. The degree of facial paralysis was evaluated based on 10 independent signs of paralysis, each of which was scored on 10 points (10 : no paralysis; 0 : complete paralysis). Two patients who had received BVDU recovered from facial paralysis so that 11-16 days after treatment they scored 50 points, and after 19-35 days they progressed to a score of 75 on 100. On the other hand, 3 patients who were not treated with BVDU needed more than 100 days to recover from the paralysis to the stage where they scored 50 points. Thus, the recovery from facial paralysis in patients treated with BVDU was faster than in the controls, who had not been treated with BVDU. In the two patients receiving BVDU, new eruptions ceased to develop within one day after BVDU treatment was started. Side effects attributable to BVDU were not revealed by any of the laboratory tests mentioned above.

Potential problems concerning the clinical use of BVDU

Laboratory parameters. The influence of oral BVDU (7.5 mg/kg/day for 7 days) on various hematological, liver and kidney functions was monitored in a normal volunteer by examining the representative laboratory parameters. As shown in Table 3, the laboratory data remained essentially unaltered during the 7-day course of BVDU treatment. Nor did the 11 herpes zoster patients and 2 patients with Ramsey Hunt disease who had been treated with oral BVDU in Fukushima show any evidence of drug toxicity for bone-marrow, liver, kidney or any other organs. None of the patients complained of any sign or symptom which might point to an adverse effect of BVDU.

Table 3. Influence of oral BVDU (7.5 mg/kg/day for 5 days) on laboratory data

Laboratory examination	Pre-administration	Post-administration (days)		
		3	5	7
WBC/mm^3 (x 10^3)	4.8	5.3	6.3	4.9
RBC/mm^3 (x 10^6)	4.95	4.84	4.85	5.05
Hb (g/dl)	16.0	16.0	16.0	16.4
Hct (%)	47.0	46.0	46.5	48.9
GOT (KU)	40	35	37	50
GPT (KU)	37	40	41	46
LDH (WrU/ml)	826	256	271	272
ALP (KAU)	6	8	5	6
T Bil (mg/dl)	1.0	1.8	0.7	1.1
D Bil (mg/dl)	0.5	1.3	0.4	0.9
LAP (GRU)	183	250	181	210
γ-GT (mU/l)	25	14	21	17
Cho-E (pH/h)	0.86	0.86	0.83	0.99
CPK (mU/l)	74	65	59	82
Urea (mg/dl)	19	14	17	14
Creatinine (mg/dl)	1.2	1.1	1.1	1.1
Uric acid (mg/dl)	9.0	8.2	7.7	7.4
Blood pressure (mmHg)	130-80	-	-	134-80

WBC: white blood cell; RBC: red blood cell; Hb: hemoglobin; hct: hematocrit; GOT: glutamic-oxaloacetic (acid) transaminase; GPT: glutamic-pyruvic (acid) transaminase; LDH: lactate dehydrogenase; ALP: alkaline phosphatase; T Bil: total bilirubin; D Bil: direct bilirubin; LAP: leucine aminopeptidase; γ-GT: γ-glutamyl transpeptidase; Cho-E: choline esterase; CPK: creatine phosphokinase.

Pharmacokinetics. A simple bioassay was devised for measuring the concentration of BVDU in the patients' serum and urine. This method was based upon the inhibitory effect of BVDU on VZV focus formation in vitro in human embryonic fibroblasts. To avoid the neutralizing effect of anti-VZV antibody on VZV focus formation, the patients' serum was diluted 1:50 and the serum samples were added to the VZV-infected cell cultures at 4 hours after VZV infection. Under these conditions the inhibitory effect of the neutralizing antibody on VZV focus formation was eliminated whereas that of BVDU was not influenced. A standard dose-response curve was established for BVDU in a 1:50 dilution of human serum (Fig. 2). When 3 healthy adult volunteers were given a single oral dose of BVDU (125 mg), the concentration of BVDU reached in the serum at 2 hr after BVDU administration was 0.6-0.7 µg/ml; it decreased to less than 0.5 µg/ml at 8 hr after BVDU administration. In the urine a concentration of 4 to 10 µg BVDU/ml was attained at 8 hr after BVDU administration.

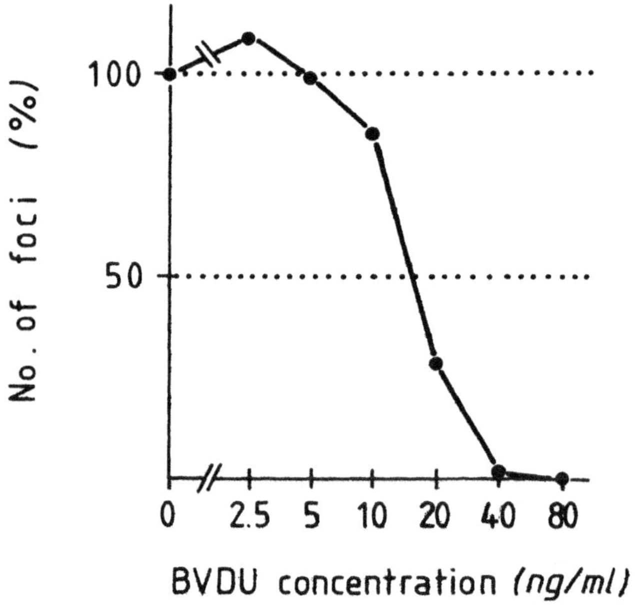

Fig. 2. Standard VZV-focus reducing curve for serial concentrations of BVDU prepared in a 1:50 dilution of human serum.

When a dose of 125 mg (2.5 mg/kg) of BVDU was administered every 8 hours, the serum BVDU concentration reached 0.64–0.8 µg/ml at 2 hr after administration of the first dose and decreased to less than 0.5 µg/ml at 8 hr, when the second dose was administered. At 10 hr, that is 2 hr after the second BVDU administration, the serum BVDU concentration increased again to a similar level as obtained at 2 hr after the first administration. At 24 hr, that is 2 hr after the fourth BVDU administration, the serum concentration of BVDU attained 0.65–1.1 µg/ml. The urinary BVDU concentrations reached an average of 15.3 µg/ml at 2 hr after the 4th BVDU administration (Fig. 3). The BVDU concentrations reached in the serum (0.65 to 1.10 µg/ml) at 2 hr after BVDU administration were 300- to 450-higher than the ID_{50} of BVDU for VZV _in vitro_ (9). Thus BVDU is effectively absorbed when given orally to humans, as has also been demonstrated in mice and monkeys (21,22).

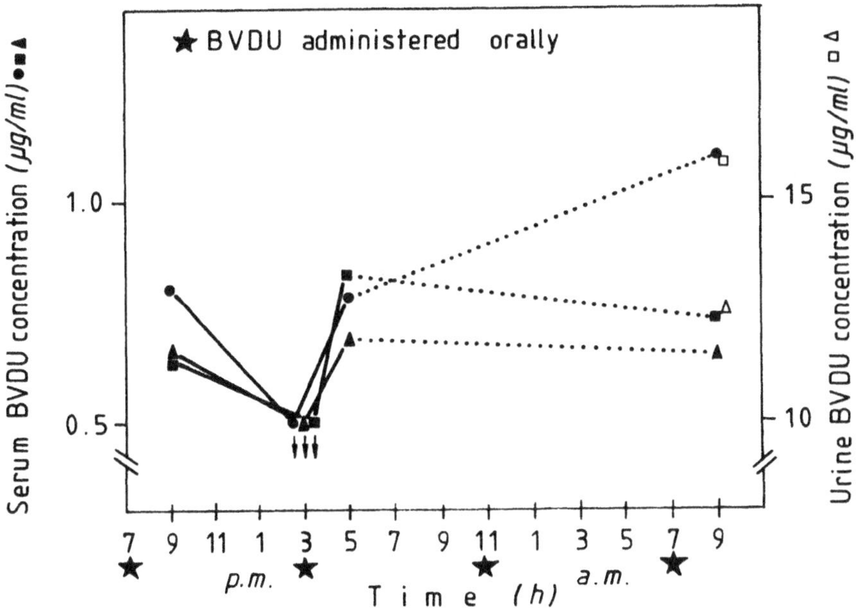

Fig. 3. Serum and urine BVDU concentrations following repeated oral administrations of BVDU (3 x 2.5 mg/kg/day).

BVDU is rapidly degraded by pyrimidine nucleoside phosphorylases, such as dThd phosphorylase, to (E)-5-(2-bromovinyl)uracil (BVU). Desgranges et al. (23) reported that BVU is cleared much more slowly from the bloodstream than BVDU. When BVDU (90 μmoles) was administered intraperitoneally (i.p.) to rats, it was completely cleared from the bloodstream 3 hours later. If at that time dThd (90 μmoles) was administered i.p., BVDU reappeared in the bloodstream (Fig. 4). Concomitantly with the reappearance of BVDU, there was a drop in the

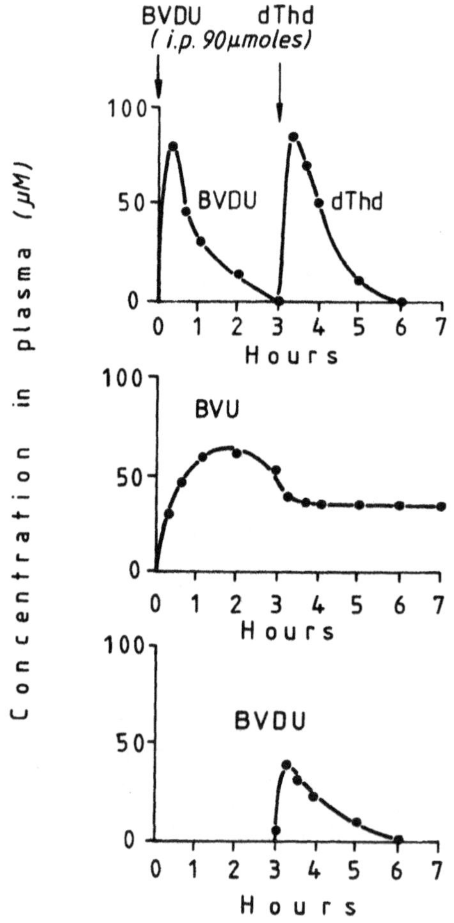

Fig. 4. Reappearance of BVDU in the bloodstream following dThd administration.

plasma concentration of BVU (Fig. 4), suggesting that BVU had reacted with dThd to give BVDU, according to the scheme : BVU + dThd ⇄ BVDU + Thy (Thy being thymine). Thus, the degradation of BVDU → BVU can be reversed, and the possibility of regenerating BVDU from BVU has obviously therapeutic implications. Furthermore, since BVU persists in the blood plasma for 24 hours, the regeneration procedure can be repeated several times until BVU has been completely cleared from the circulation.

Problems of virus-drug resistance. BVDU is phosphorylated by VZV-TK but not cellular TK. TK-negative variants of VZV are not able to phosphorylate BVDU although BVDU triphosphate is inhibitory to the DNA polymerase of TK-negative VZV strains Thus, TK-negative VZV variants are resistant to BVDU as well as other anti-herpes drugs depending for their anti-VZV activity on a specific phosphorylation by the VZV TK. So far, there have been no reports on the isolation of TK-negative or BVDU-resistant VZV strains from patients. When several clinical VZV isolates were evaluated for their susceptibility against various anti-herpes compounds, including BVDU, none proved resistant to BVDU (25). Akin to TK-negative HSV strains (26), TK-negative VZV may be less pathogenic to humans, although the lack of an appropriate animal model for VZV infection makes the evaluation of pathogenicity of VZV strains rather difficult. Recently, we have isolated a VZV variant from a patient with chickenpox (27). The TK induced by this virus had a low substrate affinity for dThd and several dThd analogues such as bromodeoxyuridine (BDU) and IDU. Hence, the VZV variant was resistant to BDU and IDU. Yet, it was still susceptible to BVDU and IVDU and the TK of this VZV variant had a high affinity for BVDU (or IVDU) as substrate (27).

CONCLUSION

BVDU is a very potent and selective anti-VZV agent, as has been demonstrated repeatedly in human embryonic fibroblasts *in vitro*. Several open clinical trials suggest that it is also efficacious against VZV infections *in vivo*. The effectiveness of BVDU in the treatment of VZV infections remains to be confirmed by double-blind controlled clinical studies. In the treatment of VZV infections BVDU could be used both systemically (i.e. orally) and topically (i.e. directly on the skin). Pharmacokinetic data indicate that upon oral administration of 3 x 2.5 mg BVDU/kg body weight per day serum BVDU levels are attained which are far in excess of those required to inhibit VZV replication *in vitro*. Whether BVDU is suitable for topical treatment of VZV infections, i.e. zoster, remains to be ascertained. To this end, a proper vehicle should be used so as to achieve optimal penetration of the drug through the skin. The importance of the vehicle for topical formulations has been repeatedly emphasized in the literature (28,29).

BVDU is particularly effective against VZV and HSV-1 infections; it is much less active against HSV-2 infection (30). To identify the causative agent of the infection and, hence, choose the most appropriate drug for treatment, rapid diagnosis is required. Immunofluorescence and enzyme immunoassays have been developed for the detection of VZV-infected cells in vesicle fluid. Early diagnosis of VZV infections may permit the prompt use of BVDU, and, when initiated sufficiently early during the course of the infection, BVDU treatment may prevent postherpetic neuralgia, one of the most feared complications of VZV infections. The beneficial effects obtained with BVDU in a number of patients, including those with Ramsey Hunt disease, indicate that prevention of postherpetic neuralgia upon oral BVDU treatment may be an achievable goal.

ACKNOWLEDGEMENTS

The original clinical data obtained in Fukushima Medical College hospital and Shirakawa Kosei hospital benefited from the enthusiastic efforts of Prof. S. Iijima, Dr. Y. Igari and Dr. T. Omata. Their contributions are gratefully acknowledged.

REFERENCES

1. Feldman, S., Hughes, W.T. and Daniel, C.B. Pediatrics 56: 388-397, 1975.
2. Locksley, R.M., Flournoy, N., Sullivan, K.M. and Meyers, J.D. J. Infect. Dis. 152 : 1172-1181, 1985.
3. Merigan, T.C., Rand, K.H., Pollard, R.B., Abdallah, P.S., Jordan, G.W. and Fried, R.P. New Engl. J. Med. 298 : 981-987, 1978.
4. Bowden, R.A., Siegel, M.S., Steele, R.W., Day, L.M. and Meyers, J.D. J. Infect. Dis. 152 : 1324-1327, 1985.
5. Groth, K.E., McCullough, J., Marker, S.C., Howard, R.J., Simmons, R.L., Najarian, J.S. and Balfour, H.H. J. Am. Med. Assoc. 239 : 1877-1879, 1978.
6. Takahashi, M., Otsuka, T., Okuno, Y., Asano, Y., Yazaki, T. and Isomura, S. Lancet ii : 1288-1290, 1974.
7. Gershon, A.A., Steinberg, S.P., Gelb, L., Galasso, G., Borkowsky, W., LaRussa, P. and Ferrara, A. J. Am. Med. Assoc. 252 : 355-362, 1984.
8. De Clercq, E., Descamps, J., Ogata, M. and Shigeta, S. Antimicrob. Agents Chemother. 21 : 33-38, 1982.
9. Shigeta, S., Yokota, T., Iwabuchi, T., Baba, M., Konno, K., Ogata, M. and De Clercq, E. J. Infect. Dis. 147 : 576-584, 1983.
10. Yokota, T., Ogata, M. and Shigeta, S. Fukushima J. Med. Sci. 29 : 101-111, 1983.
11. Yokota, T., Konno, K., Shigeta, S. and De Clercq, E. Mol. Pharmacol. 26 : 376-380, 1984.

12. Yokota, T., Konno, K., Shigeta, S. and De Clercq, E. In: Recent Advances in Chemotherapy. Proceedings of the 14th International Congress of Chemotherapy, Koyoto, 1985 (Ed. J. Ishigami), University of Tokyo Press, Tokyo, 1985, pp. 1969-1970.
13. Myers, M.G., Duer, H.L. and Hausler, C.K. J. Infect. Dis. 142 : 414-420, 1980.
14. Myers, M.G., Stanberry, L.R. and Edmond, B.J. J. Infect. Dis. 151 : 106-113, 1985.
15. Yamanishi, K., Matsunaga, Y., Otsuka, T. and Takahashi, M. Biken J. 23 : 53-55, 1980.
16. Maudgal, P.C. and De Clercq, E. In : Current Eye Research, Proceedings of the International Conference on Herpetic Eye Diseases, September 6-8, 1986, North Lake Tahoe, California, USA. In press.
17. Benoit, Y., Laureys, G., Delbeke, M.-J. and De Clercq, E. Eur. J. Pediatr. 143 : 198-202, 1985.
18. De Clercq, E., Descamps, J., De Somer, P., Barr, P.J., Jones, A.S. and Walker, R.T. Antimicrob. Agents Chemother. 16 : 234-236, 1979.
19. Wildiers, J. and De Clercq, E. Eur. J. Cancer Clin. Oncol. 20 : 471-476, 1984.
20. Tricot, G., De Clercq, E., Boogaerts, M.A. and Verwilghen, R.L. J. Med. Virol. 18 : 11-20, 1986.
21. De Clercq, E., Zhang, Z.-X. and Sim, I.S. Antimicrob. Agents Chemother. 22 : 421-425, 1982.
22. Soike, K.F., Gibson, S. and Gerone, P.J. Antiviral Res. 1 : 325-337, 1981.
23. Desgranges, C., Razaka, G., Drouillet, F., Bricaud, H., Herdewijn, P. and De Clercq, E. Nucleic Acids Res. 12 : 2081-2090, 1984.
24. De Clercq, E. Eur. J. Clin. Microbiol. 3 : 96-107, 1984.
25. Baba, M., Konno, K., Shigeta, S. and De Clercq, E. Tohoku J. Exp. Med. 148 : 275-283, 1986.
26. Field, H. and Wildy, P. J. Hyg. 81 : 267-277, 1978.
27. Shigeta, S., Mori, S., Yokota, T., Konno, K. and De Clercq, E. Antimicrob. Agents Chemother. 29 : 1053-1058, 1986.
28. Spruance, S.L., McKeough, M.B. and Cardinal, J.R. Antimicrob. Agents Chemother. 25 : 10-15, 1984.
29. De Clercq, E. Antimicrob. Agents Chemother. 26 : 155-159, 1984.
30. De Clercq, E. and Zhang, Z.-X. J. Infect. Dis. 145 : 130, 1982.

11

THERAPY AND PREVENTION OF CYTOMEGALOVIRUS INFECTIONS

Paul R. Skolnik, M.D. and Martin S. Hirsch, M.D.

Infectious Disease Unit, Department of Medicine, Massachusetts General Hospital, Boston, Massachusetts. PRS is a recipient of PHS grant CA08014-02.

ABSTRACT

Cytomegalovirus (CMV) is a major cause of morbidity and mortality in immunocompromised patients. Effective means of prevention now exist in certain groups of patients. Suppressive therapy is available for several CMV-related syndromes. This chapter will review the in-vitro, animal, and human data which suggest present indications for prophylaxis and therapy of CMV infections. Additional studies are needed to further delineate and extend present indications to other patient populations. Also needed are new agents with greater activity against CMV which also have enhanced oral bioavailability, thus making long term suppressive therapy more satisfactory.

INTRODUCTION

Cytomegalovirus, a DNA herpesvirus, may cause disease in both immunologically intact and immunocompromised individuals. In the intact host, mononucleosis characterized by fever, malaise, fatigue, and an atypical lymphocytosis is the most common manifestation of clinically apparent infection with CMV

(1-3). This illness may occur spontaneously or secondary to blood transfusion (4). Other syndromes which occur infrequently in the intact host include granulomatous hepatitis, aseptic meningitis, hemolytic anemia, pericarditis, peripheral neuropathy, acute erosive esophagitis (5), urinary retention (6), gastrointestinal ulceration, and proctitis (7-9). In the immunocompromised host, such as renal (10,11), liver (12), and bone marrow (13) transplant recipients, certain patients with malignancy, and patients with the Acquired Immunodeficiency Syndrome (AIDS), more severe illness occurs. Disease syndromes which are seen in this setting include encephalitis, retinitis (14,15), esophagitis, pneumonia, adrenalitis, gangrenous cholecystitis (often in association with cryptosporidium)(16), and colitis. These illnesses can be life-threatening, and CMV infection is a leading cause of death in both bone marrow transplant (BMT) recipients and patients with AIDS. CMV is also the most common cause of serious congenital virus infection with possible devastating consequences (9,17,18). It is for these immunocompromised patient populations that prevention, prophylaxis, rapid diagnosis, and effective therapy for CMV infection becomes crucial. This chapter will focus on antiviral chemotherapy of CMV infections in the clinical setting. Relevant in-vitro and animal data will be reviewed. Immunoprophylaxis with drug and immunoglobulin preparations will also be discussed. Prevention by vaccination will not be addressed.

Therapy of CMV Infections

DHPG

9-(1,3-dihydroxy-2-propoxymethyl)guanine, also referred to as 2'-NDG, BIOLF-62, BW759U, and hereafter DHPG, has substantial activity against CMV. The ED_{50} (that concentration of drug which causes a 50% reduction in viral effects) in various studies ranges from 0.04 to 11 uM depending on the exact method used (19-23). DHPG's congener, acyclovir, is about 10-100 times less potent against CMV (24). Peak serum levels of DHPG are usually between 20 and 75 uM with a dosage of 15 mg/kg intravenously (i.v.) and between 19 and 25 uM with 7.5 mg/kg i.v. (23,25). Oral doses of 10 mg/kg give peak plasma levels ranging from 0.90-1.93 uM (26). The half-life of DHPG is 3.4 hours (27). The mechanism of action of DHPG has not been determined with certainty. Phosphorylation to the triphosphate appears necessary for drug activity (20,28,29) but does not involve viral thymidine kinase in CMV infections (30). Interaction with a cytosol deoxyguanosine kinase which is induced in CMV-infected cells may be an important step (31). DHPG blocks the production of virus progeny but not immediate early and early gene products (24).

Studies involving murine CMV infection indicate both a decreased mortality (30) and reduced viral titers in lung and salivary tissue (32) in animals treated intraperitoneally with DHPG. Aerosolized administration has also been utilized with success in mice (33). Interestingly, latent infection does not seem to be prevented by DHPG; immunosuppression induced with rabbit antilymphocyte serum and cortisone acetate or explantation of splenic tissue following CMV infection and DHPG treatment of mice results in reactivation of CMV (34).

Several uncontrolled human trials with DHPG in immunocompromised hosts suggest a beneficial response to DHPG therapy in CMV-infected patients. In 26 patients, 22 of whom had AIDS, 17 stabilized or improved at a dose of 5 mg/kg i.v. at 8 hour intervals for a mean of 14 days. CMV retinitis, esophagitis, and colitis (35,36) were most amenable to therapy whereas CMV pneumonia responded poorly; four of 7 patients with pneumonia died before completion of 14 days of therapy. Unfortunately, both clinical and virologic relapses were common (37) occurring in 79% of the patients. A similar refractoriness of CMV pneumonia to DHPG therapy has also been noted in BMT patients despite a decrease in viral titers in lung tissue

(25). The addition of corticosteroids did not alter this outcome (38). Apparent successful therapy of CMV pneumonia in two AIDS patients has been described (36), although spontaneous resolution cannot be excluded.

Several reports describe successful suppression of CMV retinitis in AIDS patients. Distinction of cotton wool spots from CMV retinitis is important in this context (39,40). With dosages between 2.5 and 7.5 mg/kg given every eight hours, partial or complete responses were observed in 6 of 7 patients (41), 2 of 2 patients (42), 3 of 3 patients (43), 30 of 37 patients (44), and 14 of 14 patients (45) in several uncontrolled studies. Relapses were common with discontinuance of DHPG; preliminary success has been reported with maintenance DHPG given at 5 mg/kg i.v. five times per week (44) and 30 mg/kg/week i.v. (45). Further definition of an appropriate maintenance regimen is needed as breakthrough retinitis has been reported despite daily therapy of 5 mg/kg (46).

The collected experience with DHPG as of April, 1986 was recently summarized (27). These data include over 400 patients. Retinitis and gastrointestinal (GI) disease due to CMV responded best to DHPG therapy; 90% of patients with retinitis and 75% of patients with GI disease experienced improvement or stabilization of previously progressive disease. CMV pneumonitis was more responsive to therapy in AIDS patients than in BMT patients, with 75% and 30% response rates respectively.

Persistent neutropenia is an important problem in patients on maintenance DHPG therapy. In-vitro and animal data indicate direct suppressive bone marrow effects (47). One fatal case of irreversible neutropenia has been reported (27), but the neutropenia is usually reversible. Mild transaminase elevations are found in 10-20% of cases, but seldom require discontinuance of therapy. Central nervous system (CNS) toxicities including headache, hallucinations, disorientation and possible sensorineural hearing loss have also been noted and required discontinuance of therapy in approximately 1% of patients.

Other toxicities include eosinophilia, thromobocytopenia, edema, phlebitis, nausea, anorexia, and rash (35,37,42,44,45). DHPG has caused azoospermia in Sprague Dawley rats and Beagle dogs treated with doses at or below those used in humans (47). It is not known whether DHPG will have similar toxic effects on spermatogenesis in man. Further data on this potentially important toxicity are needed. DHPG appears to be the most promising drug studied to date for therapy of human CMV infections. The impressions gleaned from the open, uncontrolled studies reported above should be confirmed in prospective, controlled trials, which are planned.

PFA and PAA

Phosphonoformate (PFA), also known as Foscarnet, and phosphonoacetate (PAA) have both shown in-vitro and possibly in-vivo efficacy against CMV although controlled clinical trials are lacking. The ED_{50} of PFA against CMV strain AD-169 ranges between 32-63 uM at a multiplicity of infection (m.o.i.) of 0.1 and 125-250 uM at an m.o.i. of 0.5 (48,49). Clinical isolates seem to be somewhat less sensitive than strain AD-169 to PFA in vitro with ED_{50}'s of 16-250 uM (49,50). PFA seems to be suppressive in tissue culture rather that virocidal; immediate early and early antigens of CMV still appear in treated cultures and full productive infection can reemerge with removal of PFA as late as 35 days after the initial addition of drug (50). PAA completely inhibits AD-169 at a m.o.i. of 1 with 20-50 ug/ml (51,52) but may be more toxic than PFA (50).

Pharmacokinetic data from one study indicate a wide range of steady state plasma levels in treated patients. In BMT recipients, mean dosage was 128 mg/kg/day with steady state plasma levels of 82 mg/l and maximum plasma levels of 163 mg/l (range 61-447 mg/l). In renal transplant patients, mean dosage was 50 mg/kg/day with mean steady state and maximum plasma levels of 127 mg/l and 193 mg/l respectively (range 44-449 mg/l). These differences are presumably due to renal dysfunction in the latter group (53). PFA is generally administered intravenously as a bolus of 9-20 mg/kg followed by a continuous infusion at 0.078-0.14 mg/kg/min adjusted for renal function. PFA and PAA act by inhibiting viral DNA polymerase to a greater extent than cellular DNA polymerase; further details of the mechanism of action need elucidation (50,51,54-58).

Studies in mice utilizing PAA have shown a protective effect against CMV infection. Mortality decreased in mice treated with 500 mg/kg/day intraperitoneally (i.p.) starting 2 hours after infection, compared with untreated mice, from 93% to 40% and days until death increased from 6.1 days to 8.1 days. Viral replication was completely inhibited in the liver. Greater success was achieved with a slightly smaller viral inoculum and treatment with 250 mg/kg/day (52). Efficacy of PFA therapy could not be demonstrated in a guinea pig model of CMV infection. Severe interstitial pneumonitis developed despite treatment; perhaps initiation of therapy at day 3 post-infection was too late in this model to demonstrate anti-viral activity (59).

Small uncontrolled trials in humans suggest some effect of PFA on the course of CMV disease in immunocompromised hosts. Controlled trials with virologic studies are needed. In 16 BMT and 19 renal tranplant patients, clinical improvement defined on a subjective basis was noted in 69%. CMV cultures became negative in 12 of 20 patients with defervescence in 15 of 30 and improvements in laboratory abnormalities or x-rays in 18 of 35. Of note, all 9 BMT recipients with pneumonitis died despite PFA therapy, in contrast to 2 of 6 renal tranplant patients with pneumonitis in this same study (60). In another report utilizing similar drug dosages, two BMT patients with CMV pneumonitis treated with PFA both improved dramatically (61). Another study looked at 13 BMT and 12 renal transplant recipients with CMV infection. Twelve of these patients died, but efficacy as defined by eradication of CMV from cultures (8 of 14), resolution of fever (11 of 22), improvement of laboratory values (13 of 23), or any of the above (17 of 24 or

70%) was reported (53). In 20 renal tranplant patients treated with PFA in another study, a "good clinical effect" was thought to occur in 12 of 14 patients with primary CMV infection and 4 of 6 with secondary CMV infection (62). Another small study of 3 BMT and 3 renal transplant recipients showed dramatic improvement of CMV encephalitis in one patient, possible improvement related to PFA in another 3 patients, and suppression of positive cultures in one patient (63). The subjective nature of these assessments presents a problem in interpretation. Finally, improvement of CMV retinitis in AIDS patients during PFA therapy has been reported (64,65).

Adverse reactions to PFA have included anemia, hypercalcemia, liver function abnormalities, and renal toxicity as manifested by increasing blood urea nitrogen (BUN) and creatinine (Cr). One uremic patient with a plasma level of >400 mg/l experienced hallucinations and tremor in conjunction with PFA treatment (53,62); intention tremor has also been reported (65). The need for continuous intravenous infusions also presents problems. Differing regimens are under investigation, including early studies with an oral formulation of PFA. However, to date oral absorption has not been sufficient to achieve satisfactory blood levels of PFA (Öberg, B., personal communication).

Vidarabine

Vidarabine (9-beta-D-arabinofuranosyladenine, adenine arabinoside, ara-A) is a purine nucleoside analog with activity against all human herpesviruses. It is less active against CMV than against herpes simplex virus (HSV) or varicella-zoster virus (VZV). In one study, an ED_{80} of 87 ug/ml was found against 4 strains of CMV (66); much higher concentrations were necessary for complete inhibition (67). Vidarabine is rapidly deaminated by adenosine deaminase to hypoxanthine arabinoside in vivo (68); this metabolite is less active than the parent compound and is rapidly cleared by the kidneys with a half-life of three and one-half hours (69,70). Plasma levels of vidarabine are 1-6 ug/ml and of hypoxanthine arabinoside 1-7 ug/ml with an infusion of 10 mg/kg over 6 hours and the use of an inhibitor of deamination (68). Approximately 35% of plasma levels can be found in the cerebrospinal fluid (CSF) (71,72). Doses of 10-15 mg/kg/day of vidarabine are usually administered at a concentration of less than 0.5 mg/ml intravenously over a 12 hour period. Within cells, vidarabine is phosphorylated to a triphosphate form which acts as a relatively selective inhibitor of herpes simplex DNA polymerase; a similar mechanism presumably is involved in its activity against CMV (72).

Vidarabine has activity against the Smith strain of murine CMV in mice with Ehrlich ascites tumor. Survival of infected mice was extended from 3 to 8 days in mice treated with 4 doses of 250 mg/kg i.p. as compared to controls (73).

Clinical trials have not documented efficacy for vidarabine in CMV infections, although the numbers of patients studied have been small. A transient decrease in urinary viral excretion was noted in some of these studies. Patients with CMV mononucleosis complicated by thrombocytopenia (2 patients), congenitally-infected infants with severe CNS disease (5 patients) and 5 immunosuppressed patients were treated with 5-20 mg/kg/day infused over 12 hours for 5 to 15 days. Suppression of urinary viral excretion was seen in most of the immunologically intact and congenitally-infected patients, but viruria returned after 1-3 weeks off therapy. Two of the adults seemed to improve clinically with one showing an increased platelet count while two of the infants were thought to have improved clinically with increased appetite and decreased liver and spleen size. In the immunocompromised patients, viruria was only decreased slightly and viremia persisted. No patient was thought to have responded clinically (74). Another study reported treatment of 9 renal transplant patients with a variety of severe CMV-related diseases including pneumonia, hepatitis, prolonged fever, and pericarditis; four of these patients were treated on an open protocol and 5 in a double-blind study with placebo. Neither virological nor clinical efficacy were seen in the treated group (75). Three additional renal transplant patients were treated by another group with 5-10 mg/kg/day, again with no apparent clinical improvement. Transient suppression of viruria was seen in one patient, but CMV was isolated from all patients after therapy. One patient died with CMV pneumonitis (76). In contrast, a 14 year old girl with acute lymphocytic leukemia and CMV pneumonia for 3 weeks was treated with vidarabine at 15 mg/kg/day with clinical improvement after 48 hours and eventual resolution of her pneumonia (77). Finally, heart and renal transplant recipients and patients with lymphoma who had CMV retinitis were treated with vidarabine and compared to a group who were left untreated. Three of 7 untreated patients improved with a decrease in their immunosuppressive regimens. Five of 7 treated patients improved with a decreased inflammatory response on fundoscopic examination and a quantitative decrease in urinary CMV excretion (78). Taken in total, the above clinical trials are inconclusive as to the efficacy of vidarabine in CMV infection; if an effect exists, it certainly is not dramatic.

Toxicities of vidarabine can be extensive. The drug is relatively insoluble and must be delivered in a large volume of fluid which can create difficulties in patients with compromised cardiac or renal function. Hematologic toxicities include anemia, leukopenia, and thrombocytopenia. Gastrointestinal toxicities of nausea, vomiting, and diarrhea have been described (79,80). CNS toxicities including myoclonus, tremors, and SIADH have also been seen in several patients (75,76,78,81,82).

Acyclovir

Acyclovir, 9-(2-hydroxyethoxymethyl)guanine (ACV), has potent activity against HSV types 1 and 2 and VZV. Its activity against CMV is much less pronounced; nonetheless, it has undergone clinical trials in patients with CMV infections with variable success.

Numerous studies have looked at the in-vitro sensitivity of CMV to ACV (reviewed in 83 and 84). The ED_{50} is dependent on the virus isolate used (laboratory or clinical), the assay methodology, and the cell line employed (85). In general, the ED_{50} for laboratory strains of CMV (e.g. AD-169) ranges from 33.2 uM to 100 uM and for clinical isolates from 0.04 uM to 90 uM with an average of 80 uM (30,49,81,86-89). Clinical isolates tend to be more sensitive to ACV than laboratory strains (90). For comparison, the ED_{50} for HSV averages 0.04-1.62 uM (30,84). Interestingly, murine CMV is much more sensitive to ACV with an ED_{50} between 0.05 and 0.48 uM (83,86,91).

Serum levels achievable after a 1 hour intravenous infusion of 10 mg/kg average 90 uM (83) with peak plasma levels after an intravenous infusion of 5 mg/kg between 30 and 40 uM (92-94). The half-life is between 2 and 4 hours with 98% of the drug excreted in the urine. CSF levels are approximately 50% of serum levels. The drug has an oral bioavailability of 15 -30%; a regimen of 400 mg p.o. every 3 hours achieves serum levels of 0.79-1.22 ug/ml at steady-state (83). Peak plasma levels of 1.4-4.0 uM (mean 2.5 uM) are reached one to two hours after a 200 mg oral dose (92,95). The mechanism of action against HSV involves phosphorylation to the monophosphate by a virally-encoded thymidine kinase (TK) and subsequent phosphorylation to the more active triphosphate form by cellular kinases. The triphosphate is a relatively selective inhibitor of viral DNA polymerase (96). CMV lacks a thymidine kinase and mechanisms involving alternate phosphorylation pathways have been proposed to explain acyclovir's activity against CMV (91). The triphosphate is thought to inhibit CMV DNA polymerase as the final step (30,97,98).

Animal studies have demonstrated efficacy of ACV against murine CMV infection. ACV was able to reduce, but not prevent, CMV replication in a model of CMV adrenalitis using athymic nude mice (99). Reduced titers and lung injury were also seen in a murine model of interstitial pneumonitis (32,100). Reduction of viral titers (101) and increased survival (102) were demonstrated in disseminated murine CMV infection. It must be remembered, however, that murine CMV is much more sensitive to ACV than human strains and may not represent a good model in this respect. In contrast to the success seen in the murine model, acyclovir was ineffective in a guinea pig model of CMV disease when given 3 to 7 days following infection in a dose of 100 mg/kg i.p. twice daily (59).

Only one trial has described possible efficacy of ACV in the treatment of CMV infection in the immunocompromised host, most of whom were renal transplant recipients. Nine of 16 immunocompromised patients received acyclovir and the remainder placebo. Improvement of CMV pneumonia was seen in three patients given ACV. Two treated patients with CMV pneumonia died, but they were semicomatose and bacteremic prior to initiation of therapy. All three placebo patients with CMV pneumonia died. As a group, the treated patients exhibited more rapid clinical improvement (7 vs 31 days) and fewer days of fever after therapy was begun (13 vs 31 days). Urine and throat cultures remained positive throughtout treatment, but viremia ceased after one day of therapy in most patients. ACV was given at 500 mg/m^2 three times daily for 7 days in this protocol (103). Case reports and small uncontrolled series (104,105) have also suggested possible efficacy in transplant recipients with CMV infection. In contrast, 8 BMT patients with CMV pneumonia enrolled in an open study and treated with between 400-1200 mg/m^2 with peak plasma levels of 47-316 uM did not benefit from ACV therapy; only one patient survived despite treatment (106). Acyclovir therapy of congenital CMV infection has been attempted without success. Four infants with congenital CMV were given 15 mg/kg/day with only a temporary decrease in urinary viral titers and no apparent clinical improvement (88). Another 3 infants with congenital CMV infection were treated with ACV. One of the infants had a possible clinical response with decreased hepatomegaly and increased platelets. This infant had high peak plasma levels of ACV (136-163 uM), but this possible response is hard to evaluate since the child received anti-CMV globulin concurrently (107).

Toxicities of ACV are few. Renal crystalluria with bolus administration has been reported (108). Hematologic toxicity is rare at therapeutic concentrations (109). Local irritation if extravasation occurs has been noted (92). More recently, CNS side-effects including lethargy, ataxia, dysarthria, myoclonus, and coma have been reported, although these reactions are uncommon (110,111).

Other Nucleoside Analogs

5-iodo-2'-deoxyuridine (IUdR) was one of the first nucleoside analogs to demonstrate activity against CMV. The ED_{50} in a plaque-reduction assay is between 0.2-1.2 ug/ml (112). IUdR appears to act, after conversion to the triphosphate, as a thymidine analog with incorporation into viral and host DNA (68).

Mouse studies indicate little efficacy for IUdR against murine CMV. Twenty-five mg/kg i.p. twice daily for 8 days did not change mortality with disseminated murine infection although splenic titers of virus were somewhat reduced (112). Another study in mice also found no change in mortality with treatment (113).

A case report of a child with congenital CMV infection described decreased urine titers of CMV and possible clinical improvement following treatment with IUdR in conjunction with novobiocin (114). A second case report noted a similar decrease in urinary viral titers during therapy and possible clinical improvement (115). It is impossible to extrapolate from these case reports and IUdR toxicity has prevented its further systemic use in man. Toxicities include thrombocytopenia (114), leukopenia, and jaundice (68).

Cytosine arabinoside (1-beta-D-arabinofuranosylcytosine or ara-C) was initially thought to hold promise for the therapy of a variety of viral agents, but has been abandoned in viral infections because of its great toxicity (116,117). It has an

ED_{80} of 2.0-4.0 ug/ml in a plaque-reduction assay against CMV (64). The drug is rapidly deaminated to arabinofuranosyl uracil which is a less active metabolite than the parent compound (118). The drug is active in the phosphorylated form, but is only incorporated into DNA to a limited extent. It appears to inhibit phosphorylation of deoxycytidine (68).

Therapy of mice with disseminated murine CMV with ara-C did not appreciably change mortality or virus recovery in lung, liver, or spleen (112).

Several investigators have studied the effect of ara-C in small series of patients with congental CMV infection in open, uncontrolled studies. Dosages ranged from 2-15 mg/kg/day i.v. In the 15 infants studied, a transient suppression of viral excretion could be identified in 5 and elimination of viral excretion in 2, both possibly due to drug. Viremia abated in three infants. Possible clinical benefit was noted in 3 of the children, but was not of a dramatic nature (119-122). Nine renal transplant patients with CMV-related disease (1 with retinitis, 2 with pneumonia, and 6 with fever) were treated with ara-C at 35 mg/m^2/day i.v. for 3-4 days with possible improvement in pneumonia in both patients, no change in the retinitis, and lysis of fever in the other patients (118). Again, the uncontrolled nature of this study makes firm conclusions difficult.

Toxicities include marrow suppression (especially thrombocytopenia), nausea, vomiting, and LFT abnormalities (118,119,121,122). Mice treated with ara-C develop irreversible retinal and brain damage (123). The unexpected toxicities found in the controlled trial alluded to above emphasize the need for well-designed, controlled trials in the evaluation of antiviral compounds (124).

Several other nucleoside analogs have been assayed for activity against CMV. One of the most promising appears to be 1-(2'-deoxy-2'-fluoro-beta-D-arabinofuranosyl)-5-iodocytosine (FIAC) which has substantial anti-CMV activity and little cytotoxicity in culture. The ED_{50} for CMV is 0.3-0.6 uM

(125-127). Pharmacokinetic studies with i.v. and oral dosing in patients with herpesvirus infections indicate good bioavailability (128). The phosphorylated compound may interact with viral DNA polymerase (126). The drug is virostatic; removal from culture leads to production of infectious virus (125).

A recent randomized, double-blind study compared the efficacy of FIAC with vidarabine in immunocompromised patients with localized and disseminated VZV infection. FIAC at 400 $mg/m^2/day$ for 7 days was superior to vidarabine in decreasing the time to last lesion formation, decreasing pain, and decreasing time to initial crusting. Toxicities included only nausea and mild LFT abnormalities (129). Controlled trials utilizing FIAC for therapy of CMV infections would seem reasonable at this time.

Other nucleoside analogs have anti-CMV activity and may find applications in the future. 1-beta-D-arabinofuranosyl-5-fluorouracil (ara-FU) has an ED_{50} of 3.8 uM against AD-169. The mechanism of action appears to involve transformation to the monophosphate by cellular enzymes which inhibits thymidilate synthetase and decreases the dTTP pool. The triphosphate can inhibit viral DNA polymerase (130). (E)-5-(2-bromovinyl)-2'-deoxyuridine (BVDU) has been effective in animal models and in-vitro against HSV. The ED_{50} against HSV-1 is .008-.03 mg/l. It is also active in higher concentrations against CMV with an ED_{50} of 3.3 mg/l (131). It acts as a DNA polymerase inhibitor (132) and has additive effects in-vitro with alpha-interferon (133). 5-bromodeoxycytidine (BCDR) can completely inhibit the AD-169 strain of CMV in-vitro at 25 ug/ml. This occurs in the absence of bromodeoxycytidine kinase activity (134). No further studies with this agent have been reported.

Trifluorothymidine (TFT) has an ED_{50} of 0.22 uM for murine CMV and 0.012 uM for human CMV (135). Six clinical isolates had a mean ED_{50} of 0.57 uM (range 0.32-0.97 uM) while AD-169 had an ED_{50} of 2.1 uM in another study (136). In the presence of host thymidine kinase, TFT inhibits DNA polymerase and thymidine synthetase and is incorporated into DNA. Despite high in-vitro activity against murine CMV, a continuous subcutaneous infusion could not protect mice against lethal infection at the dosage used (135). A new compound, (S)-9-(3-hydroxy-2-phosphonyl-methoxypropyl)adenine ([S]-HPMPA), with potent and selective activity against a broad range of DNA viruses has recently been reported (137). Two laboratory strains of CMV were inhibited at 0.3 ug/ml in-vitro. No effects on host cell metabolism were seen at concentrations $\geq$ 100 ug/ml. (S)-HPMPA is efficacious in several animal models of HSV infection. The antiviral effect of this compound is not dependent on viral thymidine kinase.

Older studies have assessed other nucleoside analogs. 5-fluorouracil-2-deoxyriboside (FUDR) was given to a neonate with severe congenital CMV infection at 0.5 mg/kg/day i.v. Complete clearing of CMV pneumonia was reported although viruria persisted and CMV was recovered from CSF after therapy (138). Ribavirin (1-beta-D-ribofuranosyl-1,2,4-triazole-3-carboxamide) has an ED_{50} of 8.9-32 ug/ml against human and murine CMV. Three mechanisms of action for ribavirin have been proposed; inhibition of 5'-cap formation of mRNA; inhibition of viral RNA polymerases; and competitive inhibition of IMP dehydrogenase thereby reducing intracellular GTP (139). Ribavirin given orally to mice had no effect on survival in a model of disseminated CMV infection and no effect on tissue titers in a murine model using a low inoculum of virus (140). Clinical trials of ribavirin in human CMV infection have not been reported.

Interferons and Interferon Inducers

Interferons have been used prophylacticly and therapeuticly for CMV infections; prophylactic applications have generally met with greater success (see below). There are three types of interferons (IFNs): alpha-IFN produced by peripheral blood lymphocytes; beta-IFN produced by fibroblasts; and gamma-IFN produced largely by T-lymphocytes. Recombinant alpha-IFN is now approved for the treatment of hairy cell leukemia in the United States. Most current data suggest that CMV is relatively sensitive to IFNs (141). The degree of sensitivity is dependent on several factors including the source of the virus, the type of indicator cell used in the assay, the m.o.i., and the timing of inoculation of virus (141-143). The various species of IFNs have variable activity against CMV (144). Interferon inducers such as polyribocytidylic-inosinic acid also have in-vitro activity against CMV (145). Alpha-IFN is cleared from the circulation following intravenous administration with a half-life of 2-3 hours; the half-life with intramuscular administration is 4-6 hours. There is little CSF penetration and hemodialysis does not remove the drug (69,146).

Animal studies have shown that pretreatment with IFN is necessary to effect the course of CMV disease in mice and rats. Murine IFN had no effect on mortality in disseminated murine CMV infection; however, pretreatment with the IFN-inducer polyinosinic-polycytidylic acid or the complex with poly-L-lysine did decrease mortality in this model (147). Pretreatment with other IFN-inducers (two pyrimidone derivatives) also afforded protection in a mouse model (148). In a nonlethal model of murine CMV infection, pretreatment with mouse fibroblast IFN reversed growth retardation and partially restored splenic blastogenic responses (149). Pretreatment with rat IFN decreased viral titers of CMV in salivary and spleen tissues in a rat model of CMV infection (150). It is important to note that none of these studies were able to demonstrate any efficacy if drug was given after infection.

The use of IFN for therapy of established CMV infection in humans has also met with little success. Several uncontrolled series have been reported. Eight BMT patients with CMV pneumonia were treated with human alpha-IFN (2×10^4-6.4×10^5 units/kg/day i.m.); all 8 patients died with virus recovered in lung tissue at autopsy (151). Five BMT transplant recipients with interstitial pneumonitis (3 CMV and 2 idiopathic) were treated with recombinant alpha-A-IFN (18 x

10^6-50 x 10^6 units/day i.m. for 5-18 days); one patient with CMV pneumonia recovered and one patient with idiopathic pneumonia improved in this small series (152). Six additional BMT patients (4 of whom had CMV pneumonia) received a similar IFN preparation; all 6 patients in this series died (153). Congenital CMV infection has also been refractory to therapy with IFN. Two studies using human alpha-IFN could only demonstrate a transient (4 patients) or complete (3 patients) decrease in viruria with no effect on viremia and no apparent clinical benefit (154,155). Finally, 4 patients with CMV retinitis (3 associated with AIDS) were treated with alpha-IFN (5 x 10^6-36 x 10^6 units/day for 19-62 days) with retinal disease progression in 3 of 4 patients associated with increased viruria in 2 patients and decreased viruria in 2 patients (156). Another study treated 5 patients with AIDS and CMV retinitis with 18 x 10^6 units of recombinant alpha-IFN i.m. three times per week for 4 weeks. Only one patient had improvement in visual acuity and retinitis. Therapy in this patient was discontinued because of neurotoxicity. No consistent effect was seen on CMV viral excretion although the one patient who responded clinically had diminished titers in oral washes and urine during therapy (157).

Toxicities of IFNs include fever, chills, local pain at the injection site, anorexia, fatigue, nausea, vomiting, liver function abnormalities, confusion and reversible (but at times dose-limiting) marrow suppression (92,152,153). At the present time, IFNs do not seem to be effective agents when used alone for the treatment of established CMV infection in congenitally-infected infants or immunocompromised adults.

Immunoglobulins

Serum immunoglobulin and "hyperimmune" immunoglobulin preparations directed against CMV have been employed in the therapy and prophylaxis of CMV infections in immunocompromised hosts. Several different preparations, which differ in their route of administration and titer of anti-CMV immunoglobulin, have been tested and described (158-160). Pharmacokinetics and dosing schedules vary with each preparation. As with IFN, treatment has generally been less efficacious than prophylaxis (see below).

Antibody administered to mice with CMV interstitial pneumonia 24 hours after infection reduced viral titers in lung tissue by >90%, but did not alter lung pathology (100). Immune serum administered 6 days after acute infection in normal mice with disseminated CMV infection did not change mortality; in mice rendered immunocompromised by treatment with rabbit antiserum to murine lymphocytes, decreased viral dissemination to organs other than salivary gland was demonstrable (161).

Several uncontrolled trials and case reports have suggested efficacy of immunoglobulin (Ig) preparations in the therapy of CMV infection. Seven renal transplant patients with CMV syndromes (5 with lung infiltrates presumably due to CMV) were treated with hyperimmune anti-CMV Ig (titer 1:900); 12.5 ml was injected intraperitoneally and 12.5 ml subcutaneously. Five of 7 patients showed a "complete and sustained" response within 24 hours of therapy (162). Another uncontrolled study gave 200 mg/kg/day i.v. of human IgG with an anti-CMV titer of 1:256 by IHA to 52 renal transplant recipients. No virologic studies were done and the description of the patient population was not detailed, but the impression was of a beneficial effect (163). Two case reports describe renal transplant recipients with CMV pneumonia who recovered in conjunction with Ig therapy (164,165). Other case reports describe improvement in an infant with transfusion-acquired CMV pneumonia and an increase in platelet count in an infant with congenital CMV infection during Ig therapy (166,167). In a somewhat larger uncontrolled series, 10 BMT patients with CMV pneumonia received CMV-specific hyperimmune globulin i.v. (200-400 mg/kg for five doses over 16 days); 6 of 9 patients improved within 12-14 days of the first infusion while 3 of 9 patients died (168). This represents a dramatic decrease in mortality for this frequently fatal complication of BMT. However, 10 BMT patients in another series given a similar dose and formulation of i.v. Ig did not fare as well; 8 of 10 of these patients died (169). The reason for these differences in outcome are not immediately clear, but highlight the need for controlled studies.

Drug Combinations

Various therapeutic combinations of antiviral agents with activity against CMV have been employed with little success. In-vitro data indicate that certain combinations have synergistic activity against CMV. In these studies, clinical isolates were typically more sensitive than laboratory strains of CMV. ACV in combination with PFA (170), vidarabine (171), TFT (136,170), or alpha-IFN (133) were all synergistic in-vitro. Two studies found ACV and alpha-IFN to be additive in-vitro (172,173). ACV and beta-IFN had additive effects as well (170). TFT was synergistic with PFA but additive with beta-IFN against clinical isolates (136). Alpha-IFN was additive in combination with vidarabine and BVDU (133). Finally, DHPG was additive with alpha-IFN but synergistic with beta-IFN (173). However, no advantage of ACV in combination with beta-IFN compared to ACV alone could be demonstrated in a mouse model of CMV pneumonia (174).

Several studies of series of BMT patients with CMV pneumonia treated with combination antiviral therapy have been reported. None of these studies have demonstrated a decrease in mortality. In 7 BMT patients treated with alpha-IFN (4×10^4-1.6×10^5 units/kg/day) and vidarabine (2.5-10 mg/kg/day), mean virus titer in lung decreased with therapy but only one patient survived. In addition, granulocytopenia and severe neurotoxicity with tremors, hallucinations, myoclonus, and seizures were seen (175). Thirteen BMT patients were treated with ACV (500-1000 mg/m^2/day) and alpha-IFN (2×10^4-40×10^4 units/kg/day) intramuscularly; 10 of thirteen died, although several had decreased viral titers in lung tissue. Toxicity was again significant. Marrow suppression, renal dysfunction, and tremors with agitation were noted (176,177). Eight BMT patients with CMV pneumonia were treated with ACV (1 g/m^2 every 8 hours) and alpha-IFN (20-40 x 10^6 units/day) intravenously. Seven of 8 patients died although 2 had decreased viral titers in lung and 2 had suppression of viremia. Toxicities were severe and included lethargy, confusion, incontinence, neutropenia, renal dysfunction and hepatic abnormalities (178). Case reports of single patients have shown possible success with various combinations. A BMT recipient with CMV pneumonia was treated with ACV (1500 mg/m^2/day) and beta-IFN (10 x 10^6 units/day) with apparent response (179). A renal transplant patient with CMV pneumonia received hyperimmune anti-CMV globulin and DHPG (5 mg/kg b.i.d., i.v.) with rapid resolution of her pneumonia (164). A 20-month old child with hypogammaglobulinemia (IgG and IgA), T-cell dysfunction, and a CMV-related syndrome of fever, rash, lymphadenopathy, and pneumonia recovered after treatment with hyperimmune plasma, alpha-IFN, and levamisole (180).

Further controlled studies with different drug and/or immune serum combinations in various patient populations are needed to determine if dual therapy is efficacious. The combinations tested thus far do not appear satisfactory.

Other Agents

Several other agents are of potential interest in the therapy of CMV infections. Arildone (4-[6-(2-chloro-4-methoxy)phenoxyl]hexyl-3,5-heptanedione) is an aryl-beta-ketone which can completely suppress CMV expression in vitro at 3 ug/ml (181) when diluted in DMSO. Immediate early antigen is produced but no late antigens, viral DNA, or progeny virus (182). Another study found 64% suppression of CMV replication with 3 ug/ml of Arildone; productive replication occurred with removal of the drug (183). The mechanism of action against HSV seems to involve inhibition of viral uncoating, but another mechanism may be important in relation to CMV (181,182).

Two polyamine biosynthetic pathway inhibitors have anti-CMV activity in vitro. DMFO (D,L-alpha-difluoromethylornithine) can suppress CMV, although low viral input titers and drug pre-treatment are necessary to demonstrate an effect (184). MGB-G, another polyamine antimetabolite, has an ED_{50} of 0.5-1.8 uM against CMV in-vitro (185).

Transfer factor, a dialyzable material obtained from human lymphocytes, has been utilized in the therapy of CMV infections. Three infants with congenital CMV infection given 2 doses of transfer factor had transient suppression of urinary viral excretion (186); however, further infants treated with this same regimen and a slightly modified regimen did not have suppression of viral excretion (187). A 4-year old boy with a 2 year history of fever, rash, abdominal pain, and arthralgias presumably due to CMV, was given oral bovine transfer factor for 6 months with resolution of symptoms and a restoration of the blastogenic response to CMV antigens (188). A 7-month old infant with fever, rash, weight loss, lymphadenopathy, and hepatosplenomegaly, again presumably due to CMV, improved with transfer factor therapy (189). These case reports are inconclusive as to the value of transfer factor in the therapy of CMV infections.

Interleukin-2 (IL-2) is important in cell-mediated cytotoxicity against CMV (190) and its possible usefulness in CMV infections has been suggested (191). Analysis of data in treatment trials of AIDS patients utilizing IL-2 might shed some light on this possibility. Preliminary analysis of one such trial indicated a possible beneficial effect in decreasing CMV titers in semen with higher doses of IL-2 (192). Similar effects on viral titers were seen with isoprinosine in this study.

Two antibiotics with antiviral activity against CMV are rifampin (193,194) and novobiocin (195,196). Mithramycin can also suppress CMV infection in-vitro (197). Many other compounds have some degree of anti-CMV activity (198), but require more detailed in vitro and in vivo analysis.

Prevention of CMV Infections

The prevention of CMV infections, especially in immunocompromised hosts, has been an area of great interest and investigation. Prophylaxis takes on added importance in light of the less than ideal therapeutic modalities which are presently available. Blood transfusion represents a major mode of transmission of CMV to immunocompromised hosts. Prevention of transmission by the screening of blood for CMV antibodies or by using leukocyte-depleted or frozen, deglycerolized blood may provide a reasonable means of decreasing the incidence of CMV-related disease in the immunocompromised host. In addition, various drugs and immunoglobulin preparations have met with variable success in the prevention of CMV disease. These modalities will be reviewed here. The prevention of CMV-related disease by vaccination represents another major mode of possible prophylaxis but will not be discussed further in this review.

Transfusion

The prevention of CMV transmission via blood products has been an area of major concern (reviewed in 199-201). At risk populations include organ tranplant recipients, preterm and low birthweight infants, and seronegative pregnant women. Granulocyte transfusions seem especially effective at transmitting CMV (202-204), but all blood products can transmit the virus. Blood products with a decreased likelihood of transmitting CMV can be obtained by donor antibody screening (205-207), leukocyte depletion, or the use of frozen, deglycerolized blood (208,209,210). These products should certainly be used whenever possible in seronegative pregnant women, seronegative transplant recipients receiving organs from seronegative donors, and seronegative preterm infants with birthweights less than 1250 grams (211,212). Some would extend these recommendations to include all newborns who will be multiply transfused and all infants less than 1300 grams (213-215). These guidelines are likely to change as more data become available in each of the high risk groups. Nonetheless, the use of these techniques represents an effective means of preventing CMV-related disease in susceptible populations.

Drug Prophylaxis

Acyclovir and alpha-IFN have proven efficacious in decreasing CMV-related infections in immunocompromised hosts when used in prophylaxis. Vidarabine has been shown to be ineffective in a placebo-controlled prophylactic trial in BMT patients (216).

Acyclovir given orally at 200 mg q6h from 8 days prior to BMT through 35 days post-transplant afforded complete protection against CMV viremia and pneumonia during drug therapy as well as complete protection against HSV infection when compared to placebo; zero of 11 vs 7 of 10 patients had viremia and 0 of 11 vs 3 of 10 had CMV pneumonia (217,218). A more recent report utilizing i.v. ACV at 500 mg/m^2 q8h from 5 days prior to BMT through 30 days post-transplant also found a decrease in CMV infection and pneumonia as well as increased survival in the treatment group. The rates for infection, pneumonia, and survival were 68% and 90%, 19% and 35%, and 72% and 46% in the treatment and control groups respectively. The control group consisted of 65 HSV seronegative patients who were CMV seropositive, while the 86 patients in the treatment group were seropositive for both viruses; the groups are therefore not strictly comparable (219). Two other studies were unable to document an effect of ACV in BMT patients when given prophylacticly. A small trial using 5 mg/kg i.v., three times per day versus 800 mg p.o., four times per day documented a comparable number of CMV infections in both groups; this study suffers from the lack of an untreated control group (220). Another placebo-controlled trial treated 20 BMT recipients with i.v. ACV 250 mg/m^2 from 5 days pre-transplant through 5 weeks post-transplant and then with p.o. ACV 400 mg t.i.d. through 6 months. Twenty-two patients received placebo. There was no statistical difference between CMV infections in the two groups (221). These discrepancies may relate to differences in transplant and treatment protocols or differences in patient populations. It appears that at least in some instances, ACV may have prophylactic value in BMT patients. Further studies to define the exact circumstances are warranted.

Prophylactic alpha-IFN in renal transplant patients can decrease the incidence of CMV syndromes and possibly the incidence of superinfection. In an initial study, renal transplant patients who were both seropositive and seronegative for CMV were randomized to receive either human alpha-IFN at 3×10^6 units/day on days 1 and 2 following transplant and then twice weekly for 6 weeks (21 patients) or placebo (20 patients). There was a delay in onset of viremia in the treated patients (7.2 vs 4.2 days) and a decreased incidence of viremia (5 of 11 vs 9 of 10). The incidence of CMV infection was correlated with administration of antithymocyte globulin (ATG) in both groups (222). In a further study which included only CMV seropositive renal transplant recipients, 20 patients received 3×10^6 units i.m. three times per week for 6 weeks and then twice weekly for 8 weeks while 22 patients were given placebo. Clinical signs of CMV infection were decreased in the treated group compared with controls (1 of 20 vs 7 of 22). Two superinfections occurred in the placebo group and none in the treated group. Dosage and species of IFN used seem important; a study using high doses of recombinant IFN-alpha (36×10^6 units three times per week) was ineffective in reducing CMV infection and, in addition, induced steroid-resistant rejection episodes (223).

Lower doses of azathiaprine and prednisone administered when ATG was employed lowered the overall incidence of CMV infection in renal transplant recipients (224). Moreover, increasing use of cyclosporine and decreasing use of antilymphocyte globulins in renal transplant recipients has further reduced CMV syndromes and, thus, the need for prophylaxis in this population.

Immunoglobulins

Mice given anti-CMV Ig passively can be protected against lethal murine infection (225). Analogous findings have been described in various human trials; many of these trials represent well-designed and controlled studies. BMT patients are the best studied group. All but one of these studies showed efficacy, especially in patients who did not receive leukocyte transfusions. Treatment with high titer anti-CMV Ig (1:2048 by CF) i.m. for three doses before transplant and weekly thereafter through day 77 post-transplant significantly decreased CMV infection compared to a control group for patients who had not received granulocyte transfusions (2 of 17 vs. 8 of 19) (226). Another study administered 20 ml/kg i.v. before conditioning and then weekly thereafter through day 120 post-transplant. This trial documented a significant decrease in interstitial pneumonia due to CMV compared to controls (3 of 18 vs 8 of 18); seven of the control patients and 2 of the treated patients with CMV pneumonia died (227). In a study with 2 control arms consisting of no therapy and CMV-deficient Ig, patients treated with 200 mg/kg of hyperimmune anti-CMV Ig i.v. (titer 1:1600 by ELISA) on days 25, 50, and 75 post-transplant experienced no episodes of interstitial pneumonitis compared to 6 and 3 episodes in the two control groups respectively (228). Hyperimmune globulin compared to pooled serum immunglobulin at 0.1 g/kg i.v. for 6 doses starting 7 days pre-transplant was more efficacious in decreasing CMV pneumonia in 49 patients with BMT (1 of 26 vs. 6 of 23) (229). In another randomized, controlled trial, symptomatic CMV infection and interstitial pneumonitis were significantly decreased in patients treated with hyperimmune Ig at 10 mg/kg i.m. before conditioning and for 7 doses after transplant, when compared with no prophylactic therapy. The effect was particularly pronounced in patients who had not received leukocyte transfusions. No change in mortality could be documented in this study (230). The one study which did not document a beneficial effect of prophylactic Ig divided patients into 4 groups; administration of only seronegative blood products with or without high titer anti-CMV Ig and no screening of blood products with or without Ig. The patients were further stratified according to seropositivity to CMV pre-transplant. Seronegative blood products given to seronegative recipients were effective in decreasing the incidence of CMV-related disease. No effect of Ig therapy could be detected in any group, but the numbers were small and the confidence intervals broad (212).

Similar studies have been conducted in renal transplant recipients. In a study using historical controls, patients receiving hyperimmune globulin i.m. seemed to have fewer and less severe CMV-related syndromes. This study suffers from the fact that different immunosuppressive regimens were employed during these time periods (231). Another study reported the preliminary results of a trial in renal transplant patients stratified by pre-transplant serology and given hyperimmune globulin i.v. compared to a control group without therapy. The initial results suggested a decrease in seroconversion and CMV-related disease in the treatment group (232). Preliminary reports of a prospective, randomized study comparing hyperimmune globulin administration (31 patients) with no prophylactic therapy (38 patients) in seronegative renal transplant recipients have also suggested efficacy. A CMV-associated syndrome was documented in 24% of the treated patients and 55% of the patients not given therapy. CMV pneumonia, retinitis, opportunistic infection, viremia, and mortality were all reduced significantly in the treatment group. Seroconversion rates were similar in the two groups (233).

The toxicities of Ig preparations seem to be limited. Symptoms compatible with serum sickness have been reported in two patients given i.v. Ig (not hyperimmune) who had undergone cardiac transplant (234), but other serious toxicities have not been reported.

In summary, intramuscular and intravenous immunoglobulin seem to prevent or attenuate serious CMV disease in BMT and renal transplant recipients. Utilization of seronegative blood for seronegative recipients also seems efficacious. Further controlled trials with larger numbers of patients stratified into seropositive and seronegative recipients with and without screening of blood products are needed to further define the effectiveness of prophylactic immunoglobulin therapy and better define side effects and dosage schedules.

Summary

Considerable progress has been made in recent years towards the development of satisfactory regimens for both prophylaxis and therapy of CMV infections in immunocompromised hosts (Table 1). Effective preventative strategies now exist in certain immunocompromised groups; these modalities need to be extended and modified for use in other high risk populations. DHPG appears to be a major advance in the treatment of CMV infections. Suppressive antiviral therapy for established infection is possible in certain groups of patients and evidence exists that occassional patients may be cured of CMV-related disease. The next major hurdles to be overcome in the control of CMV infections are the development and controlled study of orally bioavailable and non-toxic regimens for both prophylaxis and therapy.

Table 1 Agents With Possible Prophylactic or Therapeutic Activity Against CMV Infections in Immunocompromised Hosts

Agent	Patient Population	References[1]
Prophylaxis		
Acyclovir	Bone Marrow Transplant	217-221
Alpha-IFN	Renal Transplant	222-224
Immunoglobulins	Bone Marrow Transplant	212,226-230
Therapy		
DHPG[2]	AIDS[3]	27,35-38,41-46
PFA[4]	Bone Marrow Transplant	25,27
	AIDS	64,65
	Bone Marrow Transplant	53,60,61,63
	Renal Transplant	53,60,62,63

1. Includes both positive and negative reports
2. DHPG = 9-(1,3-dihydroxy-2-propoxymethyl)guanine
3. AIDS = Acquired Immunodeficiency Syndrome
4. PFA = phosphonoformate

REFERENCES

1. Klemola, E., and Kääriäinen, L. Br. Med. J. 2: 1099-1102, 1965.
2. Horwitz, C. A., Henle, W., Henle, G., Snover, D., Rudnick, H., Balfour, H. H., Mazur, M. H., Watson, R., Schwartz, B., and Muller, N. Medicine 65: 124-134, 1986.
3. Pannuti, C. S., Boas, L. S. V., Angelo, M. J. O., Neto, V. A., Levi, G. C., De Mendonç, J. S., and De Godoy, C. V. F. Scand. J. Infect. Dis. 17: 153-156, 1985.
4. Lang, D. J., Scolnick, E. M., and Willerson, J.T. N Engl J Med. 278: 1147-1149, 1968.
5. Villar, L. A., Massanari, R. M., and Mitros, F. A. Am. J. Med. 76: 924-928, 1984.
6. Michaelson, R. A., Benson, G. S., and Friedman, H. M. Am. J. Med. 74: 526-528, 1983.
7. Cohen, J. I., and Corey, G. R. Medicine 64: 100-114, 1985.
8. Weller, T. H. N. Engl. J. Med. 285: 267-274, 1971.
9. Ho, M. Cytomegalovirus: Biology and Infection. Plenum Publishing Corporation: New York, N.Y. 1982 pp131-204.
10. Rubin, R. H., Cosimi, A. B., Tolkoff-Rubin, N. E., Russell, P. S., and Hirsch, M. S. Tranplantation 24: 458-464, 1977.
11. Glenn, J. Rev. Infect. Dis. 3: 1151-1178, 1981.

12. Esquivel, C. O., Jaffe, R., Gordon, R. D., Iwatsuki, S., Shaw, B. W., Jr., and Starzl, T. E. Semin. Liver Dis. 5: 369-374, 1985.
13. Betts, R. F. Prog. Med. Virol. 28:44-64, 1982.
14. Murray, H. W., Knox, D. L., Green, W. R., and Susel, R. M. Am J. Med. 63:574-584, 1977.
15. Egbert, P. R., Pollard, R. B., Gallagher, J. G., and Merigan, T. C. Ann. Intern. Med. 93:664-670, 1980.
16. Blumberg, R. S., Kelsey, P., Perrone, T., Dickersin, R., Laguaglia, M., and Ferruci, J. Am. J. Med. 76:1118-1123, 1984.
17. Yow, M. D., and Taber, L. H. Clin. Obstet. Gynecol. 15: 993-1003, 1972.
18. Reyonolds, D. W., Stagno, S., and Alford, C. A. Teratology 17: 179-182, 1978.
19. Cheng, Y.-C., Huang, E.-S., Lin, J.-C., Mar, E.-C., Pagano, J. S., Dutschman, G. E., and Grill, S. P. Proc. Natl. Acad. Sci. (USA) 80:2767-2770, 1983.
20. Mar, E.-C., Cheng, Y.-C., and Huang, E.-S. Antimicrob. Agents Chemother. 24:518-521, 1983.
21. Tyms, A. S., Davis, J. M., Jeffries, D. J., and Meyers, J. D. Lancet ii: 924-925, 1984.
22. Prisbe, E. J., Martin, J. C., McGee, D. P. C., Barker, M. F., Smee, D. F., Duke, A. E., Matthews, T. R., and Verheyden, J. P. H. J. Med. Chem. 29: 671-675, 1986.
23. Plotkin, S. A., Drew, W. L., Felsenstein, D., and Hirsch, M. S. J. Infect. Dis. 152: 833-834, 1985.
24. Tocci, M. J., Livelli, T. J., Perry, H. C., Crumpacker, C. S., and Field, A. K. Antimicrob. Agents Chemother. 25: 247-252, 1984.
25. Shepp, D. H., Dandliker, P. S., De Miranda, P., Burnette, T. C., Cederberg, D. M., Kirk, L. E., and Meyers, J. D. Ann Intern. Med. 103:368-373, 1985.
26. De Miranda, P., Burnette, T., Cederberg, D., Blum, M. R., Brodie, H. R., and Mills, J. 26th Interscience Conference on Antimicrobial Agents and Chemotherapy. p. 200 (abstract 566), 1986.
27. Mills, J. In: Antiviral Chemotherapy: New Directions for Clinical Application and Research (Eds J. Mills and L. Corey), Elsevier, N. Y., 1986, pp 195-203.
28. Field, A. K., Davies, M. E., DeWitt, C., Perry, H. C., Liou, R., Germershausen, J., Karkas, J.D., Ashton, W.T., Johnston, D. B. R., and Tolman, R. L. Proc. Natl. Acad. Sci. (USA) 80: 4139-4143, 1983.
29. Smee, D. F., Martin, J. C., Verheyden, J. P. H., and Matthews, T. R. Antimicrob. Agents Chemotherapy 23:676-682, 1983.
30. Freitas, V. R. Smee, D. F., Chernow, M., Boehme, R., and Matthews, T. R. Antimicrob. Agents Chemotherapy. 28: 240-245, 1985.
31. Smee, D. F., Mol. Cell. Biochem. 69:75-81, 1985.
32. Shanley, J. D., Morningstar, J., and Jordan, M. C. Antimicrob. Agents Chemother. 28: 172-175, 1985.
33. Debs, R., Brunette, E. Papahadjopoulos, D., Debruin, M., and Shanley, J. D., 26th Interscience Conference Antimicrob. Agents and Chemother. p. 231 (abstract 730), 1986.
34. Katzenstein, D. A., Crane, R. T., and Jordan, M. C. J. Lab Clin. Med. 108: 155-160, 1986.
35. Reed, E. C., Shepp, D. H., Dandliker, P. S., and Meyers, J. D. 26th Interscience Conference on Antimicrobial Agents and Chemotherapy p.200 (abstract 567), 1986.
36. Pollard, R. B., Matzke, D. S., Ramsey, K. M., Bissett, J. D., and Merigan, T. C. Clin. Res. 33: 416A, 1985.
37. Collaborative DHPG Treatment Study Group N. Engl. J. Med. 314: 801-805, 1986.
38. Reed, E. C., Dandliker, P. S., and Meyers, J. D. Ann. Intern. Med. 105: 214-216, 1986.
39. Akula, S. K., Mansell, P. W. A., and Ruiz, R. Ann. Intern. Med. 104: 726-727, 1986.
40. Teich, S. A., and Orellana, J. Ann. Intern. Med. 104: 132, 1986.

41. Masur, H., Lane, H. C., Palestine, A., Smith, P. D., Manischewitz, J., Stevens, G., Fujikawa, L., Macher, A. M., Nussenblatt, R., Baird, B., Megill, M., Wittek, A., Quinnan, G. V., Parrillo, J. E., Rook, A. H., Eron, L. J., Poretz, D. M., Goldenberg, R. I., Fauci, A. S., and Gelman E. P. Ann. Intern. Med. 104: 41-44, 1986.
42. Felsenstein, D., D'Amico, D. J., Hirsch, M. S., Neumeyer, A., Cederberg, D. M., De Miranda, P., and Schooley, R. T. Ann. Intern. Med. 103: 377-380, 1985.
43. Bach, M. C., Bagwell, S. P., Knapp, N. P., Davis, K. M., and Hedstrom, P. S. Ann. Intern. Med. 103: 381-384, 1985.
44. Cederberg, D., Laskin, O. L., Mills, J., Creagh-Kirk, T., and the Collaborative BW B759U Study Group. 26th Interscience Conference on Antimicrobial Agents and Chemotherapy p. 200 (abstract 565), 1986.
45. Drew, W. L., Buhles, W., Busch, D., Mills, J., Follansbee, S., and Merigan, T. C. 26th Interscience Conference on Antimicrobial Agents and Chemotherapy p. 200 (abstract 569), 1986.
46. Bach, M. C. Ann. Intern Med. 104: 587, 1986.
47. Szczeck, G.M. In: Antiviral Chemotherapy: New Directions for Clinical Application and Research (Eds J. Mills and L. Corey), Elsevier, N.Y., 1986, pp 204-225.
48. Wahren, B., and Öberg, B. Intervirology 12: 335-339, 1979.
49. Gadler, H. Antimicrob. Agents Chemother. 24: 370-374, 1983.
50. Wahren, B., and Öberg, B. Intervirology 14: 7-15, 1980.
51. Huang, E.-S., Huang, C.-H., Huong, S.-M., and Selgrade, M. Yale J. Biol. Med. 49: 93-98, 1976.
52. Overall, J. C., Jr., Kern, E. R., and Glasgow, L. A. J. Infect. Dis. 133: A237-A244, 1976.
53. Ringdén, O., Lönnqvist, B., Paulin, T., Ahlmén, J., Klintmalm, G., Wahren, B., and Lernestedt, J.-O. J. Antimicrob. Chemother. 17: 373-387, 1986.
54. Leinbach, S. S., Reno, J. M., Lee, L. F., Isbell, A. F., and Boezi, J. A. Biochemistry 15: 426-430, 1976.
55. Helgstrand, E., Eriksson, B., Johansson, N. G., Lannerö, B., Larsson, A., Misiorny, A., Norén, J. O., Sjöberg, B., Stenberg, K., Stening, G., Stridh, S., Öberg, B., Alenius, S., and Philipson, L. Science 201: 819-821, 1978.
56. Reno, J. M., Lee, L. F., and Boezi, J. A. Antimicrob. Agents Chemother. 13: 188-192, 1978.
57. Huang, E.-S. J. Virol. 16: 1560-1565, 1975.
58. Öberg, B. Pharmacol. Ther. 19: 387-415, 1983.
59. Lucia, H. L., Griffith, B. P., and Hsiung, G. D. Intervirology 21: 141-149, 1984.
60. Ringdén, O., Wilczek, H., Lönnqvist, B., Gahrton, G., Wahren, B., and Lernestedt, J.-O. Lancet i: 1503-1504, 1985.
61. Apperley, J. F., Marcus, R. E., Goldman, J. M., Wardle, D. G., Gravett, P. J., and Chanas, A. Lancet i: 1151, 1985.
62. Ahlmén, J., Wijnween, A.-C., Brynger, H., and Lycke, E. Scand. J. Urol. Nephrol. 92(Suppl): 41-44, 1985.
63. Klintmalm, G., Lönnqvist, B., Öberg, B., Gahrtron, G., Lernestedt, J.-O., Lundgren, G., Ringdén, O., Robert, K.-H., Wahren, B., and Groth, C.-G. Scand. J. Infect. Dis. 17:157-163, 1985.
64. Singer, D. R. J., Fallon, T. J., Schulenburg, W. E., Williams, G., and Cohen, J. Ann. Intern. Med. 103: 962, 1985.
65. Walmsley, S., Chew, E., Fanning, M. M., Coates, R. A., Salit, I. S., Shepherd, F. A., Rachlis, A., and Read, S. E. 26th Interscience Conference on Antimicrobial Agents and Chemotherapy p. 200 (abstract 568), 1986.
66. Fiala, M., Chow, A. W., Miyasaki, K., and Guze, L. B. J. Infect. Dis. 129: 82-85, 1974.
67. Gephart, J. F., and Lerner, A. M. Antimicrob. Agents Chemother. 19: 170-178, 1981.
68. Luby, J. P., Johnson, M. T., and Jones, S. R. Annu. Rev. Med. 25: 251-267, 1974.
69. Wong, K. K., and Hirsch, M. S. Am. J. Med. 76: 464-478, 1984.

70. Whitley, R. J., Tucker, B. C., Kinkel, A. W., Barton, N.H., Pass, R. F., Whelchel, J. D., Cobbs, C. G., Diethelm, A.G., and Buchanan, R. A. Antimicrob. Agents Chemother, 18: 709-715, 1980.
71. Aronoff, G. R., Szwed, J. J., Nelson, R. L., Marcus, E. L., and Kleit, S. A. Antimicrob. Agents Chemother. 18: 212-214, 1980.
72. Whitley, R., Alford, C., Hess, F., and Buchanan, R. Drugs 20: 267-282, 1980.
73. Schmidt-Ruppin, K. H. Chemotherapy 16: 130-143, 1971.
74. Ch'ien, L., Cannon, N. J., Whitley, R. J., Diethelm, A. G., Dismukes, W. E., Scott, C. W., Buchanan, R. A., and Alford, C. A., Jr. J. Infect. Dis. 130: 32-39, 1974.
75. Marker, S. C., Howard, R. J., Groth, K. E., Mastri, A. R., Simmons, R. L., and Balfour, H. H., Jr. Arch. Intern. Med. 140: 1441-1444, 1980.
76. Rytel, M. W., and Kauffman, H. M. J. Infect. Dis. 133: 202-205, 1976.
77. Aronson, M. D., Phillips, C. F., Gump, D. W., Albertini, R. J., and Phillips, C. A. JAMA 235:1339-1342, 1976.
78. Pollard, R. B., Egbert, P. R., Gallagher, J. G., and Merigan, T. C. Ann. Intern. Med. 93: 655-664, 1980.
79. Ross, A. H., Julila, A., and Balakrishnan, C. J. Infect. Dis. 133(supplement): A192-A198, 1976.
80. Sacks, S. L., Smith, J. L., Pollard, R. B., Sawhney, V., Mahol, A. S., Gregory, P., Merigan, T. C., and Robinson, W.S. JAMA 241: 28-29, 1979.
81. Friedman, H. M., and Grasela, T. N. Engl. J. Med. 304: 423, 1981.
82. Ramos, E., Timmons, R. F., and Schimpff, S. C. Antimicrob. Agents Chemother. 15: 142-144, 1979.
83. Richards, D. M., Carmine, A. A., Brogden, R. N., Heel, R. C., Speight, T. M., and Avery, G. S. Drugs 26: 378-438, 1983.
84. Collins, P. J. Antimicrob. Chemother. 12(Suppl. B): 19-27, 1983.
85. Harmenberg, J., Wahren, B., and Öberg, B. Intervirology 14:239-244, 1980.
86. Tyms, A. S., Scamans, E. M., and Naim, H. M. J. Antimicrob. Chemother. 8: 65-72, 1981.
87. Mar, E.-C., Patel, P. C., and Huang, E.-S. Am. J. Med. 73(1A): 82-85, 1982.
88. Plotkin, S. A., Starr, S. E., and Bryan, C. K. Am. J. Med. 73(1A): 257-261, 1982.
89. Crumpacker, C. S., Schnipper, L. E., Zaia, J. A., and Levin, M. J. Antimicrob. Agents Chemother. 15: 642-645, 1979.
90. Lang, D. J., and Cheung, K.-S. Am. J. Med. 73(1A): 49-53, 1982.
91. Burns, W. H., Wingard, J. R., Bender, W. J., and Saral, R.J. Virol. 39: 889-893, 1981.
92. Hirsch, M. S., and Schooley, R. T. N. Engl. J. Med. 309: 963-970 and 1034-1039, 1983.
93. Laskin, O. L., Longstreth, J. A., Saral, R., De Miranda, P., Keeney, R., and Lietman, P. S. Antimicrob. Agents Chemother. 21: 393-398, 1982.
94. Whitley, R. J., Blum, M. R., Barton, N., and De Miranda, P. Am. J. Med. 73(1A): 165-171, 1982.
95. Van Dyke, R. B., Connor, J. D., Wyborny, C., Hintz, M., and Keeney, R. E. Am. J. Med. 73(1A): 172-175, 1982.
96. Elion, G. B. Am. J. Med. 73(1A): 7-13, 1982.
97. Burns, W. H., Wingard, J. R., Sandford, G. R., Bender, W. J., and Saral, R. Am. J. Med. 73(1A): 118-124, 1982.
98. Sandford, G. R., Wingard, J. R., Simons, J. W., Staal, S. P., Saral, R., and Burns, W. H. J. Virol. 54: 104-113, 1985.
99. Shanley, J. D., and Pesanti, E. L. Arch. Virol. 88: 27-35, 1986.
100. Shanley, J. D., and Pesanti, E. L. J. Infect. Dis. 151: 454-458, 1985.
101. Glasgow, L. A., Richards, J. T., and Kern, E. R. Am. J. Med. 73(1A): 132-137, 1982.

102. Wingard, J. R., Bender, W. J., Saral, R., and Burns, W. H. Antimicrob. Agents Chemother. 20: 275-278, 1981.
103. Balfour, H. H., Jr., Bean, B., Mitchell, C. D., Sachs, G. W., Boen, J. R., and Edelman, C. K. Am. J. Med. 73(1A): 241-248, 1982.
104. Ashraf, M. H., Campalani, G. C., Qureshi, S. A., Froud, D. J., and Yacoub, M. H. Lancet i: 173-174, 1982.
105. Nunan, T. O., King, M., Bull, P., Banatvala, J. E., Jones, N. F., and Hilton, P. J. Clin. Nephrol. 22: 28-31, 1984.
106. Wade, J. C., Hintz, M., McGuffin, R. W., Springmeyer, S. C., Connor, J. D., and Meyers, J. D. Am. J. Med. 73(1A): 249-256, 1982.
107. Yeager, A. S. Am. J. Med. 73(1A): 205-209, 1982.
108. Peterslund, N. A., Black, F. T., and Tauris, P. Lancet i: 243-244, 1983.
109. McGuffin, R. W., Shiota, F. M., and Meyers, J. D. Antimicrob. Agents. Chemother. 10: 471-473, 1980.
110. Johnson, R., Douglas, J., Corey, L., and Krasney, H. Ann. Intern. Med. 103: 962-963, 1985.
111. Spiegal, D. M., and Lau, K. JAMA 255: 1882-1883, 1986.
112. Kelsey, D. K., Kern, E. R., Overall, J. C., Jr., and Glasgow, L. A. Antimicrob. Agents Chemother. 9: 458-464, 1976.
113. Steffenhagen, K. A., Easterday, B. C., Galasso, G. J., and Editors. J. Infect. Dis. 133: 603-612, 1976.
114. House, R. F., Jr., Person, D. A., Smith, T. F., and Harris, L. E. Lancet i: 39-40, 1973.
115. Conchie, A. F., Barton, B. W., and Tobin, J. O. Br. Med. J. ii: 162-163, 1968.
116. Karchmer, A. W., and Hirsch, M. S. N. Engl. J. Med. 289: 912-913, 1973.
117. Lauter, C. B., Bailey, E. J., and Lerner, A. M. Antimicrob. Agents Chemother. 6: 598-602, 1974.
118. Chatterjee, S. N., Fiala, M., and Myles, R. A. Am. J. Surg. 133: 719-722, 1977.
119. Plotkin, S. A., and Stetler, H. Antimicrob. Agents Chemother. (Ed Hobby, G. L). pp 372-379, 1969.
120. Emödi, G., Sartorius, J., Just, M., Rohner, F., and Buhler, U. Helv. Paediat. Acta 27: 557-564, 1972.
121. McCracken, G. H., Jr., and Luby, J. P. J. Pediatr. 80: 488-495, 1972.
122. Kraybill, E. N., Sever, J. L., Avery, G. B., and Movassaghi, N. J. Pediatr. 80: 485-487, 1972.
123. Brunell, P. J. Pediatr. 86: 317-318, 1975.
124. Stevens, D. A., Jordan, G. W., Waddell, T. F., and Merigan, T. C. N. Engl. J. Med. 289: 873-878, 1973.
125. Mar, E.-C., Patel, P. C., Cheng, Y.-C., Fox, J. J., Watanabe, K. A., and Huang, E.-S. J. Gen. Virol. 65: 47-53, 1984.
126. Colacino, J. M., and Lopez, C. Antimicrob. Agents Chemother. 28: 252-258, 1985.
127. Colacino, J. M., and Lopez, C. Antimicrob. Agents Chemother. 24: 505-508, 1983.
128. Feinberg, A., Leyland-Jones, B., Fanucchi, M. P., Hancock, C., Fox, J. J., Watanabe, K. A., Vidal, P. M., Williams, L., Young, C. W., and Philips, F. S. Antimicrob. Agents Chemother. 27: 733-738, 1985.
129. Leyland-Jones, B., Donnelly, H., Groshen, S., Myskowski, P., Donner, A. L., Fanucchi, M., Fox, J., and The Memorial Sloan-Kettering Antiviral Working Group. J. Infect. Dis. 154: 430-436, 1986.
130. Suzuki, S., Saneyoshi, M., Nakayama, C., Nishiyama, Y., and Yoshida, S. Antimicrob. Agents Chemother. 28: 326-330, 1985.
131. De Clercq, E. J. Antimicrob. Chemother. 14(Suppl. A): 85-95, 1984.
132. Allaudeen, H. S., Kozarich, J. W., Bertino, J. R., and De Clercq, E. Proc. Natl. Acad. Sci. (USA) 78: 2698-2702, 1981.
133. Smith, C. A., Wigdahl, B., and Rapp, F. Antimicrob. Agents Chemother. 24: 325-332, 1983.
134. Jerkofsky, M., Dobersen, M. J., and Greer, S. Intervirology 4: 233-238, 1980.

135. Wingard, J. R., Stuart, R. K., Saral, R., and Burns, W. H. Antimicrob. Agents Chemother. 20: 286-290, 1981.
136. Spector, S. A., Tyndall, M., and Kelley, E. Antimicrob. Agents Chemother. 23: 113-118, 1983.
137. De Clercq, E., Holy, A., Rosenberg, I., Sakuma, T., Balzarini, J., and Maudgal, P.C. Nature 323:464-467, 1986.
138. Feigin, R. D., Shackelford, P. G., DeVivo, D. C., and Haymond, M. W. Pediatrics 48:318-322, 1971.
139. Gilbert B.E., and Knight, V. Antimicrob. Agents Chemother. 30:201-205, 1986.
140. Dowling, J. N., Postic, B., and Guevarra, L. O. Antimicrob. Agents Chemother. 10: 809-813, 1976.
141. Hirsch, M. S. Tex. Rep. Biol. Med. 41: 566-570, 1981-1982.
142. Holmes, A. R., Rasmussen, L., and Merigan, T. C. Intervirology 9: 48-55, 1978.
143. Postic, B., and Dowling, J. N. Antimicrob. Agents Chemother. 11: 656-660, 1977.
144. Rasmussen, L. E., Chen, P. T., and Merigan, T. C. Antimicrob. Agents Chemother. 26:599-600, 1984.
145. Rabson, A. S., Tyrrell, S. A., and Levy, H. Proc. Soc. Exp. Biol. Med. 131: 495-497, 1969.
146. Hirsch, M. S., Tolkoff-Rubin, N. E., Kelly, A. P., and Rubin, R. H. J. Infect. Dis. 148: 335, 1983.
147. Kern, E. R., Olsen, G. A., Overall, J. C., Jr., and Glasgow, L. A. Antimicrob. Agents Chemother. 13: 344-346, 1978.
148. Brideau, R. J., and Wolcott, J. A. Antimicrob. Agents Chemother. 28: 485-488, 1985.
149. Cruz, J. R., Dammin, G. J., and Waner, J. L. Infect. Immun. 32: 332-342, 1981.
150. Bruggeman, C. A., Schellekens, H., Grauls, G., Debie, W. M. H., and Van Boven, C. P. A. Antiviral Res. 3: 315-324, 1983.
151. Meyers, J. D., McGuffin, R. W., Neiman, P. E., Singer, J. W., and Thomas, E. D. J. Infect. Dis. 141: 555-562, 1980.
152. Winston, D. J., Ho, W. G., Schroff, R. W., Champlin, R. E., and Gale, R. P. Antimicrob. Agents Chemother. 23: 846-851, 1983.
153. Meyers, J. D., Day, L. M., Lum, L. G., and Sullivan, K. M. J. Infect. Dis. 148: 551-556, 1983.
154. Arvin, A. M., Yeager, A. S., and Merigan, T. C. J. Infect. Dis. 133: A205-A210, 1976.
155. Emödi, G., O'Reilly, R., Muller, A., Everson, L. K., Binswanger, U., and Just, M. J. Infect. Dis. 133(Suppl): A199-A204, 1976.
156. Chou, S. W., Dylewski, J. S., Gaynon, M. W., Egbert, P. R., and Merigan, T. C. Antimicrob. Agents Chemother. 25: 25-28, 1984.
157. Hirsch, M.S. and Schooley, R.T. In: The Biology of the Interferon System (Eds. H. Kirchner and H. Schellekens), Elsevier, Amsterdam (in press).
158. Meyers, J. D. Infection 12: 143-150, 1984.
159. Snydman, D. R., McIver, J., Leszczynski, J., Cho, S. I., Werner, B. G., Berardi, V. P., LoGerfo, F., Heinze-Lacey, B., and Grady, G. F. Transplantation 38: 553-557, 1984.
160. Zaia, J. A., Levin, M. J., Leszczynski, J., Wright, G. G., and Grady, G. F. Transplantation 27: 66-67, 1979.
161. Shanley, J. D., Jordan, M. C., and Stevens, J. G. J. Infect. Dis. 143: 231-237, 1981.
162. Nicholls, A. J., Brown, C. B., Edward, N., Cuthbertson, B., Yap, P. L., and McClelland, D. B. L. Lancet i: 532-533, 1983.
163. Condie, R. M., Hall, B. L., Howard, R. J., Fryd, D., Simmons, R. L., and Najarian, J. S. Transplant. Proc. 11: 66-68, 1979.
164. Creasy, T. S., Flower, A. J. E., and Veitch, P. S. Lancet i: 675, 1986.
165. Dijkmans, B. A. C., Versteeg, J., Kauffmann, R. H., Van Den Broek, P. J., Eernisse, J. G., Van Zanten, J. J., Bakker, W., Kalff, M. W., and Van Hooff, J. P. Lancet i:

820-821, 1979.
166. Smith, M. J., Cooke, R. W. I., Hood, N., Hart, C. A., and Yap, P. L. Lancet i: 447-448, 1984.
167. Winiarski, J., Kreuger, A., Ejderhamn, J., and Holm, G. Scand. J. Haematol. 31: 342-348, 1983.
168. Blacklock, H. A., Griffiths, P., Stirk, P., and Prentice, H. G. Lancet ii: 152-153, 1985.
169. Reed, E. C., Bowden, R. A., Dandliker, P. S., and Meyers, J. D. 26th Interscience Conference Antimicrob. Agents and Chemother. p231 (abstract 731), 1986.
170. Spector, S. A., Tyndall, M., and Kelley, E. Am. J. Med. 73(1A): 36-39, 1982.
171. Spector, S. A., and Kelley, E. Antimicrob. Agents Chemother. 27: 600-604, 1985.
172. Levin, M. J., and Leary, P. L. Infect. Immun. 32: 995-999, 1981.
173. Rasmussen, L., Chen, P. T., Mullenax, J. G., and Merigan T. C. Antimicrob Agents Chemother. 26: 441-445, 1984.
174. Rose, R. M., Crumpacker, C., Waner, J. L., and Brain, J. D. Am. Rev. Respir. Dis. 127: 198-203, 1983.
175. Meyers, J. D., McGuffin, R. W., Bryson, Y. J., Cantell, K., and Thomas, E. D. J. Infect. Dis. 146: 80-84, 1982.
176. Wade, J. D., McGuffin, R. W., Springmeyer, S. C., Newton, B., Singer, J. W., and Meyers, J. D. J. Infect. Dis. 148: 557-562, 1983.
177. Meyers, J. D., Wade, J. C., McGuffin, R. W., Springmeyer, S. C., and Thomas, E. D. J. Antimicrob. Chemother. 12(Suppl. B): 181-193, 1983.
178. Shepp, D. H., Newton, B. A., and Meyers, J. D. J. Infect. Dis. 150: 776-777, 1984.
179. Gill, M. J., Russel, J., Burgess, K., and Tan, Y. H. Ann. Intern. Med. 104: 129-130, 1986.
180. Pahwa, S., Kirkpatrick, D., Ching, C., Lopez, C., Pahwa, R., Smithwick, E., O'Reilly, R., August, C., Pasquariello, P., and Good, R. A. Clin. Immunol. Immunopathol. 28: 77-89, 1983.
181. Jeffries, D. J., and Tyms, A. S. Lancet i: 1214-1215, 1983.
182. Tyms, A. S., Stevens, R. J., Mobberley, M. A., Ryder, T. A., and Jeffries, D. J. J. Gen. Virol. 65: 2129-2139, 1984.
183. Kim, K. S., Sapienza, V. J., and Carp, R. I. Antimicrob. Agents Chemother. 18: 276-280, 1980.
184. Gibson, W., Van Breeman, R., Fields, A., LaFemina, R., and Irmiere, A. J. Virol. 50: 145-154, 1984.
185. Tyms, A. S., and Williamson, J. D. Nature 297: 690-691, 1982.
186. Thomas, I. T., Hawkins, G. T., Soothill, J. F., and Marshall, W. C. Lancet ii: 1056-1057, 1977.
187. Paganelli, R., Soothill, J. F., Marshall, W. C., and Hamblin, A. S. Lancet i: 273-274, 1981.
188. Jones, J. F., Jeter, W. S., Fulginiti, V. A., Minnich, L. L., Pritchett, R. F., and Wedgwood, R. J. Lancet ii: 122-124, 1981.
189. Nkrumah, F., Pizza, G., Viza, D., Phillips, J., DeVinci, C., and Levine, P. Lymphokine Res. 4: 237-241, 1985.
190. Rook, A. H., Masur, H., Lane, H. C., Frederick, W., Kasahara, T., Macher, A. M., Djeu, J. Y., Manischewitz, J. F., Jackson, L., Fauci, A. S., and Quinnan, G. V., Jr. J. Clin. Invest. 72: 398-403, 1983.
191. Seigel, J. P., Rook, A. H., Djeu, J. Y., and Quinnan, G. V., Jr. Infection 12: 298-301, 1984.
192. Drew, W. L., Brodie, H., Conant, M. Volberding, P., and Rumack, J. In: Antiviral Chemotherapy: New Directions for Clinical Application and Research (Eds. J. Mills and L. Corey), Elsevier, N. Y., 1986, pp 280-288.
193. Furukawa, T., Tanaka, S., and Plotkin, S. A. J. Gen. Virol. 28:355-362, 1975.
194. Halstead, C. C., Minnefor, A. B., and Lietman, P. S. J. Infect. Dis. 125: 552-555, 1972.
195. Landini, M. P., and Baldassarri, B. J. Antimicrob. Chemother. 10: 533-537, 1982.

196. Furlini, G., Coppolecchia, P., Re, M. C., Baldassarri, B., Ripalti, A., and Landini, M. P. J. Antimicrob. Chemother. 12: 503-506, 1983.
197. Smith, R. D., Henson, D., Gehrke, J., and Barton, J. R. Proc. Soc. Exp. Biol. Med. 121: 209-211, 1966.
198. Sidwell, R. W., Arnett, G., and Schabel, F. M., Jr., Chemotherapy 17: 259-282, 1972.
199. Tegtmeier, G. E. Prog. Clin. Biol. Res. 182: 175-199, 1985.
200. Vox Sang. 46: 387-414, 1984.
201. Adler, S. P. Rev. Infect. Dis. 5: 977-993, 1983.
202. Winston, D. J., Ho, W. G., Howell, C. L., Miller, M. J., Mickey, R., Martin, W. J., Lin, C.-H., and Gale, R. P. Ann. Intern. Med. 93: 671-675, 1980.
203. Verdonck, L. F., Middeldorp, J. M., Kreeft, H. A. J. G., The, T. H., Hekker, A., and De Gast, G. C. Transplantation 39: 455-457, 1985.
204. Tegtmeier, G. E. Semin. Liver Dis. 6:82-95, 1986.
205. Benson, J. W. T., Bodden, S. J., and Tobin, J. O. Arch. Dis. Child. 54: 538-541, 1979.
206. Kurtz, J. B., and Barlow, M. E. Lancet ii: 294-295, 1984.
207. Ringdén, O., Paulin, T., Pihlstedt, P., Lönnqvist, B., and Wahren, B. Lancet ii: 1044, 1984.
208. Tolkoff-Rubin, N. E., Rubin, R. H., Keller, E. E., Baker, G. P. Stewart, J. A., and Hirsch, M. S. Ann. Intern. Med. 89: 625-628, 1978.
209. Taylor, B. J., Jacobs, R. F., Baker, R. L., Moses, E. B., McSwain, B. E., and Shulman, G. Pediatr. Infect. Dis. 5: 188-191, 1986.
210. Brady, M. T., Milam, J. D., Anderson, D. C., Hawkins, E. P., Speer, M. E., Seavy, D., Bijou, H., and Yow, M.D. J. Infect. Dis. 150: 334-339, 1984.
211. Adler, S. P., Lawrence, L. T., Baggett, J., Biro, V., and Sharp, D. E. Transfusion 24: 333-335, 1984.
212. Bowden, R. A., Sayers, M., Flournoy, N., Newton, B., Banaji, M., Thomas, E. D., and Meyers, J. D. N. Engl. J. Med. 314: 1006-1010, 1986.
213. Adler, S. P. Pediatr. Infect. Dis. 5: 239-246, 1986.
214. Primhak, R. A., Duffy, P., and Becker, L. E. J. Pediatr. 102: 1013, 1983.
215. Yeager, A. S., Grumet, F. C., Hafleigh, E. B., Arvin, A. M., Bradley, J. S., and Prober, C. G. J. Pediatr. 98: 281-287, 1981.
216. Kraemer, K. G., Neiman, P. E., Reeves, W. C., and Thomas, E. D. Transplant. Proc. 10: 237-240, 1978.
217. Gluckman, E., Lotsberg, J., Devergie, A., Zhao, X. M., Melo, R., Gomez-Morales, M., Mazeron, M. C., and Perol, Y. J. Antimicrob. Chemother. 12 (Suppl. B): 161-167, 1983.
218. Gluckman, E., Devergie, A., Melo, R., Nebout, T., Lotsberg, J., Zhao, X. M., Gomez-Morales, M., Mazeron, M. C., and Perol, Y. Lancet ii: 706-708, 1983.
219. Meyers., J. D., Reed, E. C., Shepp, D. H., Flournoy, N., Dandliker, P., Vicary, C., Kirk, L. E., and Balfour, H. H., Jr. 26th Interscience Conference Antimicrob. Agents and Chemother. p. 231(abstract 732), 1986.
220. Selby, P., Powles, R., Stolle, K., and Stern, H. Br. Med. J. 289: 253, 1984.
221. Lundgren, G., Wilczek, H., Lönnqvist, B., Lindholm, A., Wahren, B., and Ringdén, O. Scand. J. Infect. Dis. 47(Suppl): 137-144, 1985.
222. Cheeseman, S. H., Rubin, R. H., Stewart, J. A., Tolkoff-Rubin, N. E., Cosimi, A. B., Cantell, K., Gilbert, J., Winkle, S., Herrin, J. T., Black, P. H., Russell, P. S., and Hirsch, M. S. N. Engl. J. Med. 300: 1345-1349, 1979.
223. Hirsch, M. S., Schooley, R. T., Cosimi, A. B., Russell, P. S., Delmonico, F. L., Tolkoff-Rubin, N. E., Herrin, J. T., Cantell, K., Farrel, M.-L., Rota, T. R., and Rubin, R. H. N. Engl. J. Med. 308: 1489-1493, 1983.

224. Weimar, W., Kramer, P., Bijnen, A. B., Jeekel, J., Rothbarth, P. H., and Masurel, N. Scand. J. Urol. Nephrol. 92(Suppl): 37-39, 1985.
225. Araullo-Cruz, T. P., Ho, M., and Armstrong, J. A. Infect. Immun. 21: 840-842, 1978.
226. Meyers, J. D., Leszczynski, J., Zaia, J. A., Flournoy, N., Newton, B., Snydman, D. R., Wright, G. G., Levin, M. J., and Thomas, E. D. Ann. Intern. Med. 98: 442-446, 1983.
227. Winston, D. J., Ho, W. G., Lin, C.-H., Budinger, M. D., Champlin, R. E., and Gale, R. P. Am. J. Med. Symposium; March 30, 1984: 128-133, 1984.
228. Condie, R. M., and O'Reilly, R. J. Am. J. Med. Symposium; March 30, 1984: 134-141, 1984.
229. Jacobsen, N., Schäfer, U., Ostendorf, P., Kubaneck, B., and Wolf, H. Tokai J. Exp. Clin. Med. 10: 193-195, 1985.
230. Winston, D. J., Pollard, R. B., Ho, W. G., Gallagher, J. G., Rasmussen, L. E., Huang, S. N-Y., Lin, C.-H., Gossett, T. G., Merigan, T. C., and Gale, R. P. Ann. Intern. Med. 97: 11-18, 1982.
231. Dreikorn, K., Doerr, H.-W., and Geursen, R. G. Scand. J. Urol. Nephrol. 92(Suppl): 15-21, 1985.
232. Fassbinder, W., Bechstein, P.-B., Scheuermann, E.-H., and Schoeppe, W. Scand. J. Urol. Nephrol. 92(Suppl): 23-28, 1985.
233. Snydman, D. R., Werner, B. G., Heinze-Lacey, B., Tilney, N. L., Kirkman, R. L., Milford, E. L., Strom, T. B., Carpenter, C.B., Levey, R. H., Harman, W. E., Zimmerman, C. E., II, Shapiro, M. E., Steinman, T., Cho, S. I., Bush, H. L. Jr., Levey, A., LoGerfo, F., Schroter, G. P. J., Levin, M. J., Berardi, V. P., McIver, J., Leszczynski, J., and Grady, G.F. American Society of Transplant Physicians, Chicago, Illinois, abstract, 1986.
234. Preiksaitis, J. K., Rosno, S., Rasmussen, L., and Merigan, T. C. J. Clin. Immunol. 2(Suppl): 36S-41S, 1982.

12

THE PROPHYLAXIS OF HERPES GROUP INFECTIONS IN PATIENTS WITH HAEMATOLOGICAL MALIGNANCY

H. G. Prentice[1] and I. M. Hann[2].

[1]Royal Free Hospital, London and [2]Royal Hospital for Sick Children, Glasgow.

INTRODUCTION

Haematological malignancy provides a model of acquired combined immunodeficiency. During intensive treatment the degree to which the patient is immunocompromised (IC) is increased and is comparable to that seen in the Acquired Immunodeficiency Syndrome (AIDS). Fortunately the patient with a haematological malignancy usually has a limited period of compromised immunity unlike the AIDS patient.

We have chosen to deal with haematological malignancy in this Chapter on the prophylaxis of herpes group infections but the studies detailed are often applicable to other IC patients, from renal transplant recipients to those with AIDS.

BACKGROUND

Herpes (DNA) viruses are ubiquitous in the human race and, as such, most individuals have been previously infected (based on serological surveys) and are therefore carriers of latent virus. The majority of patients who become immunocompromised, for whatever reason, will experience a reactivation of certain of the Herpes group viruses. These are, most commonly, Herpes Simplex (type I) (HSV), Varicella Zoster Virus (VZV) and Cytomegalovirus (CMV). The extent to which patients with latent Epstein-Barr Virus (EBV) reactivate and develop infection with this virus is presently not quantified but appears to be less frequent than with the other viruses. HSV type II reactivations are equally frequent in those (the minority) who have latent virus.

The incidence of reactivation illustrated by the BMT recipient is staggeringly high at around 50-80% for HSV type I (1,2,3,4,5).

The BMT patient illustrates another fascinating feature of the Herpes family. Young and others (6,7,8) have documented a remarkably consistent temporal pattern of reactivation following BMT. HSV in the first few weeks followed by CMV between 30 and 90 days (mainly) with VZV reactivation occurring usually from 3 up to 18 months after the procedure. Not only is this observation of biological interest, but it also introduces an important logistic complication: How do we provide long term effective prophylaxis in this setting?

Of course the IC patient may also acquire a de novo (primary) infection but in numerical terms this is of lesser importance, although the possibilities for prophylaxis are probably greater.

Not only are the Herpes group infections frequent in the IC patient but what in the normal patient is frequently a "trivial" infection will often cause considerable morbidity and even mortality. HSV reactivation is the most frequent infection in leukaemic and BMT patients and it was associated with a 4% mortality in BMT recipients prior to the availability of effective treatment or prophylaxis (9). CMV reactivations often lead to CMV pneumonitis in the BMT patient and after leukaemic relapse are now the most frequent single cause of death. In allogeneic BMT 22% get interstitial pneumonitis and in 40% this is associated with CMV with case fatality in the IBMTR in excess of 80% (10).

Prophylaxis of Herpes Simplex infections

Following the observations of impressive in-vitro and in-vivo activity of acyclovir (ACV) against HSV including treatment in IC patients (12) a randomised placebo controlled multicentre study in the U.S.A. demonstrated a highly significant reduction in the period of viral excretion (13). Both the Johns Hopkins and Royal Free groups elected to use acyclovir as prophylaxis of HSV infection in BMT (1) and chemotherapy plus BMT patients (2). In the latter EBV excretion was also studied.

Saral and co-workers gave 250mg/m² ACV 8 hourly by 1 hour I.V. infusion starting 3 days prior to transplant for 18 days. Patients who were seropositive (≥1;8) to HSV were randomised to receive ACV or placebo. Five of 10 patients having placebo developed an HSV mucocutaneous reactivation against none of 10 who received ACV.

In the Royal Free study (2) ACV (5mg/kg) was given by one hour infusion every 12 hours. All seropositive (≥1:8) consecutive patients having chemotherapy (n = 39) or BMT (n = 20) were randomised by stratification to receive ACV or placebo within 24 hours of starting the chemotherapy or receiving the BMT. In this study total protection was again confirmed in the BMT group with a 50% failure rate in the placebo arm ($p = 0.033$). In the chemotherapy stratification the failure rate on ACV was 2/19 against 10/20 in the placebo arm ($p = 0.018$).

The Royal Free study showed additional benefits. First, prevention of HSV infection in the BMT stratification was associated with a significant reduction in the period of neutropenia ($<1.0 \times 10^9/l$) ($p = 0.001$) (one CMV failure excluded). This was an unexpected observation but was subsequently confirmed by the Seattle group (Meyers - Personal Communication), whilst previous studies had demonstrated transient neutropenia in chickenpox (14). Second, the number of febrile days was reduced in the BMT/ACV group ($p = 0.047$) and this was mirrored by a reduction in the number of days of antibiotic usage ($p = 0.06$). Subsequent studies from Sweden using oral ACV (see below) showed a statistical significance for a reduction in the risk of bacteraemia (15) in a placebo controlled study ($p = <0.05$). The reduced rate of bacteraemia has been attributed to a protective effect on mucosal integrity when HSV reactivation is prevented (16). The major bacteraemic pathogen in this study was Streptococcus viridans considered to have been derived from the oral cavity.

Follow-up of patients in the BMT arm of the Royal Free study showed delayed reactivation of HSV in those who had received ACV and it was concluded that the period of protection should be extended to cover the period of major risk. That is, until immune function was considered adequate to at least protect against dissemination of HSV. The consensus is that this could be for up to 6 months after BMT. Clearly I.V. prophylaxis is not practical and we and others therefore embarked upon a series of studies utilising oral agents.

Our own open study of oral ACV 400mg 6 hourly for 6 weeks after BMT resulted in an approximate 50% reduction in HSV reactivation (to 25%) and was considered a failure (17). By contrast the Seattle group (18) showed total protection where patient compliance was at least 40% with a prescribed dose of 400mg five times daily in seropositive BMT patients. Gluckman et al (19) showed total protection with only 200mg 6 hourly. Although ACV absorption was relatively poor in the Royal Free study (twice the dose was administered), the resulting plasma levels were identical to those of the French group. Thus, the difference in protection remains unexplained.

Attempts have been made to improve the oral absorption of ACV by preparing prodrugs. The first of these BW134U (Aminoacyclovir) [2,6-diamino-9-(2-hydroxyethoxymethyl)purine] was shown in animal studies to be well absorbed (20). Following absorption BW134U is rapidly deaminated to ACV (approximately 65%). Higher levels of ACV were generated in normal subjects compared with oral ACV and this was confirmed in BMT recipients. An equivalent dose of BW134U and ACV will result in an approximate doubling of the plasma ACV level with the former (21). This drug was recently withdrawn in the light of preclinical toxicology studies but, in any event, has also been superseded.

A515U (Deoxyacyclovir/amino(hydroxyethoxymethyl)purine) - a deoxy cogener of ACV- is almost completely absorbed in normal volunteers (22). It is then converted by xanthine oxidase to ACV. To date only the pharmacokinetic profile has been determined in immunocompromised hosts. Selby et al in (23) gave A515U at a dose of 250mgs 6 hourly to patients with acute leukaemia undergoing remission induction therapy. The peak plasma level achieved was 23.7μmol (mean). No toxicity attributable to the A515U or ACV was noted.

To eleven consecutive HSV seropositive BMT recipients we gave I.V. ACV (5mg/kg 12 hourly) for 14 days followed by oral A515U (250mg 6 hourly) for 7 days and then oral ACV 400mg 6 hourly for 7-21 days. The pharmacokinetic profile was obtained during each phase of the study (for A515U on days 4 or 5 of treatment). Again excellent plasma levels were obtained (8 of 11 adequately studied). These were approximately 4 fold higher than those obtained with oral ACV. The mean peak plasma level achieved was 16.4μM at 2 hours post dosing for A515U and 4.59μM for oral ACV. The ACV level for A515U was well in excess of the ID_{50} for HSV 1 & 2 (and VZV - see below). Residual A515U with a peak plasma level at 1 hour post-dose of 3.63μM was also found. Again no toxicities attributable to either ACV or A515U were seen. It seems likely that current studies will confirm the suitability of A515U for long term prophylaxis.

Other drugs with proven in-vitro activity against HSV have, to date, only been tested in open therapeutic studies but may have clinical utility in prophylaxis. These include Bromovinyldeoxyuridine (BVDU) (24), Vidarabine (Ara-A - Adenine Arabinoside [although this drug must be given by slow I.V. infusion and is myelotoxic]) and Interferon.

Prophylaxis of Varicella Zoster Virus Infection

Control of VZV infection is a mainly cellular immune phenomenon but investigations in patients with Hodgkin's disease have shown that those with the lowest antibody titres are most susceptible to infection on exposure (25,26). The value of the use of zoster immune globulin (ZIG) in children with acute lymphoblastic leukaemia who are immunised within 72 hours of exposure, is very well documented (27). However, anecdotally, we have for some years treated contacts with oral acyclovir for 14 days (400mg p.o. 6 hourly in adults) with no case of overt infection.

There have now been a large number of studies of a live varicella vaccine, nearly all amongst children with leukaemia, in Japan (28). Care has been taken to vaccinate the children after a period of at least 6 months in continuous remission and when they had positive reactions to skin test antigens. In the study of Kamiya 54 children were vaccinated and 21 followed up for over 5 years. Seroconversion occurred in all cases but seroregression was observed in 7 (13%). Thirty three definite exposures to varicella occurred but only 3 developed an infection (6%) which was mild in all cases. Herpes Zoster occurred at a rate of 15.4% which was similar to the rate in patients who had natural VZV infection before the onset of disease.

For the future, double-blind placebo-controlled studies of oral acyclovir, probably as the prodrug A515U and Bromovinyldeoxyuridine (BVDU) need to be undertaken in high risk groups (i.e. marrow transplant patients) throughout the prolonged period of risk. However, it remains to be seen whether this approach will be logistically or economically viable and an early intervention approach with intravenous acyclovir, along with treatment of contacts, will probably continue to be the mainstay of management. Future developments in vaccine technology should also become available, that is the use of VZV recombinant DNA vaccines. In marrow transplant recipients the Royal Free BMT group have recently demonstrated the adoptive transfer of B cell immunity from donor to recipient in T lymphocyte depleted BMTs (29). That it might have clinical utility is anticipated.

Cytomegalovirus

Acyclovir has only minimal activity against CMV in-vitro, presumably because this virus does not code for Thymidine Kinase (TK). Despite this a randomised study in Paris (19) showed apparent protection. In the group receiving oral ACV (200mg 6 hourly) given from day -8 to day +35 in BMT recipients there were no instances of CMV infection during the period of therapy (0/20) against 7/19 receiving placebo. Following treatment (35 - 100 days) there were 7 episodes in the ACV group and 11 in the placebo arm. Whilst this was a surprising result the timing of the CMV infections following BMT in the placebo arm was also unusual. These infections are unusual before 30 days after BMT in the experience of most groups. At the Royal Free Hospital we have observed CMV infection occurring during I.V. ACV prophylaxis. A large multicentre European group is studying this question at present.

A recent multicentre study from the U.S. (30) has reported a reduced risk of CMV infection and mortality in CMV seropositive recipients receiving high dose I.V. ACV (500mgs/m^2). The therapy was allocated such that patients who were HSV seropositive received ACV whilst the others had placebo. This important observation may require confirmation with a placebo controlled study where randomisation is made in all CMV seropositive patients since HSV serological status may have influenced the outcome.

Many drugs with activity against CMV both in-vitro and in-vivo have been tested therapeutically but none has, as yet, been used for prophylaxis. The following drugs with anti-CMV activity are of interest; FIAC (2′-fluoro-5-iodoarabinosyl)cytosine (31), Foscarnet (Phosphonoformate) (32) and BWB759U/DHPG(9-(1,3,-dihydroxy-2-propoxymethyl)guanine. The latter, an ACV derivative, has particularly impressive in-vitro activity (33). In-vitro studies in established CMV interstitial pneumonitis from the Seattle group (34) confirms this activity with up to a 4 log reduction in viral titre in serial lung biopsies as well as cessation of viraemia and viruria. Unfortunately this effect had no influence on survival (9/10 succumbed). This observation supports the possibility that the lethality of CMV pneumonitis in BMT recipients is not entirely attributable to the virus and may, in part, be caused by the host response.

Several groups have now published their results on the use of CMV immune plasma or "hyperimmune" globulin, with conflicting results in BMT recipients. Winston et al (35) used immune plasma with complement fixing antibody titres of 1:64+. BMT (for severe aplasia or leukaemia) recipients were randomised to receive this plasma or no therapy intermittently for approximately 4 months. There were 39 CMV negative recipients and 9 who had titres of 1:8 or greater. The leukaemia group were conditioned with total body irradiation, a known risk factor for the development of CMV I.pn. Patients who did not receive granulocyte transfusions developed CMV infections in 9 of 18 controls and 7 of 17 who received the immune plasma. Severe CMV infections were less frequent in those having the immune plasma (1/17) compared to (8/18) controls ($p = 0.03$) as was CMV pneumonitis (6/18 versus 6/18) ($p = 0.02$). In seropositive recipients 5/5 in the control arm developed CMV infection compared with 1 of 4 receiving the plasma ($p = 0.048$). Survival was similar in the two groups.

The Sloan-Kettering group (36) in a randomised study compared CMV hyperimmune globulin with globulin deficient in CMV antibody or no therapy. Treatment was administered 25, 50 and 70 days post-BMT. Most of the patients were CMV seronegative. Of 17 patients who received the hyperimmune globulin none developed any evidence of CMV infection whereas 6 of 18 and 10 of 20 developed infections on, respectively, CMV deficient globulin and no therapy ($p = 0.019$ and $p = 0.001$). Four of 20 patients in the control arm suffered CMV pneumonitis whilst none of 17 receiving CMV globulin did ($p = 0.15$).

In a study using I.M. CMV globulin the Seattle team (37) showed no benefit compared with controls if CMV seropositive donors were used for granulocyte support. In this study, where treatment was given pre-transplant and for up to 77 days post-transplant, 2 of 17 having globulin developed a CMV infection (not pneumonitis) against 8 of 19 in the control group (p= 0.05), but only one infection resulted in pneumonitis.

Another study suggesting a protective effect of CMV hyperimmune globulin (Cytotect® [Biotest]) was from Kubanek at al (38). Compared with a non-specific globulin preparation (both given up to day +93) there was a reduction in CMV infection and pneumonitis. Some control patients received granulocytes in this study which entered patients irrespective of CMV serological status.

On the other hand two recent studies failed to show a benefit. Blacklock et al (39) in a multicentre study entering only seropositive patients used the same preparations as Kubanek but included both a no treatment control arm and an arm having intraglobin (Biotest®). The Biotest hyperimmune globulin with an anti-CMV titre of >1:1,250,000 (radioimmune assay) was given at 1g/5kg on day -2 (or -1) and 1g/10kg weekly for 5 weeks, then every 2 weeks for the remainer of a 3 month post-transplant period. After the entry of only 29 patients the study was deemed to have failed and was discontinued. There were a total of 5 infections on Cytotect (1 asymptomatic), Intraglobin 6 (one asymptomatic), untreated control 7 (one asymptomatic). CMV pneumonitis was seen respectively in 2 (2 fatal), 3 (1 fatal) and one non-fatal in the no treatment control arm.

In the study of Bowden et al (40) from the Seattle Transplant Centre a comparative trial with I.V. immunoglobulin was reported. They were unable to confirm that CMV Ig prevented CMV infection or reduced the severity of CMV disease. CMV seronegative blood products had a significant protective effect ($p < 0.007$) in seronegative patients with seronegative donors, but not when the marrow donor was seropositive. They have followed 112 seronegative patients all with seronegative donors and all of whom received seronegative blood products. Only 2 patients developed CMV infection and they had received seropositive blood products in error. Thus, this programme was highly effective but they pointed out the logistic problems of providing over 5,500 units of platelets and nearly a 1,000 pints of blood over this period.

Epstein-Barr Virus (EBV)

The spectrum of infection with EBV in immunocompromised patients is at present not well-defined. There have, in fact, been very few reports of serious infection (41) and the only serious sequela has been the rare association with B cell lymphomas in bone marrow transplant recipients. It may be the case that severe infections occur infrequently because of the presence of activated natural killer (NK) cells which appear rapidly after marrow transplantation and have effective killing activity on EBV-infected targets (42).

The only published study on EBV infection in marrow transplant patients (2) receiving acyclovir prophylaxis was inconclusive. Clearly, more studies are needed which document the pattern of EBV infection in immunocompromised patients. One study has showed reduced EBV culture positivity following the use of interferon in renal transplant patients (43). This prophylactic approach may be worth pursuing. Unpublished studies performed in Baltimore show that the vast majority of patients who undergo marrow transplantation have latent EBV prior to the procedure and over 60% show reactivation of this virus.

CONCLUSIONS

At this time, prevention of CMV infection in immunocompromised patients is the major challenge which faces clincians. It would seem that oral acyclovir and CMV hyperimmune globulin are not effective but in seronegative patients we can usually prevent infection if carefully-screened CMV antibody negative blood products are used. Further studies are needed to determine whether intravenous acyclovir or BWB759U or foscarnet will be of prophylactic value.

Epstein-Barr virus does not appear to be a major threat and, in any case, it seems unlikely that current or proposed prophylactic strategies would have any prophylactic value. Primary VZV infections can usually be effectively treated and short courses of oral acyclovir and/or ZIG following infectious contact make this a rare problem nowadays. Reactivation of VZV is a more difficult question. Prophylaxis is feasible with ACV I.V. but given the, often, very prolonged period of risk in the IC patient, long-term prophylaxis is less easy. Oral ACV *may* be effective but the plasma levels achieved are often below the ID_{50} for this virus. Some hope can be held out for A515U, but we suspect that financial considerations will play an important role in therapeutic decisions. If vaccines prove ineffective then, perhaps, we will prefer to use rapid therapeutic intervention.

HSV has become much less of a problem in this situation where the patients have relatively short-lived susceptibility to severe infection. We expect that oral acyclovir, or its prodrug A515U, will become standard prophylaxis for patients at high risk, especially those who are seropositive and receive a bone marrow transplant.

REFERENCES

1. Saral, R., Burns, W.H., Laskin, O.O. et al. New Eng. J. Med. 305: 63-67, 1981.
2. Hann, I.M., Prentice, H.G., Blacklock, H.A., Ross, M.G.R., Brigden, D., Rosling, A.E., Burke, C., Crawford, D.H., Brumfitt, W. and Hoffbrand, A.V. Br. Med. J. 287: 375-382, 1983.
3. Wade, J.C., Newton, B., McLaren, C. et al. Ann. Int. Med. 96: 265-269, 1982.
4. Atkinson, K., Meyers, J.D., Storb, R. et al. Transplantation 29: 47-50, 1980.
5. Meyers, J.D., Fournoy, N. and Thomas, E.D. J. Infect. Dis. 153; 3: 478-488, 1986.
6. Young, L.S. In: Proceedings of the Second International Symposium on the Compromised Host (Eds. H.S. Gaya & C. Easmon), Academic Press, London, 1983, pp. 27-35.
7. Winston, D.J., Gale, R.P., Meyer, D.V. et al. Medicine 58: 1-21, 1979.
8. Meyers, J.D. & Thomas, E.D. In: Clinical Approach to Infection in the Immunocompromised Host (Eds. L.S. Young & R.H. Rubin), Plenum Press, New York, 1981, pp. 507-551.
9. Buckner, C.D., Clift, R.A. et al. In: Bone Marrow Transplantation. Recent Advances in Haematology (Ed. A.V. Hoffbrand) Churchill Livingstone. 1982, pp. 144-159,
10. Weiner, R.S., Bortin, M. and Rimm, A. In: Journal of Cellular Biochemistry, Alan R. Liss, New York, 1986, Suppl. 10d, p.214.
11. Elion, G.B., Furman, P.A., Fyfe, J.A. et al. Proc. Nat. Acad. Sci. 74: 5716-5720, 1977.
12. Selby, P.J., Powles, R.L., Jameson, B. et al. Lancet ii: 1267-1270, 1979.
13. Meyers, J.D., Wade, J.C., Mitchell, C.D. et al. Am. J. Med. 73: 229-235, 1982.
14. Holbrook, A.A. Arch. Int. Med. 68: 294-300, 1941.
15. Lönqvist, B., Palmblad, J., Gimfors, G. et al. In: 14th International Chemotherapy Congress, Kyoto, Japan, 1985, p. 140, Abst. S22.11.

16. Heimdahl, A., Zetterqvist, L., Lönqvist, B. and Ringden, O. Infections from the oral cavity in bone marrow transplantation. In: 4th International Symposium on Infections in the Immunocompromised Host, Sweden, 1986, Abst. 44.
17. Prentice, H.G. J. Antimicrob. Chemother. 12B: 153-159, 1983.
18. Wade, J.C., Newton, B., Flournoy, N. & Meyers, J.D. 22nd ICAAC, Miami, Florida, 1982.
19. Gluckman, E., Lotsberg, J., Devergie, A. et al. J. Antimicrob. Chemother. 12B: 161-167, 1983.
20. Good, S.S., Krasny, H.C., Elion, G.B. and de Miranda, P. J. Pharmacol. Exp. Ther. 227: 644-651. 1983.
21. Prentice, H.G., Robbins, G., Hann, I.M., Blacklock, H.A., Ross, M., Graphakos, S., Brigden, D., Rees, P.J., Collins, P., Hoffbrand, A.V. and Griffiths, P. (In preparation).
22. Selby, R., Powles, R.L., Blake, S. et al. Lancet ii, 1428-1430, 1984.
23. Whiteman, P.D. Eur. J. Clin. Pharmacol. 27: 471-475, 1984.
24. DeClercq, E., Descamps, J., Ogata, M. and Shigeta, S. Antimicrob. Agents Chemother. 21: 622, 1986.
25. Schimpff, S.C., Serpick, A., Bloch, J. et al. Ann. Int. Med. 76: 241-254, 1972.
26. Ruckdeschel, J.C., Schimpff, S.C., Smyth, A.C. et al. Am. J. Med. 62: 77-85, 1977.
27. Brunell, P.A., Ross, A., Miller, L.H. et al. New Eng. J. Med. 280: 1191-1194, 1969.
28. Kamiya, H., Kitamura, K., Ohta, Y. et al. In: 4th International Symposium on Infections in the Immunocompromised Host, Sweden, 1986, Abst. 131.
29. Wimperis, J.Z., Brenner, M.K., Prentice, H.G., Reittie, J.E., Karayiannis, P., Griffiths, P.D. and Hoffbrand, A.V. Lancet i: 339-343, 1986.
30. Meyers, J.D., Reed, E.C., Shepp, D.H., Flournoy, N., Dandliker, P., Vicary, C., Kirk, L.E. and Balfour, Jr. H.H. Abstract of the ICAAC, p. 231, 1986.
31. Lopez, C., Watanabe, K. and Fox, J.J. Antimicrob. Agents Chemother. 17: 803-806, 1980.
32. Ringden, O., Lönnqvist, B., Paulin, T., Ahlmen, J., Klintmalm, G., Wahren, B. and Lernestedt, J-O. J. Antimicrob. Chemother. 17: 373-387, 1986.
33. Mar, E-C., Cheng, Y-C. and Huang, E.S. Antimicrob. Agents Chemother. 24: 518-521, 1983.
34. Shepp, D.H., Dandliker, P.S., deMiranda, P., Burnett, T.C., Cederberg, D.M., Kirk, L.E. and Meyers, J.D. Ann. Int. Med. 103: 368-373, 1985.
35. Winston, D.J., Pollard, R.B., Ho, W.G. et al. Ann. Int. Med. 97: 11-18, 1982.
36. O'Reilly, R.J., Reich, L., Gold, J. et al. Transpl. Proc. IV: 1405-1411, 1983.
37. Meyers, J.D., Leszczynski, J., Zaia, J.A. et al. Ann. Int. Med. 98: 442-446, 1983.
38. Kubanek, B., Ernst, P., Ostendorf, P. et al. Transp. Proc. 17: 468-469, 1985.
39. Blacklock, H.A., Griffiths, P.D., Kay, H.E. et al. Bone Marrow Transplantation, 1986, (Submitted for publication).

40. Bowden, R.A., Sayers, M., Flournoy, N. et al. 4th International Symposium on Infection in the Immunocompromised Host, Sweden, 1986, Abstr. 28.
41. Lange, B., Henle, W., Meyers, J.D. et al. Int. J. Cancer 26: 151-157, 1980.
42. Rooney, C.M., Wimperis, J.Z., Brenner, M.K., Patterson, J., Hoffbrand, A.V. and Prentice, H.G. Brit. J. Haematol 62: 413-420, 1986.
43. Cheeseman, J.H., Henle, W., Rubin, R.H. et al. Ann. Int. Med. 93: 39-42, 1980.

13

RESISTANCE OF HERPES VIRUSES TO NUCLEOSIDE ANALOGUES - MECHANISMS AND CLINICAL IMPORTANCE

C. CRUMPACKER

Beth Israel Hospital, Harvard Medical School, Boston, Massachusetts 02215, U.S.A.

INTRODUCTION

The study of resistance to antiviral drugs provides an important method to determine mechanisms of drug action. A mutation in the gene for a viral enzyme which enables a virus to replicate in the presence of the active form of a drug provides evidence that the mechanism of action involves the viral gene product. In the study of DNA replication, Kornberg observed that mutants possessing temperature sensitive and drug resistant mutations were the most important classes of mutants for defining precise mechanisms of DNA replication (1). A key feature in showing that the current class of antiviral drugs are specific for herpes viruses has been related to the fact that they readily select for resistance in herpes viruses and the resistance markers can be mapped to viral specific functions. Another important reason to study resistance of herpes viruses to antiviral drugs is that resistance can occur in the course of clinical treatment of herpes virus infections and development of resistance may account for failure of antiviral therapy. Resistant mutants of herpes viruses may also possess altered pathogenetic properties and produce unusual disease manifestations. In this review, the mechanisms of resistance to various nucleoside analogues will be outlined, and the cases where clinical resistance has been observed will be summarized.

MECHANISMS OF ACTION OF ANTIVIRAL DRUGS FOR HERPES VIRUSES

The deoxyguanosine analogue, acyclovir, is the prototype of a group of anti-herpes drugs which utilize a virus-specific enzyme to form an active inhibitor of viral DNA replication. Acyclovir is a preferred substrate for herpes virus thymidine kinase (TK) which phosphorylates acyclovir to acyclovir monophosphate (2). Cellular enzymes, particularly guanydylate kinase, lead to the formation of acyclovir diphosphate which is then converted to its triphosphate (3). Acyclovir triphosphate is a substrate for herpes DNA polymerase activity, and is a chain terminator for elongating herpes virus DNA (4,5). Acy-

DeClercq, E. (ed.), *CLINICAL USE OF ANTIVIRAL DRUGS.*

clovir triphosphate also appears to be a suicide inhibitor of herpes simplex virus DNA polymerase, binding irreversibly to the polymerase enzyme and preventing the polymerase from dissociating from the template-primer to engage in replication of other viral DNA molecules (6). When cells are infected with HSV-1 in the presence of acyclovir for 20 hours and then assayed for functional HSV DNA polymerase, very little enzyme can be detected, whereas infection with an acyclovir resistant HSV-1/HSV-2 recombinant virus for 20 hours results in a large amount of HSV DNA polymerase being detected (7). This indicates that the recombinant virus possesses an altered viral DNA polymerase which may not be bound irreversibly by the ACV-TP-template-primer complex and is free to direct HSV DNA synthesis. Acyclovir triphosphate inhibits herpes virus DNA polymerase 30 times more preferentially than mammalian cellular polymerases.

A close derivative of acyclovir, containing an additional CH_2OH group is 9-(1,3-dihydroxy-2-propoxymethyl)guanine (known as 2'NDG, DHPG, or BW759), and also requires a viral thymidine kinase for activation. This compound, 2'NDG, is a much better substrate for HSV thymidine kinase than is acyclovir and HSV infected cells produce about 7 fold higher amounts of 2'NDG triphosphate than of ACV-triphosphate (8). The affinity of 2'NDG-triphosphate for the HSV DNA polymerase is less than that of ACV-triphosphate and less ACV-triphosphate on a μg/ml basis is required to inhibit the HSV-DNA polymerase. The 2'NDG triphosphate acts as an alternative substrate to deoxyguanosine triphosphate (dGTP) for the HSV DNA polymerase, and 2'NDG is incorporated into DNA and results in a slowing down of the rate of DNA synthesis (9). Unlike ACV which acts as a chain terminator, DHPG gets incorporated at both internal and terminal linkages.

The main importance of this close derivative of acyclovir, 9-(1,3-dihydroxy-2-propoxymethyl)guanine (2'NDG, DHPG, BW759), is that it is an effective inhibitor of CMV replication and in clinical trials has provided the first successful treatment of serious CMV infections in immunocompromised patients (10-12). The DHPG-triphosphate and ACV-triphosphate successfully inhibit CMV DNA polymerase (10) which indicates that the viral polymerase remains sensitive to the active triphosphate inhibitor of these 2 compounds if adequate amounts of the triphosphate can be generated in infected cells. Since human cytomegalovirus does not induce a viral thymidine kinase enzyme, CMV-infected cells treated with acyclovir generate very little acyclovir triphosphate and CMV replication is not effectively inhibited by ACV. The use of ACV in clinical trials to treat serious CMV infection has been disappointing. With DHPG, however, effective inhibition of human CMV does occur and the mechanism by which DHPG is converted to DHPG triphosphate in CMV-infected cells must be clearly different from the activation of ACV in these cells. An enzyme has been found in calf thymus cells, a mitochondrial deoxyguanosine kinase, which has been partially purified and shown to comigrate with DHPG phosphorylating activity on DEAE-cellulose chromatography. The levels of this DHPG kinase are markedly elevated in CMV-infected cells and it has been suggested that this cellular enzyme is induced by CMV infection and results in high levels of DHPG triphosphate (13). This evidence supports the concept that mitochondrial deoxyguanosine kinase is the enzyme phosphorylating DHPG. Alternative evidence derived from passing a sensitive isolate of CMV in the presence of DHPG indicates that CMV can become resistant to DHPG. The cells infected with the resistant CMV fail to phosphorylate DHPG to form high levels of DHPG-triphosphate. This suggests that CMV-infected cells may possibly possess another viral induced kinase which can phosphorylate DHPG and when resistance develops, this kinase is no longer able to phosphorylate DHPG (14). An alternative explanation could be that CMV infection greatly induces the mitochondrial deoxyguanosine kinase activity and the resistant CMV fails to induce the activity of this enzyme. The existence of an as yet unidentified CMV kinase needs to be established. Additional indirect evidence which suggests that the steps in the cellular pathways for deoxyguanosine metabolism are important for the activity of DHPG can be obtained from experiments which examine the ability of natural deoxyribonucleosides to overcome the toxic effects of 2'NDG or DHPG on mouse cells growing in culture. Mouse LM cells or LM (TK^-) cells exhibit significant reversal of 2'NDG toxicity when exposed to increasing deoxyguanosine concentrations indicating that metabolic pathways for deoxyguanosine are important in overcoming the toxic effects of 2'NDG (15). A postulated mechanism to explain this observation could be that increasing concentrations of deoxyguanosine compete with DHPG for phosphorylation by deoxyguanosine kinase resulting in lower levels of DHPG triphosphate formed and improved cell survival.

The drug (E)-5-(2-bromovinyl)-2'-deoxyuridine (BVDU) is a thymidine analogue which also requires HSV thymidine kinase for phosphorylation. BVDU triphosphate is an effective inhibitor of viral DNA polymerase and BVDU gets incorporated into elongating DNA. BVDU is an effective inhibitor of HSV-1 and varicella zoster virus, but is less efficient in inhibiting HSV-2 (16). This difference in sensitivity between HSV-1 and HSV-2 is primarily due to the ability of the herpes thymidine kinase enzyme from HSV-1 to efficiently form a BVDU monophosphate and diphosphate whereas the thymidine kinase of HSV-2 efficiently forms the monophosphate of BVDU but this does not result in high intracellular levels of the di- and triphosphate of BVDU. This is probably due to the inability of the thymidylate kinase activity associated with the HSV-2 thymidine kinase enzyme to effectively phosphorylate the BVDU monophosphate whereas the HSV-1 TK enzyme efficiently adds a second phosphate group and leads to formation of large amounts of BVDU triphosphate, the active inhibitor of herpes DNA replication. BVDU triphosphate has been shown to be an effective competitive inhibitor of HSV-DNA polymerase and resistance to BVDU has been mapped to the HSV DNA polymerase gene (17). In addition, BVDU also inhibits the growth of thymidine kinase deficient cells and it appears to inhibit glycosylation of cellular membranes. This may provide an additional mechanism of action for BVDU in inhibiting cellular function (18).

The antiherpes drug adenine arabinoside (Ara-A) is phosphorylated directly to Ara-ATP by cellular enzymes (mainly adenosine kinase). Ara-ATP is a potent competitive inhibitor of HSV DNA polymerase and is incorporated by internucleotide linkage into DNA (19). It does not require specific activation by viral enzymes. Ara-A monophosphate becomes incorporated into primer terminus and produces a significant decrease in the rate of primer elongation. The 3'-terminal Ara-AMP residues are removed by action of the HSV-1 DNA polymerase associated exonuclease activity (20). The HSV exonuclease, however, does not remove terminally incorporated ACV-TP, and does not appear to possess an editing function for this nucleoside analogue.

Two closely related compounds phosphonoacetic acid (PAA) and phosphonoformic acid (PFA) are inhibitors of pyrophosphate exchange and act directly as non-competitive inhibitors of viral DNA polymerase. These compounds are not incorporated into viral DNA. All of the triphosphates of the nucleoside analogues inhibit viral DNA synthesis directly as competitive inhibitors of viral DNA polymerase, chain termination of viral DNA elongation, or by incorporation into viral DNA.

MECHANISMS OF RESISTANCE TO ANTIVIRAL DRUGS

Three mechanisms have been associated with the development of resistance of HSV to currently active antiviral drugs (Table 1). Two mechanisms involve the viral thymidine kinase for drugs which employ a viral thymidine kinase (vTK) to initially phosphorylate the nucleoside analogues (ACV, DHPG, BVDU, FIAC). When a sensitive strain of HSV-1 is passed in tissue culture in the presence of ACV, for example, a mutant is selected which is deficient in viral TK activity (Fig. 1) (21). The selected HSV population induces very little TK activity. The emphasis is on selection because these mutants are probably pre-

Table 1. Mechanisms of resistance to acyclovir.

1.	Selection of thymidine kinase-deficient mutants
2.	Selection of mutants possessing a thymidine kinase enzyme possessing an altered substrate specificity
3.	Selection of a mutant with an altered DNA polymerase

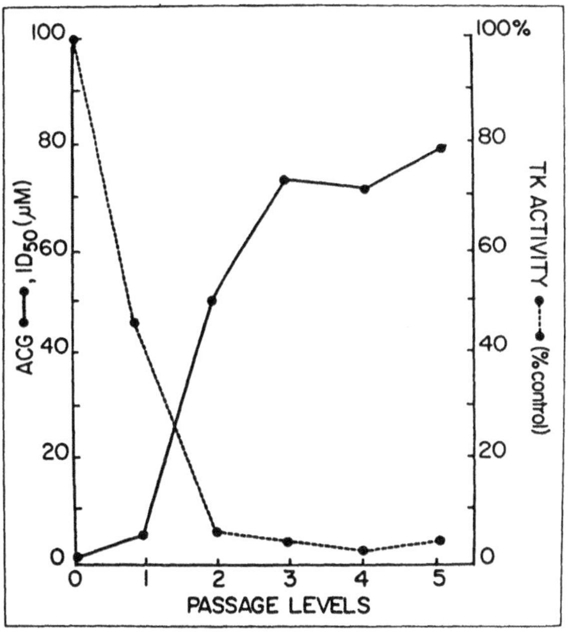

Fig. 1. Resistance to ACV in HSV-1 by passage in 10 μM ACV. Concentration of drug to inhibit plaque formation (ID_{50}) increases and virus pool fails to induce a HSV-specific TK (reprinted with permission from ref. 21).

sent in the population of wild-type HSV, and ACV treatment enables this mutant to become the majority of a population. This mechanism is by far the most common way by which resistance to ACV develops and it is the most important mechanism for the development of resistance following clinical treatment with acyclovir. Resistance to ACV following clinical treatment in well documented examples has always been by the mechanism of selection for TK^- virus.

Another mechanism for resistance involving the TK enzyme is the selection of viral mutants which possess an altered thymidine kinase which phosphorylates thymidine but is no longer able to phosphorylate acyclovir (22). Mutants possessing an altered TK were selected by Darby and co-workers by passing a wild type strain of HSV-1 in resting cells in the presence of ACV. The resting cells were produced by growing cells in the absence of serum. Under these conditions cells cease to divide and cellular thymidine kinase decreases markedly. In order to replicate in these cells, HSV needs to make its own thymidine kinase enzyme and mutants possessing an altered enzyme are selected by ACV. A similar mutant possessing an altered TK can be selected by passing in the presence of bromovinyldeoxyuridine (BVDU) which also requires a viral thymidine kinase for activation (23).

A mutant possessing an altered TK by selection in ACV fails to efficiently phosphorylate ACV and BVDU. The inability to phosphorylate these nucleoside analogues has been shown to be associated with a single amino acid change in the TK protein in a region of the protein defining a putative nucleotide binding site for ATP (24). With varicella zoster virus, passage in the presence of ACV can also select for ACV resistant mutants which are markedly deficient in thymidine kinase activity (25).

An altered HSV DNA polymerase activity is also associated with resistance to acyclovir and other nucleoside analogues. The ability to map resistance of all of the currently effective nucleoside analogues for herpes viruses to the viral DNA polymerase locus establishes that inhibition of viral DNA polymerase activity is a common mechanism by which these drugs stop viral DNA synthesis. Alterations in viral DNA polymerase function were first shown to be associated with resistance to acyclovir by demonstrating that mutants resistant to phosphonoacetic acid, PAA, a known inhibitor of viral DNA polymerase, and temperature-sensitive mutants with thermolabile lesions in the polymerase gene were also resistant to acyclovir (21,26). Resistance mutations of Ara-A mapped within the HSV polymerase gene and conferred resistance to Ara-A (17,27). At present no clear documentation of a DNA polymerase mutant being selected during clinical treatment of herpes infections with acyclovir or other nucleoside analogue has occurred.

Isolates of HSV from brain biopsy specimens of patients who were not treated with any antiviral drug have been reported to demonstrate resistance to PAA, but resistance to acyclovir was not described (28).

RESISTANCE TO ANTIVIRAL DRUGS AND THE STRUCTURE OF THE HERPES SIMPLEX DNA POLYMERASE

Since all of the clinically useful antiviral drugs for herpes viruses work by interfering with the herpes DNA polymerase, the interaction of the drugs with the polymerase enzyme can reveal a great deal concerning the structure of viral DNA polymerase. In fact, resistance to antiviral drugs is a powerful marker for polymerase function. The technique of marker rescue of temperature sensitive mutations by the use of HSV-1/HSV-2 intertypic recombinant viruses was employed to show that resistance of herpes DNA polymerase to PAA could be mapped within a 2.6 kilobase-pair (kbp) region of DNA sequences around map unit 0.40 in the long unique region of HSV-1 (strain 17) (29). The resistance of HSV-2 (strain HG52) to PAA was also mapped to a 1.3 kbp region of sequences which overlapped with the sequences for PAA resistance in HSV-1, and showed that the polymerase genes of HSV-1 or HSV-2 were colinear on the HSV genome. Physical mapping of acyclovir resistance and Ara-A resistance by the use of HSV-1/HSV-2 intertypic recombinants revealed that resistance to these two nucleoside analogues was determined by the same region of DNA sequences conferring resistance to PAA, establishing that these three drugs of dissimilar structure were able to interact with the same part of the viral DNA polymerase enzyme (30). The viral DNA polymerase enzyme induced by these recombinants was purified on glycerol gradients and the resistant polymerases were found to have a decreased binding affinity for PAA and all 4 of the dideoxynucleotide triphosphates (7). On further purification through phosphocellulose and DNA cellulose columns, the purified DNA polymerase from a resistant recombinant (R6-34) exhibited higher K_i to ACV-TP, Ara-ATP and BVDU-TP indicating that a mutation within the DNA sequences involving the polymerase gene could produce a polymerase which was able to function in the presence of high concentrations of the nucleoside analogue triphosphates (Allaudeen, Schnipper and Crumpacker, personal communication). The DNA sequences which contained the mutations conferring resistance to these inhibitors were found within the limits of the viral DNA polymerase gene and suggested that the mutations occurred in the region of DNA sequences encoding the carboxyl terminal one third of the viral polymerase enzyme (7). The mapping of independently derived mutations in the polymerase gene conferring resistance to PAA, Ara-A, and ACV would also suggest that mutations within a region of DNA encoding the carboxyl terminal one half of the polymerase enzyme determines resistance to nucleoside analogues of different structures and suggests that this region of the polymerase enzyme determines an important catalytic domain (31). The mutants which were found to be resistant to PAA and ACV were uniformly sensitive to 2'NDG and physical studies with the HSV-1/HSV-2 intertypic recombinants revealed that the resistance to 2'NDG was closely linked to the mutation for a temperature-sensitive mutant ts6 of HSV-2 (strain HG52) within a 2.2 kilobase-pair region of DNA sequences which was physically separate from the region for the $\underline{paa}^r$ and $\underline{acv}^r$ mutations (32). In studies on mutant polymerase enzymes the determinants of nucleotide binding specificity of DHPG-TP and ACV-TP were different, suggesting that DHPG-TP and ACV-TP interacted with different regions of the polymerase enzyme (33). It was also noted that there was no correlation between the interaction of virus DNA polymerase with DHPG-TP and inhibition of virus replication by DHPG. This was very different from the observations with ACV-TP where a strong correlation with the antiviral activity of ACV has been observed (34). These physical mapping studies suggest that a region of 3.5 kilobase-pairs contained all of the drug resistance markers and that the entire polymerase polypeptide may be able to contribute to interaction with nucleotides of different structure by tertiary folding of the polypeptide (32).

The complete nucleotide sequence of the DNA polymerase gene of HSV-1 (strain 17) (35), HSV-1 (strain KOS) (31), and HSV-1 (strain Angelotti) (36) has been determined and the gene consists of a 3705 base-pair open reading frame to the right of the origin of replication in the long unique region of DNA sequences. The amino terminal end of the polymerase gene begins with an ATG located 154 base pairs to the right of the Bam HI v-r site and the carboxyl end of the coding region has a 63 base non-coding region in the messenger RNA. Two abundant RNA transcripts of 4.2 and 4.3 kbps encode the DNA polymerase and these RNAs appear to be unspliced (37). The DNA sequence of the polymerase gene reveals no splice donor or acceptor sites. All of the drug resistant mutations and temperature sensitive mutations mapped previously by HSV-1/HSV-2 intertypic recombinants lie within the nucleotide sequences defining the open reading frame of the pol gene (29,30,32,38). The mutations mapped by marker transfer using cloned fragments of the HSV genome also map within this open reading frame (39). The 1235 amino acids of the HSV-1 pol enzyme contain a region of 16 amino acids extending from amino acid residue at 880 to 896 which shares strong sequence homology with the DNA polymerases of EBV and Ad2 (35). The CMV DNA polymerase gene has also been identified and sequenced and shown to share strong sequence homology in this region with the DNA polymerases of HSV-1, EBV and Ad2 (40,41). A region of 6 amino acids extending from Met-Gly-Asp-Thre-Asp-Ser was found to be absolutely conserved in all 4 polymerases. This highly conserved region of amino acids is probably a common functional element in the DNA polymerase of these four viruses. The 16 amino acids of the HSV-1 DNA polymerase sharing strong homology with EBV, CMV and Ad2 polymerases lie within the carboxyl terminal one third of the polymerase enzyme, the region previously designated from physical mapping of drug resistant mutations as defining an active center on the polymerase polypeptide (7) (Fig. 2). This region of the carboxyl terminal one third to one half of the polymerase enzyme appears likely to define a catalytic domain on the polymerase enzyme. The nucleotide sequence comparison of the HSV-1 Angelotti strain and a PAA resistant mutant of this strain reveals that a single base change is able to confer an amino acid alteration at position 719 in the polymerase enzyme and results in PAA resistance of the purified polymerase (37). In another acyclovir-resistance mutant sequenced by Larder and Darby a change of asparagine to serine at position 815 was found to be associated with ACV resistance (42). The asparagine at position 815 is highly conserved in the polymerases of HSV, CMV, EBV, Ad2 and phage 29 (41) and this conservation suggests important func-

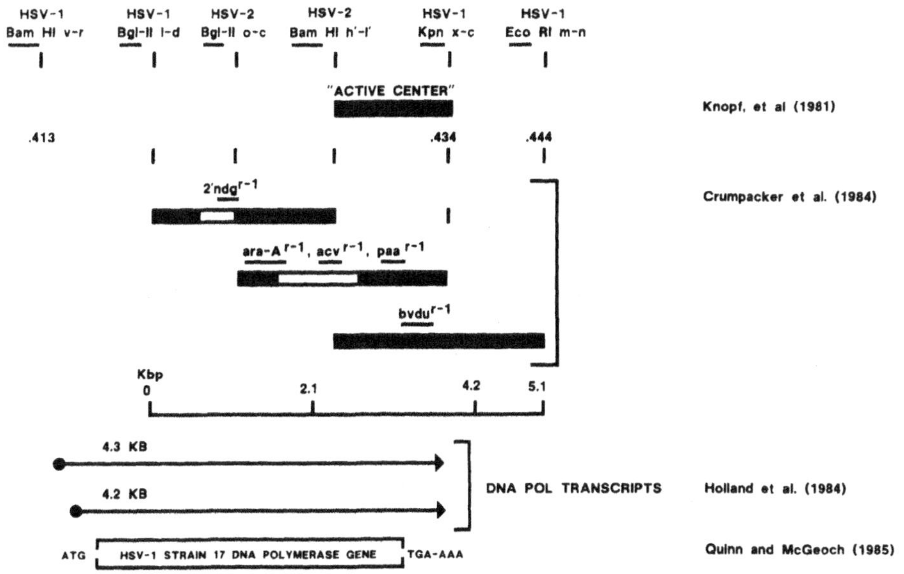

Fig. 2. Diagram of the HSV pol gene showing limits of DNA sequence, RNA transcripts, mapping of resistance markers, and the region of DNA defining active center or catalytic domain of the pol enzyme on the map of the HSV genome.

tional significance. These studies on the physical mapping of drug resistant mutations within the polymerase gene of HSV indicate the importance of drug resistant markers in defining functional and structural features of the HSV polymerase. They also predict that the clarification of the function of the highly conserved region of HSV-1 polymerase will have important consequences for Ad2, CMV and EBV as well. The remarkable agreement of physical mapping limits of drug resistance markers with the open reading frame of nucleotide sequences for the polymerase gene is gratifying.

CLINICAL SIGNIFICANCE OF RESISTANCE TO NUCLEOSIDE ANALOGUES

The development of resistance of herpes viruses to nucleoside analogues with clinical use is important for three primary reasons : 1) resistance may explain the failure of an antiviral drug to produce healing of a serious herpes infection; 2) resistant strains of herpes viruses may produce disease in unexpected ways; 3) resistance isolates may spread widely through a human population and make effective antiviral therapy difficult or impossible. In the clinical use of acyclovir, resistance has been observed almost exclusively in patients with severe immunodeficiency disorders who have been treated with several courses of intravenous or oral acyclovir (43-46; Table 2).

In all of the clinical cases of resistance to acyclovir, the mechanism has been based on the selection of thymidine kinase-deficient (TK^-) mutants. In a well documented case, resistance to acyclovir occurred in a 7 year old child with severe combined immunodeficiency syndrome who received three courses of intravenous acyclovir (43). The initial HSV-1 isolate from the child was sensitive to acyclovir and dramatic healing occurred with the first treatment episode. By the third treatment course, however, healing did not occur and virus persisted in the oral cavity of the child. The isolate from the third treatment episode was markedly deficient in thymidine kinase and resistant to acyclovir and another drug requiring TK for activation, bromovinyldeoxyuridine (BVDU). The resistant isolate was identical to the sensitive HSV-1 isolate as revealed by restriction endonuclease analysis (47). The sensitive and resistant isolates had not undergone alterations in the DNA polymerase gene of the virus and the isolate remained sensitive to drugs which act directly on the viral DNA polymerase, Ara-A and PAA. The clinical use of acyclovir selected for a mutant virus which was markedly deficient in TK activity and was unable to produce the TK protein of 43,000 daltons or the associated proteins of 39K and 38K (47). Other early reports of resistant isolates of HSV mentioned resistance occurring in a bone marrow transplant recipient (45) and

Table 2. Resistance to acyclovir in herpes simplex virus following treatment.

Study	Journal	Comment
Crumpacker *et al.*	New Engl. J. Med., 1982	Resistance in an immunodeficient child with 3 courses of ACV
Sibrack *et al.*	J. Infect. Dis., 1982	Resistance and pathogenesis of a TK-virus from a child treated with ACV
Burns *et al.*	Lancet, 1982	Resistance in a marrow transplant patient following ACV
Wade *et al.*	Ann. Intern. Med., 1982	Summary of the incidence of ACV resistance in a marrow transplant unit
Straus *et al.*	New Engl. J. Med., 1984	Resistance in a normal patient
Schinazi *et al.*	J. Antimicrob. Agents Chemother., 1986	Resistance and shedding in a patient with severe genital herpes
Svennerholm *et al.*	Scand. J. Infect. Dis., 1985	Resistance in a patient with severe genital herpes

in another immunodeficient child (44). These reports of resistance to acyclovir occurred in immunodeficient patients who received multiple courses of intravenous drug. The clinical course associated with the isolation of resistant virus showed slow healing but no widespread dissemination. The most severe clinical consequences associated with resistant virus have been observed in two immunocompromised patients with extensive genital herpes associated with HSV-2 (48,49). Resistant virus developed following prolonged acyclovir therapy and was associated with deep extensive genital herpes lesions which failed to heal.

A careful analysis of the pathogenicity of an acyclovir resistant isolate of HSV from an immunodeficient child treated with acyclovir was carried out in mice (44). The acyclovir resistant isolate was present in the cerebrospinal fluid of the patient but was associated with an absence of inflammation or perivascular infiltration. The virus induced a decreased amount of viral thymidine kinase and intracerebral inoculation in mice was associated with a 1000-

fold decrease in neurovirulence and death due to encephalitis. Cutaneous lesions induced by this resistant mutant were slow to heal and chronic infection resulted. Some lesions eventually healed, but others tended to persist for prolonged periods with the ultimate death of the mouse. In several reports it has been well documented that thymidine kinase-deficient mutants of herpes simplex are less efficient at establishing infection in the central nervous system (50-52). Thymidine kinase-deficient mutants of HSV are capable of establishing latent infection but at a decreased frequency. The possibility exists that thymidine kinase deficient mutants can revert back to a wild type phenotype with expression of TK and an increase in neurovirulence.

There is only one report of resistant isolates of HSV occurring in immunologically normal patients treated with prolonged oral acyclovir to suppress recurrent genital herpes (53). These isolates were obtained during an outbreak of genital herpes while a patient was taking oral ACV for suppression, were not associated with severe disease, and the recurrence following cessation of ACV exhibited a return to the wild type sensitivity of HSV for ACV. The mechanism of resistance developing in clinical situations has been by the selection of TK-deficient virus which is unable to phosphorylate ACV (43). One of these mutants selected with ACV treatment has been well characterized and reveals a complete cessation of the viral TK polypeptide (47). There has been a single report of a mutant with a thymidine kinase possessing an altered substrate specificity being isolated but this mutant has not been well characterized. A mutant possessing altered DNA polymerase activity has not been selected in any patient treated with acyclovir or other antiviral drug. It has been suggested that naturally occurring acyclovir-resistant mutants containing an altered DNA polymerase may exist (28). A mutant exhibiting an altered viral DNA polymerase conferring resistance to acyclovir triphosphate may exhibit resistance to several drugs such as phosphonoacetic acid, adenine arabinoside and bromovinyldeoxyuridine (32,59).

Although mutants with an altered polymerase have not been isolated from patients following treatment with acyclovir, mutants prepared in vitro with an altered polymerase exhibiting resistance to antiviral drugs produce unusual disease manifestations when inoculated intracerebrally into mice (54). The mice were inoculated intracerebrally with PAA and Ara-A resistant mutants possessing an altered polymerase and exhibiting multiple drug resistance. The mice did not die of encephalitis but a high proportion of the mice developed cataracts and blindness. It should be stressed that intracerebral inoculation is a very unnatural route of infection, but this study does point out that mutants possessing an altered polymerase may produce disease in unexpected ways.

Although the development of resistance has not been associated with severe pathogenicity or dissemination of virus in most cases in immunocompromised patients, two recent reports of an ACV resistant strain of HSV-2 producing genital lesions have been associated with the continued presence of virus and severe genital ulcers which did not heal in two severely immunocompromised patients (48,49). This represents an example of moderately severe disease associated with ACV-resistant virus and suggests that clinicians should be alert to the presence of disease which fails to heal in a normal manner in patients treated with acyclovir. Although resistance has been well documented following the clinical use of acyclovir, a case of well documented resistance following treatment of HSV and VZV with Ara-A or treatment of CMV with DHPG has not been reported. Possible resistant viruses to these other antiviral drugs are being actively sought for.

PREVENTION AND ALTERNATE TREATMENT FOR RESISTANT VIRUSES

Since the most common manner in which resistant HSV has developed has been in association with multiple courses of intravenous acyclovir can be very effective in immunocompromised patients. When resistance develops by selection of a thymidine kinase-deficient mutant, then all of the nucleosides which employ viral TK for activation become ineffective (Fig. 2; 43,59). An important goal of antiviral therapy for herpes viruses will be the development of drugs which do not require viral TK for activation. One such drug is the cyclic phosphate derivative of 2'NDG known as 2'nor cGMP (55). This compound easily gets across cell membranes in large amounts, does not require viral thymidine kinase for activation, and is an effective inhibitor of TK-deficient mutants of HSV (55,15) (Fig. 2). This drug provides an example of a compound which may prove useful for treating serious infections associated with resistant HSV.

Another antiviral compound which does not require a virus-induced thymidine kinase for activity is designated (S)-9-(3-hydroxy-2-phosphonylmethoxypropyl)adenine [(S)-HPMPA] (56). This drug has activity against many DNA viruses and is also active against TK^- mutants of HSV and VZV.

In normal patients, resistance of HSV to acyclovir has been observed infrequently. In a study of topical 5 % and 10 % acyclovir ointment for facial oral herpes or oral ACV for genital herpes, no change in sensitivity of HSV before and after ACV treatment was observed (57). In this study of breakthrough recurrences employing a sensitive plaque reduction assay, a change in sensitivity to ACV was not detected in the isolate obtained from a breakthrough (57). In a larger study employing a less sensitive dye uptake assay, the *in vitro* sensitivities of 183 virus isolates from 107 patients which were obtained before, during, and after acyclovir therapy showed no change with therapy. In particular, six isolates from breakthrough recurrences obtained during acyclovir therapy showed no change in sensitivity. This study concluded that acyclovir-resistant strains were not commonly recovered during acyclovir therapy and a high frequency of resistance was not observed after 4 months of chronic suppressive therapy (58). These results suggest that development of resistance is not a simple explanation for breakthrough recurrences and resistance is not a common problem in normal patients using oral acyclovir or topical acyclovir ointment.

Finally, it appears that in patients with normal immunological defenses, the emergence of resistant viruses will not be a common problem with the use of acyclovir for a short period. The full importance of resistance to nucleoside analogues will only become clear with more widespread use of these drugs, appreciation of the problems associated with their use, and more careful monitoring for the emergence of resistance.

REFERENCES

1. Kornberg, A. In: DNA Replication. Chapter 13. Freeman, New York, 1980.
2. Elion, G.B., Furman, P.A., Fyfe, J.A., de Miranda, P., Beauchamp, L. and Schaeffer, H.J. Proc. Natl. Acad. Sci. USA 74 : 5716-5720, 1977.
3. Miller, W.H. and Miller, R.L. J. Biol. Chem. 255 : 7204-7207, 1980.
4. Derse, D., Cheng, Y.-C., Furman, P.A., St. Clair, M.H. and Elion, G.B. J. Biol. Chem. 256 : 11447-11451, 1981.
5. Furman, P.A., St. Clair, M.H., Fyfe, J.A., Rideout, J.L., Keller, P.M. and Elion, G.B. J. Virol. 32 : 72-77, 1979.
6. Furman, P.A., St. Clair, M.H. and Spector, T. J. Biol. Chem. 259 : 9575-9579, 1984.
7. Knopf, K.W., Kaufman, E.R. and Crumpacker, C. J. Virol. 39 : 746-757, 1981.
8. Field, A.K., Davies, M.E., DeWitt, C., Perry, H.C., Liou, R., Germershausen, J., Karkas, J.D., Ashton, W.T., Johnston, D.B.R. and Tolman, R.L. Proc. Natl. Acad. Sci. USA 80 : 4139-4143, 1983.
9. Cheng, Y.-C., Grill, S.P., Dutschman, G.E., Nakayama, K. and Bastow, K.F. J. Biol. Chem. 258 : 12460-12464, 1983.
10. Mar, E.-C., Chiou, J.-F., Cheng, Y.-C. and Huang, E.-S. J. Virol. 53 : 776-780, 1985.
11. Tocci, M.J., Livelli, T.J., Perry, H.C., Crumpacker, C.S. and Field, A.K. Antimicrob. Agents Chemother. 25 : 247-252, 1984.
12. Collaborative DHPG Treatment Study Group. New Engl. J. Med. 314 : 801-806, 1986.
13. Smee, D.F. Mol. Cell. Biochem. 69 : 75-81, 1985.
14. Biron, K.K., Fyfe, J.A., Stanat, S.C., Leslie, L.K., Sorrell, J.B., Lambe, C.U. and Coen, D.M. Proc. Natl. Acad. Sci. USA 83 : 8769-8773, 1986.
15. Oliver, S., Bubley, G. and Crumpacker, C. Virology 145 : 84-93, 1985.

16. De Clercq, E., Descamps, J., De Somer, P., Barr, P.J., Jones, A.S. and Walker, R.T. Proc. Natl. Acad. Sci. USA 76 : 2947-2951, 1979.
17. Crumpacker, C.S., Schnipper, L.E., Kowalsky, P.N. and Sherman, D.M. J. Infect. Dis. 146 : 167-172, 1982.
18. Balzarini, J., De Clercq, E., Verbruggen, A., Crumpacker, C., Ayusawa, D. and Seno, T. Anticancer Res. 6 : 1077-1084, 1986.
19. Pelling, J.C., Drach, J.C. and Shipman, C. Jr. Virology 109 : 323-335, 1981.
20. Derse, D. and Cheng, Y.-C. J. Biol. Chem. 256 : 8525-8530, 1981.
21. Schnipper, L.E. and Crumpacker, C.S. Proc. Natl. Acad. Sci. USA 77 : 2270-2273, 1980.
22. Darby, G., Field, H.J. and Salisbury, S.A. Nature 289 : 81-83, 1981.
23. Larder, B.A., Derse, D., Cheng, Y.-C. and Darby, G. J. Biol. Chem. 258 : 2027-2033, 1983.
24. Larder, B.A., Cheng, Y.-C. and Darby, G. J. Gen. Virol. 64 : 523-532, 1983.
25. Shigeta, S., Yokota, T., Iwabuchi, T., Baba, M., Konno, K., Ogata, M. and De Clercq, E. J. Infect. Dis. 147 : 576-584, 1983.
26. Coen, D.M. and Schaffer, P.A. Proc. Natl. Acad. Sci. USA 77 : 2265-2269, 1980.
27. Coen, D.M., Furman, P.A., Gelep, P.T. and Schaffer, P.A. J. Virol. 41 : 909-918, 1982.
28. Parris, D.S. and Harrington, J.E. Antimicrob. Agents Chemother. 22 : 71-77, 1982.
29. Chartrand, P., Stow, N.D., Timbury, M.C. and Wilkie, N.M. J. Virol. 31 : 265-276, 1979.
30. Crumpacker, C.S., Chartrand, P., Subak-Sharpe, J.H. and Wilkie, N.M. Virology 105 : 171-184, 1980.
31. Gibbs, J.S., Chiou, H.C., Hall, J.D., Mount, D.W., Retondo, M.J., Weller, S.K. and Coen, D.M. Proc. Natl. Acad. Sci. USA 82 : 7969-7973, 1985.
32. Crumpacker, C.S., Kowalsky, P.N., Oliver, S.A., Schnipper, L.E. and Field, A.K. Proc. Natl. Acad. Sci. USA 81 : 1556-1560, 1984.
33. Frank, K.B., Chiou, J.-F. and Cheng, Y.-C. J. Biol. Chem. 259 : 1566-1569, 1984.
34. Furman, P.A., Coen, D.M., St. Clair, M.H. and Schaffer, P.A. J. Virol. 40 : 936-941, 1981.
35. Quinn, J.P. and McGeoch, D.J. Nucleic Acids Res. 13 : 8143-8163, 1985.
36. Knopf, C.W. Nucleic Acids Res. 14 : 8225-8226, 1986.
37. Holland, L.E., Sandri-Goldin, R.M., Goldin, A.L., Glorioso, J.C. and Levine, M. J. Virol. 49 : 947-959, 1984.
38. Chartrand, P., Crumpacker, C.S., Schaffer, P.A. and Wilkie, N.M. Virology 103 : 311-326, 1980.
39. Coen, D.M., Aschman, D.P., Gelep, P.T., Retondo, M.J., Weller, S.K. and Schaffer, P.A. J. Virol. 49 : 236-247, 1984.
40. Kouzarides, T., Bankier, A.T., Satchwell, S.C., Weston, K., Tomlinson, P. and Barrell, B.G. J. Virol. 61 : 125-133, 1987.
41. Reinhold, R. et al. J. Virol. 61, in press, 1987.
42. Larder, B.A., Kemp, S.D. and Darby, G. EMBO J. 6 : 169-175, 1987.
43. Crumpacker, C.S., Schnipper, L.E., Marlowe, S.I., Kowalsky, P.N., Hershey, B.J. and Levin, M.J. New Engl. J. Med. 306 : 343-346, 1982.
44. Sibrack, D., Gutman, L.T., Wilfert, C.M., McLaren, C., St. Clair, M.H., Keller, P.M. and Barry, D.W. J. Infect. Dis. 146 : 673-682, 1982.
45. Burns, W.H., Saral, R., Santos, G.W., Laskin, O.L, Lietman, P.S., McLaren, C. and Barry, D.W. Lancet 1 : 421-423, 1982.
46. Wade, J.C., Newton, B., McLaren, C., Flournoy, N., Keeney, R.E. and Meyers, J.D. Ann. Intern. Med. 96 : 265-269, 1982.

47. Chatis, P., Germershausen, J., Field, A.K. and Crumpacker, C.S. Abstract of the 11th International Herpes Workshop, 1985.
48. Schinazi, R.F., del Bene, V., Scott, R.T. and Dudley-Thorpe, J.-B. J. Antimicrob. Agents Chemother. 18 (Suppl. B) : 127-134, 1986.
49. Svennerholm, E., Vahlne, A., Löwhagen, G.B., Widell, A. and Lycke, E. Scand. J. Infect. Dis., Suppl. 47 : 149-154, 1985.
50. Field, H.J. and Wildy, P. J. Hygiene 81 : 267-277, 1978.
51. Tenser, R.B. and Dunstan, M.E. Virology 99 : 417-422, 1979.
52. Tenser, R.B., Miller, R.L. and Rapp, F. Science 205 : 915-917, 1979.
53. Straus, S.E., Takiff, H.E., Seidlin, M., Bachrach, S., Lininger, L., Di-Giovanna, J.J., Western, K.A., Smith, H.A., Nusinoff Lehrman, S., Creagh-Kirk, T. and Alling, D.W. New Engl. J. Med. 310 : 1545-1550, 1984.
54. Field, H.J. and Coen, D.M. J. Virol. 60 : 286-289, 1986.
55. Tolman, R.L., Field, A.K., Karkas, J.D., Wagner, A.F., Germershausen, J., Crumpacker, C. and Scolnick, E.M. Biochem. Biophys. Res. Commun. 128 : 1329-1335, 1985.
56. De Clercq, E., Holý, A., Rosenberg, I., Sakuma, T., Balzarini, J. and Maudgal, P.C. Nature 323 : 464-467, 1986.
57. Marlowe, S.I., Kowalsky, P.A., Douglas, J., Corey, C. and Crumpacker, C.S. Abstract of the 24th Interscience Conference on Antimicrobial Agents and Chemotherapy, 1984.
58. Nusinoff Lehrman, S., Douglas, J.M., Corey, L. and Barry, D.W. Ann. Intern. Med. 104 : 786-790, 1986.
59. Field, H., McMillan, A. and Darby, G. J. Infect. Dis. 143 : 281-285, 1981.

14

CLINICAL USE OF FOSCARNET (PHOSPHONOFORMATE)

B. ÖBERG[1], S. BEHRNETZ[2], B. ERIKSSON[2], H. JOZWIAK[2], A. LARSSON[2], J. O. LERNESTEDT[2] and V. LINDSÖ ÅBERG[2].

[1]Department of Antiviral Chemotherapy and [2]Department of Clinical Research, ASTRA ALAB AB, S-151 85 Södertälje, Sweden

ABSTRACT

Foscarnet is a selective inhibitor of several viral DNA polymerases. When used topically, foscarnet has a limited effect on recurrent mucocutaneous herpes, possibly due to the rapid course of virus replication. In immunocompromised patients with severe CMV infections, parenteral foscarnet has beneficial effects especially against CMV retinitis in AIDS patients. Preliminary data indicate that parenteral foscarnet decreases HIV replication in vivo. Following iv treatment with foscarnet for two weeks or longer and with median daily doses of 6 to 10 g in different subgroups of patients, transient renal impairment, decreases in haemoglobin concentration and calcium have been observed. Effects on the central nervous system, e.g. hallucinations and tremor, may be associated with high plasma foscarnet levels.

INTRODUCTION

Viral RNA and DNA polymerases have been major targets for antiviral drugs. Most of these have been substrate analogues e.g. nucleoside analogues competing, as triphosphates, with the normal substrates. One product analogue, trisodium phosphonoformate (INN: foscarnet sodium), has been found to inhibit all the human herpesvirus DNA polymerases, hepatitis B DNA polymerase and reverse transcriptases, notably the reverse transcriptase of human immunodeficiency virus (HIV). The mechanism of inhibition is non- or un-competitive and cellular polymerases are not affected at concentrations inhibitory to the viral enzymes. The selective antiviral activities of foscarnet are outlined in Fig. 1.

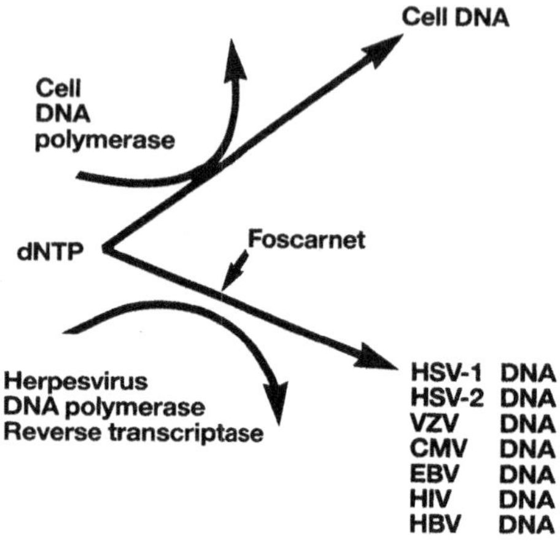

Fig. 1. Mechanism of action of foscarnet.

The inhibition of viral polymerases by foscarnet results in an inhibition of virus multiplication in cell-cultures and in animals. The antiviral properties of foscarnet were first described by Helgstrand et al. (1) and have been reviewed by Helgstrand et al. (2), Öberg (3), Eriksson and Öberg (4), Öberg (5, 6).

The present review describes the clinical results of using topical and parenteral foscarnet against viral diseases.

In these studies foscarnet has been used as a cream in the treatment of labial and genital herpes and as an iv treatment against severe virus infections. The iv use of foscarnet has mainly been directed against cytomegalovirus (CMV) infections. Recently attempts to treat HIV infections in AIDS and ARC patients have started and a few patients with fulminant hepatitis B have also been treated with iv foscarnet.

LABIAL HERPES

Ten clinical studies, comprising about 1000 patients, on recurrent cutaneous herpes using a 3 % (w/w) foscarnet cream have been carried out. As presented elsewhere in this book by S. L. Spruance, a small clinical benefit has been observed in patients who initiated treatment very early.

GENITAL HERPES

The effect of topically applied foscarnet on recurrent genital herpes has been investigated using 0.3 % and 1 % foscarnet cream. Several clinical trials have been completed but the results have been conflicting.

In a double-blind placebo-controlled study, Wallin et al. (7) followed 86 patients given 0.3 % foscarnet cream or placebo cream for a total of 129 episodes of recurrent genital herpes. The cream was applied every second hour during the first day and subsequently 6 times a day for another 4 days. When the results were analyzed in a parallel group manner and all patients were considered, a significantly shorter time to healing (4 days) was observed in the foscarnet group as compared to the placebo group (5 days).

Lassus et al. (8) evaluated 1 % foscarnet cream against genital herpes in a double-blind, randomized, placebo-controlled study with a cross-over design of two consecutive episodes. Patients were randomized to the treatment sequence foscarnet - placebo, or placebo - foscarnet. Among 118 patients valid for efficacy evaluation in the first of the two consecutive episodes, the median time to healing was not reduced in foscarnet-treated men but reduced from 5 to 4 days ($p < 0.001$) in foscarnet-treated women as compared to placebo.

Sacks et al. (9) have evaluated 0.3 % foscarnet cream in men and 1 % foscarnet cream in women as a treatment of recurrent genital herpes. In this multicenter double-blind, parallel group trial, 314 patients with clinically active herpes, and a duration of 6 hours or less after symptom onset, were randomly assigned to foscarnet cream or placebo cream treatment which was initiated at the clinic. Medication was applied every 2 hours during the first day and then every 4 hours for 4 more days.

Foscarnet did not shorten the time to healing or decline of overall symptoms but did result in a higher proportion of symptom-free patients after 1 day of treatment. Foscarnet-treated patients had a reduced duration of virus shedding which was significant in men. The duration of virus excretion and lesion evolution in culture-confirmed episodes are shown in Table 1. The culture-positive episodes healed significantly slower than the culture-negative episodes. Foscarnet and placebo cream were well tolerated.

Table 1. Effect of foscarnet on duration of virus excretion and lesion evolution in culture-confirmed episodes of recurrent genital herpes.

Sex, treatment group		Duration of viral excretion	Healing time of lesions
Men			
0.3 %	Foscarnet	2.0	5.8
	Placebo	3.0	6.0
	P value	0.014	NS
Women			
1 %	Foscarnet	2.0	4.9
	Placebo	3.0	4.6
	P value	NS	NS

Note: Durations are expressed as the median number of days.
Modified from Sachs et al. (9).

Three other multicenter placebo-controlled double-blind studies on recurrent genital herpes were performed similar to the study by Sacks et al. (9) using 0.3 % foscarnet cream for men and 1 % foscarnet cream for women (10, 11, Astra Alab, data on file).

In the study by Barton et al. (10) there was no difference between 253 foscarnet or placebo-treated patients in time to healing for either men or women. The median time to healing was 5 days in men and 4 days for women in both groups. The development of new lesions was significantly less common in patients given foscarnet.

Gundersen et al. (11) evaluated the effect of foscarnet cream in 119 men and 72 women. A significant reduction in time to healing (4 versus 5 days) was seen in men and a reduction of positive HSV cultures in women (17 % versus 40 % positive isolations) was observed in the foscarnet treatment group when compared to placebo. However, there was a tendency to longer time for healing in the foscarnet treatment group in women.

In a multicenter study evaluating 0.3 % foscarnet in 128 men in South East Asia (Astra Alab, data on file) no difference in time to healing (median 4 days) was observed between foscarnet and placebo-treated lesions.

The overall impression of the clinical use of foscarnet cream against recurrent genital herpes is that a slight reduction in time to healing, compared to placebo, can be achieved if treatment is initiated very early before vesicles appear and that time for virus shedding is consistently reduced. A compilation of data for all patients from the studies by Sachs et al. (9), Barton et al. (10), Gundersen et al. (11) and Astra Alab, data on file, does show a significant reduction of number of days with sores from 1.6 to 1.0 ($p = 0.02$) in 190 patients who started their treatment in the prevesicular stage.

The evaluation of topical foscarnet against recurrent herpes is no longer being pursued at Astra.

CYTOMEGALOVIRUS INFECTIONS

The effect of foscarnet against CMV replication offers a possibility to treat patients with severe CMV infections which is a considerable clinical problem, especially in transplant recipients and, at a rapidly increasing rate, in patients with an immunodeficiency caused by HIV infection. To date more than 600 patients have been given foscarnet by the iv route on a compassionate basis for severe CMV infections. Some results from the compassionate use of iv foscarnet have been published (12-27).

Generally, when administered by the iv route as a 2-2.4 % solution, foscarnet treatment has been initiated with a bolus injection of 9-20 mg/kg followed by continuous infusion of 0.015-0.20 mg/kg/min depending on the renal function as assessed by the s-creatinine. Treatment duration has as a rule been 2 weeks but has occasionally been extended several weeks depending on the clinical situation.

CMV INFECTIONS IN RENAL TRANSPLANT PATIENTS

Ahlmén et al. (13) treated 20 renal transplant patients with iv foscarnet for CMV infections. All patients had continuous or spiking fever before treatment. Leukopenia, pathological liver tests, suspected pneumonitis on pulmonary x-ray investigations and unspecific symptoms as arthralgia, muscular tenderness, and headache were also noted in this group of patients. The clinical effect of foscarnet was judged as good in 12 of 14 patients with primary CMV infection and as good in 4 of 6 patients with reactivated CMV infection. No clinical side effects were observed. A slight increase, within the normal limits, in s-calcium was noted in a few patients and a decrease in s-calcium from 2.3 to 1.9 mmol/l in one patient. These values returned to normal within a week after treatment. Four of the 20 patients died during 5 - 79 days after the end of foscarnet treatment. Deaths were not considered attributable to foscarnet treatment.

The effect of foscarnet treatment in 12 renal patients with CMV infections have been reported by Ringdén et al. (23). Results from two of these patients have been reported in detail by Klintmalm et al. (16). Only one of the 4 renal transplant patients with pneumonitis died. Five of 10 assessable patients became afebrile and laboratory abnormalities improved in 7. In 2 of 4 assessable patients, CMV was eradicated during therapy. Overall clinical improvement was seen following foscarnet therapy in 8 of 11 assessable patients. One patient developed adverse symptoms such as hallucinations and flapping tremor. This occurred at inadvertently high plasma levels (1490 µM) of foscarnet. The symptoms disappeared when foscarnet infusion was discontinued.

A potentially lethal CMV pneumonitis after renal transplantation was successfully treated with a combination of i.v. foscarnet, CMV antibodies and transplantectomy (22). These measures rapidly normalized the pulmonary function which had been severely impaired.

CMV INFECTIONS IN BONE MARROW TRANSPLANT PATIENTS

Apperley et al. (14) reported on foscarnet treatment of CMV pneumonitis in two bone marrow transplant patients. In the first patient bronchial lavage revealed CMV and anti-CMV hyperimmune globulin was given. However, this did not affect the fever, and the pulmonary function deteriorated. Clinical symptoms improved and the patient was subsequently discharged from hospital. One year later the chest x-ray was almost entirely normal. The second patient was dyspnoeic, had a cough with mucoid sputum and bronchoscopy showed CMV. Because of a raised serum creatinine he was given a reduced dosage of foscarnet. At the end of treatment the patient was afebrile with obvious improvements in his symptoms and chest signs. The clinical improvement was maintained after his discharge from hospital.

Ringdén et al. (23) evaluated the efficacy of foscarnet treatment in 13 bone marrow patients with severe CMV infections. These patients have partly been described also in other reports (12, 16, 18-19) and Figure 2 shows the results for one of these patients. The treatment was the same as in the renal transplant patients described above but the serum levels of foscarnet in the bone marrow transplant patients were on the average 75 mg/l as compared to 162 mg/l in the renal transplant patients (23).

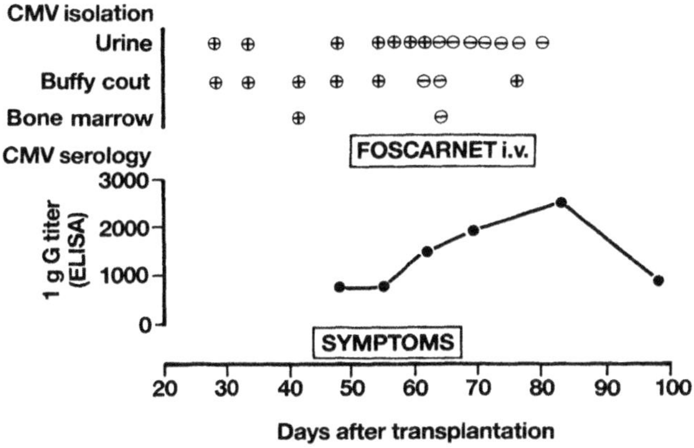

Fig. 2. Serological findings and isolation of CMV in a bone marrow recipient during treatment with iv foscarnet. From Klintmalm et al. (16) with permission.

Overall clinical improvement was seen in 9 of 13 patients. However, all 8 patients with pneumonitis and one patient with pancytopenia died. The latter patient had a severe acute graft versus host disease (GVHD) and died of septicemia. CMV was considered the principal cause of mortality in 4 of the 8 patients with pneumonitis and contributed together with other disorders in another 3. Two patients died of other causes. Further evaluation by Ringdén et al. (24) of the previous 13 patients and another 7 patients demonstrated in the total material of 20 patients, clinical improvement, eradication of CMV, afebrility, improvements of laboratory abnormalities or chest x-ray in 13 patients. Although chest x-ray improved in 3 patients all 14 patients with pneumonitis died. Eight of these had other complicating illnesses such as serious bacterial and fungal infections, severe GVHD or veno-occlusive liver disease. Among the remaining 6 of the total of 20 patients one died of graft versus host disease and 5 improved.

CMV INFECTIONS IN AIDS PATIENTS

Several AIDS patients have been treated with iv foscarnet for CMV infections. A few of these cases have been reported. Singer et al. (25, 28) treated a CMV retinitis in an AIDS patient, who also had received a kidney transplant, with foscarnet. The patient, who was blind in the left eye due to a previous CMV retinitis, became affected in the right eye, with retinal haemorrhages and necrosis in the fundus; visual acuity was 6/18. He was given CMV hyperimmune globulin, but the retinitis progressed. After 12 days of foscarnet treatment the retinitis had decreased and visual acuity had improved to 6/12. CMV excretion in urine ceased at 24 days and in saliva at 32 days after the initiation of foscarnet treatment. Mean foscarnet plasma level during treatment was 65.6±10.2 mg/l. Remission was sustained for 3 months, from the onset of improvement without maintenance treatment. Retinitis recurred 3 months after the first foscarnet treatment and further foscarnet treatment for two weeks, without hyperimmune globulin, resulted in remission with no relapse on follow up 1 month later.

Three patients with CMV colitis and one with a CMV ulceration of the oesophagus were given a continuous iv infusion of 0.08 mg/kg/min for 14 days (26). One patient showed a prompt response to therapy on two occasions but CMV colitis relapsed after each course. The second patient had a symptomatic response only, and the third colitis patient died of a non-CMV opportunistic infection while on treatment. The fourth patient was treated for 10 days but died of over-whelming pneumonitis on the 12th day. CMV was isolated in high titre from a post mortem lung biopsy.

Michon et al. (21 and pers. comm.) have reported the administration of 21 courses of iv foscarnet to 17 AIDS patients with CMV infections. The clinical outcome is summarized in Table 2 and the virological results in Table 3. These indicate beneficial effects of foscarnet on clinical outcome and viremia. Relapses occurred within two months in 8 of 9 treatments for retinitis. The findings suggest that foscarnet can inhibit the progression of CMV infections in AIDS patients but relapses are common.

Table 2. Clinical outcome after treatment of CMV infections in AIDS patients.

	N	Not assessable	Ineffective	Partial Improvement	Complete improvement
Retinitis	14	1	1	1	11
Colitis	3			3	
Pneumonitis	2		1	1	
Oesophagitis	1			1	
Cholangitis	1	1			
Total	21	2	2	6	11

Michon et al., pers. comm.

Similar results were reported by Le Huang et al. (29) and Robinet et al. (30) who found that 10 of 13 AIDS patients with CMV retinitis responded to iv foscarnet treatment. In 7 of these cases relapses of CMV retinitis were observed within 4 weeks after discontinuation of therapy.

Treatment of five AIDS patients with CMV retinitis with 14 days of foscarnet infusion has been reported by Walmsley et al. (31) who found stabilization or improvement of the disease. Urine cultures for CMV became negative during treatment in these patients.

Farthing et al. (32) have studied the outcome of treating 8 AIDS patients with iv foscarnet for CMV pneumonitis. Foscarnet was given as a continuous infusion for a minimum of 8 days. As shown in Table 4 all patients improved, 3 showing complete resolution of symptoms. The adverse effects noted were reversible and changes in haemoglobin concentration and serum creatinine returned to habitual levels after administration of foscarnet was discontinued.

Table 3. Virological results in AIDS patients with CMV treated with foscarnet.

	Number of samples	Viremia -	Viremia +
Before treatment	21	1	20
During treatment	25	24	1
First month after treatment	14	3	11

Michon et al., pers. comm.

Table 4. Treatment of CMV pneumonitis with iv foscarnet in AIDS patients

Patient No.	Treatment outcome: End of treatment	Treatment outcome: Follow-up	Adverse events
1	Improved	Died of untreated recurrence of CMV symptoms 5 months later	Hb decrease by 3.1 g/l
2	Improved	Worse after 2 weeks with development of PCP	Hb decrease by 1.4 g/l
3	Improved	Died of neurological disease 4 months later	Hb decrease by 2.6 g/l. Creatinine increase by 99 μmol/l
4	Improved	Died from intraabdominal sepsis 21 days later	Phlebitis
5	Pneumonia resolved	Well after 166 days	Thrombophlebitis
6	Pneumonia resolved	Well after 140 days	Thrombophlebitis
7	Pneumonia resolved	Well after 120 days	None
8	Improved	Died of untreated recurrence of CMV symptoms 21 days later	Thrombophlebitis Creatinine increase by 53 μmol/l

From Farthing et al. (32), with permission.

GENERAL ASPECTS ON THE TREATMENT OF CMV INFECTIONS WITH FOSCARNET

The clinical experience available on the efficacy of iv foscarnet against CMV infections is so far based on compassionate use in severely ill patients. The first patients given iv foscarnet were critically ill and an earlier institution of treatment might have been more favourable. However, the overall impression clearly suggests a therapeutic effect. The immunodepression is less severe in renal transplant patients and the best permanent effects seem to have been achieved among these patients. In more immunosuppressed patients, and especially in AIDS patients with CMV retinitis, relapses seem to be frequent. In bone-marrow transplant recipients, who are heavily immunosuppressed, the response to foscarnet therapy has been varying and difficult to interpret, especially in the CMV pneumonitis cases. In these patients the irradiation damage to the lungs, and immune reactions are likely to be important for the symptoms, rather than a damage due to lytic replication of CMV in the lungs. In contrast, AIDS patients with CMV pneumonitis have responded to foscarnet treatment (Table 4). Prolonged treatment until the immune system has recovered at least partly, combination with hyper-immune globulin, and intermittent prophylactic treatment might be ways to decrease the frequency of relapses.

Evaluation of efficacy of foscarnet in the CMV patients has been complicated due to the presence of several other infectious agents concomitantly but both clinical and virological data suggest beneficial effects.

Measurement of cellular deoxythymidine kinase activity in serum has been used as a marker for an ongoing CMV infection (33). A rapid decrease in serum thymidine kinase during foscarnet treatment has been observed in both renal and bone-marrow transplant recipients with CMV infections (15, 17). The decrease in TK suggests that foscarnet prevents a CMV-induced destruction of cells with release of cellular thymidine kinase into serum. It also suggests a rapid onset of the therapeutic effect.

The renal transplant patients have, on average, had higher serum levels of foscarnet than other patients. The foscarnet concentration necessary to inhibit most CMV isolates in cell culture, when using a high multiplicity of infection, is 150 μg/ml (500 μM) (27). In recently treated patients this level has been aimed at and the dosing of foscarnet has been increased. It is unclear how endogenous interferon will affect the *in vivo* requirements for antiviral activity but cell culture data indicate a synergistic effect. Dose-response studies *in vivo* are required to settle this issue and animal studies using rats infected with rat CMV might be helpful in determining the optimal dose regimen. Controlled clinical trials using iv foscarnet against CMV infections in immunosuppressed patients have been started.

INFECTIONS WITH HUMAN IMMUNODEFICIENCY VIRUS (HIV)

The inhibitory effect of foscarnet on HIV in cell culture has made it important to evaluate its clinical effect against HIV infections. Results from controlled studies are not yet available but 40 patients have been treated in open uncontrolled studies. Farthing et al. (34, 35) gave iv foscarnet for 3 weeks (bolus 20 mg/kg and then continuous infusion 0.05-0.11 mg/kg/min adjusted according to serum creatinine) to 11 patients with AIDS or AIDS-related complex (ARC). Virus isolations were performed at regular intervals up to 3 months post infusion on treated patients as well as on 4 untreated control patients. Virus production was registered as reverse transcriptase activity and syncytial induction in lymphocytes. Virus was recovered post treatment in 20 % of attempted isolations in foscarnet-treated patients and in 79 % of attempted occasions in the 4 untreated patients. This indicates an inhibitory effect of foscarnet on the possibility to isolate HIV, which could represent the amount of HIV present in lymphocytes. No significant trend was noted in OKT4 positive lymphocyte counts. This is, however, not surprising when considering the relatively short duration of treatment. One patient developed an axillary vein trombosis within 4 days of treatment, which was then discontinued. A further patient had a reversible renal dysfunction at the end of 3 weeks treatment.

Very similar results with respect to virus isolations have been seen by Bergdahl et al. (36) in 14 patients, mainly ARC, further indicating that foscarnet *in vivo* might reduce the amount of virus. These patients also experienced a decrease in the frequency of mucocutaneous herpes simplex infections, a decrease in night sweats and a decrease in the frequency of diarrhea for 4-16 weeks after foscarnet treatment. The decrease in frequency of herpes simplex symptoms is not surprising in view of the inhibitory effect of foscarnet against herpesviruses.

Gaub et al. (37) have studied the effect of iv foscarnet given for 6-21 days to 15 AIDS patients. In contrast to the results from Farthing et al. (32) and Bergdahl et al. (36), Gaub et al. (37) did not observe any decrease in the frequency of positive HIV isolations in relation to foscarnet treatment. However, 8 patients had detectable free HIV antigen in serum before therapy, and in 5 of these HIV antigen disappeared during therapy, but reappeared 3-23 weeks after therapy. No patient lost HIV antigen except during foscarnet therapy, and no patient became HIV antigen positive during foscarnet therapy. Foscarnet treatment was accompanied by improvement in some probably CMV related symptoms and the occurrence of positive CMV cultures decreased during therapy. Renal functional impairment was seen in 12 patients, possibly due to a reversible tubular damage. In patients where follow up data were available, s-creatinine normalized post treatment. Severe renal function impairment was only seen in patients who at the start of foscarnet therapy were chronically affected by the disease indicating that treatment should be attempted in less severe manifestations of HIV infection. The mean steady state serum concentration of foscarnet was 260 μM which is about twice the 50 % inhibitory concentration for CMV and at least 20 times the 50 % inhibitory concentration for HIV in cell culture (6).

Since the expected effect of foscarnet would be a prevention of HIV infecting new cells, a long-term foscarnet treatment seems necessary. This might be more effective if combined with interferon or other compounds with activity against HIV and possibly also if combined with immunomodulating compounds. These could both increase the rate of lytic destruction of HIV-infected cells and improve the cellular immunity. The effect of treatment with foscarnet could also depend on the state of HIV infection, whether asymptomatic, ARC or AIDS. Considering the efficacy of foscarnet against HIV in cell culture (38, B. Wahren, pers. comm.) the use of lower serum levels of foscarnet than those used above may be sufficiently effective and should be tried. Controlled studies on foscarnet against HIV infections are under way.

HEPATITIS B VIRUS INFECTIONS

There is yet no effective treatment for HBV infections, whether acute, fulminant or chronic. Five patients with fulminant hepatitis B and fulminant hepatitis B and D coinfection have recently been treated with foscarnet as reported by Hedin et al. (39). All patients received a bolus dose of 20 mg/kg BW over 30 min followed by a continuous infusion of 0.16 mg/kg/min for 2-10 days. An improvement in liver function and disappearance of coma was closely correlated to institution of foscarnet treatment. Although all patients recovered and the disease is known to have a grave prognosis with conventional treatment the number of patients in this study is still too small to make any conclusions about efficacy of treatment.

A severe, but reversible, increase in s-creatinine during foscarnet treatment was seen in two patients. It is known that HBV can cause glomerulonephritis and that the few patients surviving a fulminant HBV disease often experience renal damage or failure (40). One patient had a kidney biopsy taken and this showed the presence of IgM complexes, possibly supporting the idea that immune complexes were involved in the renal damage in this patient.

The successful treatment, with foscarnet, of a patient with fulminant hepatitis B has recently also been reported by Price et al. (41) and a few additional patients with this severe condition have responded well to foscarnet treatment (unpublished observations). The positive results from treating some patients with fulminant hepatitis B and the severity of this disease makes further evaluation of iv foscarnet against fulminant hepatitis B warranted. The reduction in duck HBV during treatment of chronically infected ducks with foscarnet (42) and the results from patients with fulminant hepatitis B (39, 41 and unpublished data) indicate that a clinical evaluation of foscarnet against chronic and acute hepatitis B might be worthwhile.

OTHER VIRAL INFECTIONS

Herpes simplex and varicella zoster virus infections in a few renal transplant recipients have been treated with iv foscarnet (13, 43). The clinical response was judged as good but the low number of patients does not permit any conclusions about efficacy.

SAFETY ASPECTS ON PARENTERAL FOSCARNET

As described above, the clinical experience of parenteral foscarnet is to the greater extent derived from the compassionate use of the drug in treatment of serious/life-threatening CMV infections in immunocompromised hosts i.e. transplant recipients and AIDS patients.

Evaluation of the safety of foscarnet per se in these patients is very complex due to several factors such as severe underlying diseases, severe viral infections, other concomitant bacterial or fungal infections, concomitant treatment with other potentially toxic drugs, dialyses, transfusions etc. and the lack of control or reference groups.

A summary has been published recently (23), specifically adressing the issue of foscarnet safety. The present summary comprises a total of 145 patients, whereof 30 bone-marrow and 40 renal transplant recipients, 62 patients with AIDS or ARC and 13 patients with other underlying immunodeficiency. The following observations, where foscarnet might have been associated, were reported. With respect to clinical chemistry, renal function impairment, evidenced by increases in s-creatinine, and in some patients progressing to severe renal dysfunction, has been observed. Fluctuations in s-calcium have also been observed, and in many patients a decrease of the haemoglobin concentration was recorded during the period of foscarnet administration. Administration of foscarnet might also have been associated with effects on the central nervous system. Local intolerance, associated with thrombophlebitis, has been reported after infusion of foscarnet stock solution (2-2.4 %) into peripheral veins. When administered in central iv lines the stock solution has been well tolerated.

Renal function impairment of varying degrees was observed in several patients. In toxicity studies in animals, foscarnet has induced morphological changes in the kidneys, which have been to a large extent reversible. Available follow-up data from patients who experienced renal function impairment in connection with foscarnet treatment, also suggest the renal function impairment in man to be of a reversible nature.

The decrease of haemoglobin concentration was often substantial but not of an acute nature. The interpretation of and association with foscarnet therapy of this observation is in many patients complicated by concomitant medication with myelosuppressive drugs, dialyses and transfusions.

With respect to s-calcium, increases as well as decreases have been reported, which in the latter case, in connection with tetany/paresthesia, necessiated administration of calcium.

The clinical observations, implying that foscarnet may have some effect on the nervous system comprise: tiredness, irritability, anxiety, hallucination, tremor, headache, meningism/encephalitis-like symptoms and status epilepticus. Some of these observations, e.g. hallucinations and tremor, have been closely related to inadvertent over-dosage resulting in extremely high plasma foscarnet levels (>400 μg/ml). Status epilepticus has been observed in a few patients. Although a concomitant HIV meningo/encephalopathy or other cerebral infections may have contributed to the status epilepticus in these patients, an association with foscarnet can not be excluded.

Among these patients the average daily foscarnet dose has ranged from 6-10 g, and the average treatment duration has been two weeks. However, in the individual patient, the variation in daily dose and treatment duration is considerable.

Although some of the above-mentioned adverse effects may have been related to the underlying condition, the concomitant infections or other medication, a relationship with foscarnet therapy can not be excluded. However, the possible adverse effects of high doses of parenteral foscarnet must be weighed against the potential benefits of the treatment in patients with severe infections due to CMV, HIV or HBV and the unfavourable prognosis of the disease expected in many of these patients.

REFERENCES

1. Helgstrand, E., Eriksson, B., Johansson, N.-G., Lannerö, B., Larsson, A., Misiorny, A., Norén, J. O., Sjöberg, B., Stenberg, K., Stening, G., Stridh, S., Öberg, B., Alenius, S. and Philipson, L. Science 201: 819-821, 1978.
2. Helgstrand, E., Alenius, S., Johansson, N.-G. and Öberg, B. In: Current Chemotherapy and Infectious Diseases 2 (Eds. J. D. Nelson and C. Grassi), American Society for Microbiology, Washington, D.C., 1980, pp. 1359-1361.
3. Öberg, B. Pharmac. Ther. 19: 387-415, 1983.
4. Eriksson, B. and Öberg, B. In: Antiviral drugs and interferon: The molecular basis of their activity (Ed. Y. Becker), Martinus Nijhoff Publishing, Boston, 1984, pp. 127-142.
5. Öberg, B. In: Human Herpesvirus Infections: Pathogenesis, Diagnosis and Treatment (Eds. C. Lopez and B. Roizman), Raven Press, New York, 1986, pp. 141-151.

6. Öberg, B. Viral Chemotherapy. Ed. D. Shugar. Pergamon Press. In press, 1987
7. Wallin, J., Lernestedt, J.-O., Ogenstad, S. and Lycke, E. Scand. J. Inf. Dis. 17: 165-172, 1985.
8. Lassus, A., Vainio, E., Kalimo, K. and Eriksson, B. Manuscript.
9. Sacks, S. L., Portnoy, J., Lawee, D., Schlech III, W., Aoki, F. Y., Tyrrell, D. L., Poisson, M., Bright, C., Kaluski, J. and the Canadian Cooperative Study Group. J. Inf. Dis. In press, 1987.
10. Barton, S. E., Munday, P. E., Kinghorn, G. R., van der Meijden, W. I., Stolz, E., Notowicz, A., Rashid, S., Schuller, J. L., Essex-Cater, A. J., Kuijpers, M. H. M. and Chanas, A. C. Genitourin. Med. 62: 247-250, 1986.
11. Gundersen, T. et al. Manuscript.
12. Lönnqvist, B., Klintmalm, G., Öberg, B., Gahrton, G., Lernestedt, J.-O., Lundgren, G., Ringdén, O., Robert, K.-H. and Wahren, B. Third International Symposium on Infections in the Immunocompromised Host, Toronto, 1984, Abstract 150.
13. Ahlmén, J., Wijnween, A.-C., Brynger, H. and Lycke, E. Scand. J. Urol. Nephrol. Suppl. 92: 41-44, 1985.
14. Apperley, J. F., Marcus, R. E., Goldman, J. M., Wardle, D. G., Gravetti, P. J. and Chanas, A. Lancet I: 1151, 1985.
15. Gronowitz, J. S., Larsson, A., Källander, C. F. R., Claesson, K., Sjöberg, O., Lernestedt, J.-O., Frödin, L. and Tufveson, G. Ann. Clin. Res. 18: 71-75, 1985.
16. Klintmalm, G., Lönnqvist, B., Öberg, B., Gahrton, G., Lernestedt, J.-O., Lundgren, G., Ringdén, O., Robert, K.-H., Wahren, B. and Groth, C.-G. Scand. J. Inf. Dis. 17: 157-163, 1985.
17. Larsson, A., Frödin, L., Tufveson, G., Larsson, E., Källander, C. F. R. and Gronowitz, J. S. Scand. J. Urol. Nephrol. 20: 75-76, 1986.
18. Ringdén, O., Wilczek, H., Lönnqvist, B., Gahrton, G., Wahren, B. and Lernestedt, J.-O. Lancet I: 1503-1504, 1985.
19. Ringdén, O., Lönnqvist, B., Paulin, T., Ahlmén, J., Bolme, P., Wahren, B., Wilczek, H. and Lernestedt, J.-O. Proc. 14th Int. Chemother. Meeting, Japan, 1985.
20. Wahren, B., Öberg, B., Wiklund, M., Lönnqvist, B., Ljungman, P., Ringdén, O., Lernestedt, J.O. and Gahrton, G. Excerpta Med. Int. Congress Series, Elsevier, 67-71, 1985.
21. Michon, C. P., Katlama, C., Le Hoang, P., Leport, C., Rowzioux, C., Matheron, S. et al. Int. Conf. on Acquired Immunodeficiency Syndrome (AIDS), Paris, 1986.
22. Modig, J., Hedstrand, U. and Tufveson, G. Manuscript.
23. Ringdén, O., Lönnqvist, B., Paulin, T., Ahlmén, J., Klintmalm, G., Wahren, B. and Lernestedt, J.-O. J. Antimicrob. Chemother. 17: 373-387, 1986.
24. Ringdén, O., Lönnqvist, B., Paulin, T., Ljungman, B. Wahren, B. and Lernestedt, J.-O. J. Cell Biochem. Suppl. D: 259, 1986.
25. Singer, D. R. J., Fallon, T. J., Schulenburg, W. W., Boyd, A. S., Cohen, J. and Williams, G. Proceedings EDTA Congress, 1986.
26. Weber, J. N., Thom, S., Unwin, R., Forster, S., Jeffries, D. J., Boylston, A. and Pinching, A. J. Gut. In press, 1987.
27. Åkesson, A., Lernestedt, J.-O., Ringdén, O., Lönnqvist, B. and Wahren, B. Bone marrow transplantation. In press, 1986.

28. Singer, D. R. J., Fallon, T. J., Schulenburg, W. E., Williams, G. and Cohen, J. Ann. Int. Med. 103: 962, 1985.
29. Le Huang, P., Dhermy, P., Robinet, M., Holgado, P. and Matheron, S. Second World Congress on Sexually Transmitted Diseases (STD), Paris, 1986. Abstract.
30. Robinet, M., Matheron, S., Katlama, C., Rozenbaum, W., Leport, C., Dhermy, P., Gaudric, A., Sterkers, M., Coulaud, J. P., Vilde, J. L., Gentilini, M. and Rousselie, F. Second World Congress on Sexually Transmitted Diseases (STD), Paris, 1986. Abstract.
31. Walmsley, S., Chew, E., Fanning, M., Coates, R. A., Salit, I. S., Shepherd, F. A., Rachlis, A. and Read, S. E. ICAAC, 1986. Abstract 568.
32. Farthing, C., Anderson, M. G., Ellis, M. E., Gazzard, B. G. and Chanas, A. C. J. Med. Virol. In press, 1987.
33. Gronowitz, J. S., Källander, C. F. R., Diderholm, H., Hagberg, H. and Pettersson, U. Int. J. Cancer 33: 5-12, 1984.
34. Farthing, C., Dalgleich, A. G., Clarke, A., McClure, M., Chanas, A. and Gazzard, B. Aids Research, submitted 1987.
35. Farthing, C. F., Dalgleish, A. G., Clarke, A. L., Chanas, A., McClure, M. and Gazzard, B. G. Int. Conf. on Acquired Immuno-deficiency Syndrome (AIDS), Paris, Abstract 87: S 14 e, 1986.
36. Bergdahl, S., Albert, J., Biberfeldt, G., Böttiger, B., Halvarsson, M., Julander, I., Lernestedt, J.O., Morfeldt-Månsson, L., Åsjö, B. and von Stedingk, L. V. Manuscript.
37. Gaub, J., Pedersen, C., Poulsen, A.-G., Mathiesen, L. R., Ulrich, K., Lindhardt, B. O., Faber, V., Gerstoft, J., Hofmann, B., Lernestedt, J.-O., Nielsen, J. O., Nielsen, C. M. and Platz, P. Manuscript.
38. Hartshorn, K. L., Sandström, E. G., Neumeyer, D., Paradis, T. J., Chou, T.-C., Scholey, R. L. and Hirsch, M. S. Antimicrob. Ag. Chemother. 30: 189-191, 1986a.
39. Hedin, G., Weiland, O., Ljunggren, K., Strömberg, A., Nordenfelt, E., Hansson, B.-G. and Öberg, B. In: The Hepatitis Delta and its Infection (Eds. M. Rizzetto et al.), A. R. Liss Inc. In press, 1986.
40. Wyszynka, T., Jung, H., Madalinski, K., Morzycka, M. Int. J. Pediatr. Nephrol. 5: 147-158, 1984.
41. Price, J. S., France, A. J., Moaven, L. D. and Welsby, P. D. Lancet II: 1273, 1986.
42. Sherker, A. H., Hirota, K., Omata, M. and Okuda, K. Gastroenterol. 91: 818-824, 1986.
43. Blohmé, I., Ahlmén, J., Jeansson, S., Ranch, T. and Brynger, H. Transplantation Proceedings XVI: 1672-1674, 1984.

15

PROPHYLAXIS AND TREATMENT OF RHINOVIRUS INFECTIONS

D. A. J. TYRRELL, W. AL-NAKIB

MRC Common Cold Unit, Harvard Hospital, Salisbury SP2 8BW UK

ABSTRACT

Considerable progress has been made in the development of some active antirhinovirus agents. Thus, various interferons (IFNs) including recombinant human IFN alpha and beta were found active and protected volunteers from experimental rhinovirus infections. However, clinical studies indicated that IFNs need to be given in large doses and over frequent intervals and therefore were only useful in short-term prophylaxis (4-7 days). Prolonged administration (21-28 days) resulted in local toxicity manifested by nasal irritation, inflammation, ulceration and bleeding.

Field studies clearly showed that IFN-alpha-2 can also be very effective in preventing naturally occurring rhinovirus infection (88% protective efficacy) among those family members who received medication as compared with those who received placebo. It remains to be seen whether the same regimen will also prove useful in protecting individuals who may be at particular risk e.g. with obstructive airways disease, from infection by rhinoviruses.

In contrast, despite progress in developing some very active synthetic antirhinovirus compounds (MIC < 0.25 µg/ml), none of these substances were found effective in preventing experimental or, in the case of the compound enviroxime , naturally occurring rhinovirus infection in man.

This chapter discusses the prospect of developing more potent and perhaps less toxic substances and the possible

DeClercq, E. (ed.), *CLINICAL USE OF ANTIVIRAL DRUGS.*

application of these in the prophylaxis and therapy of rhinovirus infections.

INTRODUCTION

This is not the place to give a long review of the course of research and development of this field but a short section is appropriate, if for no other reason than that it is well known that acute respiratory infections, mainly common colds, occur frequently throughout the world - the average individual has from 2 to 5 each year, the exact count depending on the method of ascertainment used and the age, sex and circumstance. Furthermore, our inability to prevent or treat these conditions is the point of many jokes and more serious complaints about modern medicine. The syndrome of the common cold is induced by infection with a range of viruses - including coronaviruses, parainfluenza and influenza viruses, respiratory syncytial virus, and others, but the most frequent cause is one of the 100 or more serotypes of rhinoviruses. In the most intensive studies rhinovirus infection has been detected in about half of the cases, so an effective drug against rhinovirus infection could have a significant impact on the whole clinical problem.

It has to be admitted that there is still no generally useful specific medication for respiratory infections - apart from amantadine for influenza A virus infections only treatment for symptoms or complications are on the market. However, in the last three decades very considerable progress has been made and certain doubts about whether drug prophylaxis or treatment of respiratory virus infection would ever be possible have been set at rest.

For instance, it is now known that there are molecules which have a powerful inhibitory effect on virus replication so that it is possible to introduce effective amounts of an antiviral drug into the respiratory tract. Furthermore, many of these molecules are remarkably non-toxic in vitro and in vivo. There are nevertheless serious problems in the delivery of these molecules. Amantadine and its congeners may be given by mouth and are concentrated in the lung and respiratory secretions but most molecules do not behave in this way: if oral treatment is impossible, local administration may get effective amounts of drug onto the respiratory mucosa - for example interferons were shown in 1973 to be effective against rhinoviruses by nasal spray and ribavirin has recently been used against RSV and influenza A as an aerosol generated by a modified Collison apparatus.

Because of the difficulties most work in the last two or three decades has aimed to prevent infection - it was felt this was the easier task and one which should be undertaken first before attempting to treat an infection which was already under way. Indeed there were those who felt that symptoms and signs developed so late in the course of common respiratory infections that even if a potent antiviral drug were to become available it would not be possible to derive any benefit from it, simply because one could never administer it soon enough. This view was disproved relatively early on when it was shown that amantadine had a measurable effect on the course of influenza A infections in patients given the drug soon after the onset of disease. Subsequently this has been confirmed, for instance by showing that inhaled ribavirin improves the course of influenza infections in adults and of RSV infections in children.

We would be the first to admit that for the average patient with a rhinovirus cold or for his clinician there is still practically nothing which can be done in the way of specific prevention or treatment. However we believe that effective and safe antirhinovirus drugs are in principle possible and for years now we have been making steady progress toward certain thresholds. When we cross these it will be appreciated that there really is some useful chemoprophylaxis and chemotherapy. Amantadine has been known to be effective for 20 years, but like the early antiherpes drugs it was not sufficiently effective or free of side effects to commend itself to practitioners. Now that better derivatives are becoming available and rapid diagnosis is easier we are approaching the point, already reached with antiherpes drugs, that their use is a self-evident part of good clinical practice. Something similar may well happen with rhinoviruses.

HISTORY

Since the early 1960s rhinoviruses have been included in screening programmes for antiviral activity in pharmaceutical research laboratories. As a result compounds have been identified which have antirhinovirus activity and are relatively harmless to cells in tissue culture. Some compounds were selected for clinical trial in prophylaxis against experimental rhinovirus infections but all the earlier trials gave negative results (1). In these early trials the *in vitro* activity as measured by the MIC was in the region of 4-10 μg/ml. (Table 1).

Table 1. The *in vitro* MIC and cytotoxicity of some of the early antirhinovirus compounds (Modified from Reed et al (5))

Compounds	MIC (μg/ml) for HRV-9	Cytotoxicity (μg/ml)*
ICI 73,602	> 4	4
SKF 40491	> 4	10
GL R9-338**	> 4	> 20
RP 19826**	> 4	> 32

* for HeLa cells

** in MRC-5, R9-338 was cytotoxic at 8 μg/ml while ICI 73,602 was cytotoxic at 16 μg/ml

The first synthetic compound to show what appeared to be clinical effectiveness was enviroxime, and it has an MIC of <1 µg/ml, which seemed unusually low at the time it was first studied. Since then compounds of higher activity have been discovered. Indeed, as can be seen from the data presented in Table 2 the majority of the more recent synthetic compounds are active at concentrations of <0.25 µg/ml.

Table 2. In vitro MIC and cytotoxicity of some of the more recent antirhinovirus compounds

Compound	MIC (µg/ml) for HRV-9	Cytotoxicity (µg/ml)
Dichloroflavan (Wellcome)	0.25	$\geq$ 12.5
Ro-09-0410 (Roche)	0.13	$\geq$ 6.25
Enviroxime (Eli Lilly)	0.06	$\geq$ 6.25
62025 (Janssen)	0.0078	$\geq$ 25

These were conducted in Ohio HeLa cells

In addition, the majority of interferons tested in our laboratory are active at concentrations of <8 IU/ml in inhibiting the replication of rhinovirus infection in cell cultures (Table 3). Early compounds had no effect on the virus particle itself, and were presumed to act by interfering with virus replication. More recently studied compounds, such as dichloroflavan and the chalcones inactivate the virus particle directly, whether or not they interfere with replication in the cell (2, 3, 4).

Table 3. *In vitro* MIC of interferon for rhinoviruses

Type of IFN	MIC* (IU/ml)
HuIFN-α2 (IFN-alfa-2b)	2
HuIFN-β	2
HuIFN-β x 401	4
HuIFN-γ **	8

* IFN assay conducted in WISH and MRC-5 cells

** MRC-5 cells are resistant to HuIFN-γ

The experiments that showed that intranasal sprays of leucocyte interferon (IFN) prevent rhinovirus colds (6) was an important landmark, both in indicating that local administration of an antiviral could be effective and also in suggesting that large amounts of antiviral activity (a total dose of 14 MU) as measured in *in vitro* units had to be introduced into the nose to produce a significant effect on the course of infection. Judged by this standard all previous and many subsequent trials had used relatively little antiviral activity.

The compounds selected for clinical trials were all screened for toxicity in tissue cultures, by systemic administration to animals, and then locally. Those that were acceptable were tested for tolerance in man. Again there has been a favourable trend in that more recent compounds show very low toxicity in tissue cultures and animals (Table 2), in other words the *in vitro* therapeutic index has improved. There has however been difficulty in preparing suitable formulations of the active non-toxic compounds. Not only was it the aim to get the maximum amount of drug into a stable solution or suspension but also to find a vehicle that was well tolerated, both for the comfort of the volunteer and also to avoid local irritation and excess nasal secretions which could be confused with the effect of the virus. With this went the difficulty of finding a way of delivering to the nose, but because of the success of interferon the general policy has been to produce a rather coarse spray which would be deposited in the nasal epithelium and would not enter the lower respiratory tract.

There were quite early hints that, as in other areas of clinical pharmacology, the tissue distribution and pharmacokinetics of compounds could be of great importance. An experiment with fusidic acid (7) showed that after oral administration the plasma contained amounts of drug that were inhibitory for coxsackie virus A21 - an enterovirus that behaves much like a rhinovirus: nevertheless it was ineffective in preventing colds due to that virus, and it was found that the concentration of drug in nasal secretion was much lower than that in the plasma - and very similar results were seen after giving dichloroflavan or chalcone by mouth (see below). The clearance of drugs from the nasal mucosa may be important and in experiments with interferon, an antihistamine was given as well as the IFN, but the effect on drug clearance or prophylactic effect was not specifically determined (8). It was known that the normal respiratory tract could clear drugs and other materials sprayed into it but it is only recently when sensitive and convenient methods of assay have become available so that the clearance of model solution (9) or drug from the nose has been measured quantitatively. In spite of this many important parameters still cannot be measured. We do not know how much drug gets close to and binds to cells, how much of the relatively insoluble drugs are in solution, and if in solution are free or protein bound and how long drug stays cell associated before it is released or metabolised.

Thus the initial emphasis of research on antivirals was on finding substances with high antiviral potency and low toxicity, but has later moved to finding other important characteristics of the molecule such as more advantageous tissue distribution and long half-life in the cell.

PRESENT METHODS OF EVALUATION

The present methods of selecting and testing potential antirhinovirus drugs are based on this historical experience and the techniques which have been adapted for the development of therapeutic agents in general.

Screening for activity

In initial experiments a candidate compound is added to the medium of tissue cultures which are rhinovirus susceptible and these are then challenged by the addition of a laboratory strain of rhinovirus. If the compound is active then virus replication is prevented and this is usually detected by inhibition of the cytopathic effect, which can be detected by microscopic examination of living cultures, or by naked eye examination of stained cultures, for example for the presence of a cell sheet in the wells of a microtitre plate or the presence of plaques in a plate assay.

Any inhibition needs to be supported by further experiments for instance using viruses of a range of serotypes, and other cell systems, in particular cells like those of the human host such as diploid fibroblast cell strains or organ cultures of human respiratory epithelium. There are examples of compounds with large differences in activity against different serotypes and others which are active in HeLa cells but not in fibroblasts or respiratory epithelium.

Animal studies

In the case of other viruses, such as influenza or herpes simplex,it would be normal to test candidate compounds in an animal model. These tests are particularly valuable as the results depend on important properties of the molecule such as how it is metabolised, what its half-life may be and whether it reached infected cells in adequate concentrations. Although experimental infections have been produced in primates and even in mice there is no satisfactory animal model of human rhinovirus infection so these experiments cannot be done. Thus it is important to make direct chemical studies of how well it is absorbed, whether it is metabolised or not and if it is, whether the metabolites are active or inactive. We recommend in particular that the concentration of compound in respiratory secretions should be studied. On the basis of present data, if the compound does not enter the secretions in inhibitory concentrations, it is unlikely to be active in man by oral administration.

Animal studies are also needed to assess the toxicity of a compound, the lethal dose, acute and chronic toxicity, target organs and carcinogenicity need to be determined and if the results are satisfactory teratology is needed if women are to be included in the volunteer studies. If the drug is to be given intranasally a preliminary test of blood tolerance in an animal is needed.

Human tolerance and pharmacokinetics

From the antiviral assays and other studies it may be decided to proceed to studies with man. In this case one needs to establish human tolerance for either intranasal or oral administration. It is usual to use an escalating design and to monitor the subjects clinically and by biochemistry, haematology, etc. Data on blood concentrations and excretion are also obtained. If the compound is to be given intranasally the intranasal pharmacokinetics can also be determined (Fig. 1) by washing out the nose at different intervals after a dose. There is evidence that some individuals may clear more rapidly than others.

Prophylaxis in human volunteers

In general the initial trials in this Unit are designed in such a way as to have the maximum chance of observing a protective effect. The compound is given for 1 or 2 days before the virus is administered. The virus is chosen to have the highest possible *in vitro* sensitivity to the compound and is given in a low dose. The trials are conducted in isolation and are placebo controlled and double blinded. Samples are collected daily for virus isolation and paired sera for antibody assay. The outcome is assessed not only by classifying individual subjects as having colds

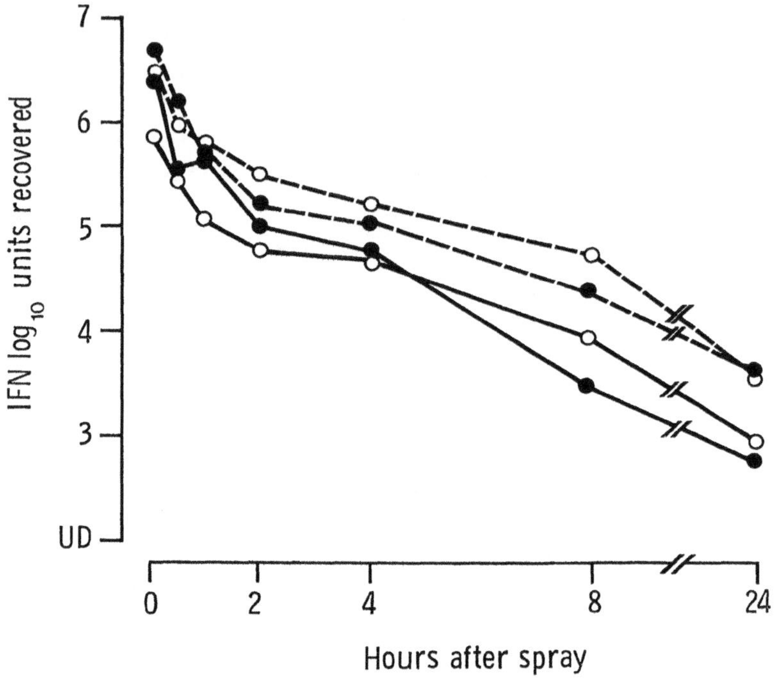

Fig. 1. Pharmacokinetics of IFN- alpha. Mean estimated (-) and immunoreactive (----) HuIFN α 2 recovered after single doses by physician- (●) and self- (O) administered sprays (five observations at each point).

or being infected, but also by calculating a score including contributions from all the clinical signs and symptoms and recording the nasal secretion weight. Nevertheless power calculations show that to demonstrate, for example, a reduction of 50% in the number of colds with $p < 0.05$ in 90% of trials one needs to have two groups of about 30 volunteers each. It is important to plan trials in such a way that a negative result can be regarded as conclusive, so in addition it is normal to use what seems to be the maximum tolerated dose of compound and a high frequency of administration. If the compound proves active, as happened for instance with IFN α, it is possible then to undertake further studies to find the minimum effective dose and the most convenient method and schedule of administration.

Therapy in human volunteers

If a compound is effective prophylactically it can then be assessed for any therapeutic effect. Groups of volunteers are inoculated and when individuals develop symptoms of a cold they are allocated at random to receive drug or placebo. The outcome is then assessed as in the prophylactic studies, but the only changes that can be anticipated are reductions in the scores and in the amount or duration of virus shed.

Community studies

If the possible value of an antiviral has been assessed in volunteers the crucial test is to discover whether it is effective in subjects living in normal families in the community and catching colds in the natural way. Such tests have so far been made only with IFNs (except for a study with enviroxime). In the first place IFN was given regularly by intranasal spray and this revealed that on prolonged administration IFNs were more toxic than was originally thought and produced unacceptable nasal symptoms. Later series showed that daily IFN or placebo sprays could be given to family members to take for a week after another member of the family developed a cold. Analysis of daily symptom records showed that the number of colds was roughly halved, but these results were further refined by doing virologic diagnostic tests on all the subjects with colds. These indicated that the drug was reducing the frequency of rhinovirus colds, but not of those due to other viruses. Details of these studies will be reviewed below.

RECENTLY STUDIED SUBSTANCES

Interferons

Studies on prophylaxis in experimental rhinovirus infection. It was in 1973 that Merigan et al (6) established that partially purified leucocyte interferon when given as repeated nasal sprays and at a total dose of 14 MU (14×10^6 IU) over four days protected some 80% of volunteers from illness as compared with those who received placebo following intranasal challenge with human rhinovirus 4 (HRV-4).

The result of this experiment was particularly significant since it demonstrated that rhinovirus colds can be prevented with interferon. However, relatively large doses of interferon had to be given and at frequent intervals. Unfortunately, this experiment could not be repeated since interferon was not available in adequate quantities. However, by 1980 a number of highly purified IFNs did become available and, with the advent of recombinant DNA technology, it was possible to clone and produce large quantities of highly purified IFNs. These were now available for further clinical evaluation in man as possible prophylaxis against colds caused by rhinoviruses.

Table 4 shows that the majority of interferons evaluated were found, in fact, to be very efficacious in reducing the incidence of rhinovirus colds, virus shedding and even the number of volunteers showing serological evidence of infection. Thus, Scott et al (10) demonstrated that human leucocyte interferon (IFN-α(Le)), purified by affinity chromatography using a monoclonal antibody, when given at a very high total dose of 90 MU over four days resulted in 100% protection against rhinovirus challenge in that none of the volunteers who received interferon developed significant colds (i.e. mild, moderate or severe).

Scott et al (11) then reported the results of another interferon trial, this time with a high-titre interferon alpha-2 (rIFN-alfa-2b) prepared by recombinant DNA technology. They used a dose similar to that previously used for the purified IFN-α(Le) (90 MU) and again they were able to demonstrate a 100% protection against illness.

Both these studies confirmed that the early antiviral effect detected with the partially purified interferon (6) was not due to some impurities but rather was due to the effect of the IFN itself since both purified IFN and that prepared by a totally different method, recombinant DNA technology, protected against challenge with a virulent rhinovirus strain.

Table 4. Summary of efficacy data on interferons in the prevention of experimental rhinovirus infections in man.

Type of IFN (Ref)	Total dose (MU)	Dose (MU/day)	% reduction relative to placebo		
			Significant colds*	Virus shedding	Seroconversion or Ab rises
IFN-α (Le) (6)	14	9 x 0.36 for 4 days	80	76	65
IFN-β (16)	<40	9 x <4 for 4 days	22.6	0	0
Purified IFN-α (Le) (10)	90	3 x 7.5 for 4 days	100	31	31
rIFN alfa-2b (11)	90	3 x 7.5 for 4 days	100	50	56
IFN-α (Le) (8)	1-4	1-2 x 1 or 3 for 1 to 2 days	38-45	0	1.2
rIFN-alfa-2a (14)	40	10 x 1 for 4 days	89	67	4.8
rIFN-alfa-2b (13)	182	4 x 11.4 for 4 days	100	78	100
	212	42.8 x 1 for 5 days	75	64	28
IFN-alfa-N1 (18)	35	3 x 2.7 for 4 days	100	25	31
rHuIFN-β (17)	26	3 x 2 for 4 days	66.6	41.7	18.2

*Classed as Mild, Moderate or Severe

In the studies of Scott et al, high doses of IFN were deliberately given since the optimal dose necessary to establish a protective effect was not known. Previously, Aoki and Crawley (9) showed that the initial half-life clearance of radiolabelled albumin given intranasally was in fact only 20 minutes. These data, therefore, suggested that substances applied topically in the nose may be removed rapidly by the mucociliary clearance mechanism of the nasal epithelium and hence large concentrations of interferon would have to be given topically and perhaps over relatively frequent intervals if sufficient antiviral activity were to be maintained in the nasal mucosa. Indeed, studies by Greenberg et al (8) demonstrated a significant prophylactic effect against rhinovirus illness and infection when low doses (1 or 3 MU) of leucocyte interferon were given as a saturated cotton pledget inserted in the nose for one hour to increase the contact time with the nasal mucosa. Volunteers in these studies were pretreated with a dose of an antihistamine (chlortrimetron, 4 mg) prior to virus challenge in order to reduce the nasal secretion of mucus.

Despite the statistically significant beneficial effect, it was clear from these studies that the degree of effectiveness was still not optimal. In order that an effective dosage regimen for prophylaxis against rhinovirus infection by intranasal administration of IFN be established, Phillpotts et al (12) conducted a detailed study to establish the quantity and interval when interferon should be given to protect against virus challenge. They found that a medication regimen of about 2 - 4 MU of recombinant IFN-alfa-2b per day was effective in protecting against a rhinovirus challenge. However, it was clear that small dosages would have to be given three times daily if protection against natural virus infection were to be maintained since the protective effect seems to wear off a few hours after a dose (12). Hayden and Gwaltney (13) also showed that 11.4 MU of recombinant IFN-alfa-2b given four times a day for four days was more effective than 42 MU per day given as a single dose in preventing rhinovirus infection, virus shedding and rhinovirus specific colds (78%, 78% and 100% vs 45%, 64% and 75% respectively). Samo et al (14) demonstrated that recombinant IFN-alfa-2a given at a dosage of 10 MU four times a day for four days protected 89% of volunteers against illness when challenged with human rhinovirus type 13 (HRV-13). Samo et al (15) later showed that a dose of 2.4 MU of recombinant IFN-alfa-2a per day for four days resulted in about 60% efficacy in protecting against rhinovirus challenge whereas a dosage regimen of 0.7 MU per day for four days did not protect against virus challenge.

All these studies have, therefore, highlighted, if anything, the difficulty of finding the correct dosage level that will offer protection against challenge with a rhinovirus infection and, at the same time, have minimal toxicity. Clearly these studies indicated that concentrations in the range of 2 - 4 MU per day are protective and that frequent application, e.g. 3 - 4 times daily, are advantageous to maintain the antiviral activity.

Attempts to assess the antiviral activity of other interferons were made by Scott et al (16). In double-blind placebo controlled trials they assessed the prophylactic efficacy of an interferon prepared from human fibroblasts (HuIFN-β). The results of these trials were negative (Table 4) but it was not clear whether this lack of protective effect was due to lack of antiviral activity *in vivo* of the interferon or whether it was due to the instability of these molecules with consequent loss of activity during the trial.

In 1986, we had the opportunity to evaluate a recombinant HuIFN-β_{ser} as prophylaxis against challenge with HRV-9 and 14 in human volunteer studies. We gave it at a total dose of 26 MU over four days and were able to show that HuIFN-β, like HuIFN-alpha, was also active *in vivo* against rhinoviruses and protected volunteers from experimental rhinovirus illness and infection (17) (Table 4).

Finally, another interferon that has recently been

evaluated is the lymphoblastoid interferon. It was administered intranasally in a total dose of 35 MU over five days and this resulted in total protection against challenge with HRV-9 and 14 (18).

Studies on prophylaxis against naturally occurring rhinovirus infection. Experiments on interferon as prophylaxis against experimental rhinovirus infection have demonstrated that various interferons are effective in preventing a rhinovirus common cold. However, they have to be given in relatively large doses and at frequent intervals. The situation with the prevention of naturally occurring rhinovirus colds is even more complex. For example, it is difficult to predict when an individual is likely to acquire a cold and hence interferons would have to be given for prolonged periods of time and probably at relatively high doses and frequent intervals, e.g. 3 - 4 times daily, in order to remain effective at preventing infection for 12 - 24 hours as suggested by Phillpotts et al (12). Indeed, when IFN-alfa-2b was administered at 5 MU once daily for 28 days (19) or 10 MU twice daily for 21 days (20) it did not prevent illness although it had a dramatic effect on preventing infection as compared with placebo (Table 5). The proportion of individuals who developed colds was in fact higher among the interferon than among the placebo treated group, but this was attributed mainly to the local toxicity, e.g. nasal irritation, stuffiness and bleeding. Hayden et al (21) conducted a tolerance study in which 52 healthy volunteers were randomly assigned IFN-alfa-2b (8.4 MU/day) or placebo for 28 days. Of those who received IFN they detected signs of mucosal irritation and/or symptoms in 23%, moderate or marked epithelial acute inflammation with ulceration in 19% and pronounced submucosal lymphocytic and mononuclear infiltrates in 58%.

Although these histological abnormalities resolved within eight weeks of terminating medication, they concluded that intranasal administration of IFN at this dosage was not a feasible strategy for prophylaxis of respiratory viral

Table 5. Prophylactic use of intranasal interferon in the prevention of naturally occurring rhinovirus common colds.

Type of IFN (Ref)	Dose (MU)	Dose (MU)	% reduction relative to placebo	
			Significant colds	Virus shedding
IFN-alfa-2b (19)	5	Once daily for 28 days	0*	95
IFN-alfa-2b (20)	10	Twice daily for 21 days	0*	100
IFN-α (Le) (22)	I 0.4	Twice daily for 28 days	0*	0
	II 1.5		0*	-
	III 4.4		0*	-
IFN-alfa-2b (24)	1	Twice daily for 28 days	0*	87
IFN-alfa-2b (23)	1.25	Twice daily - trial discontinued after 12 days	0*	-
IFN-alfa-2b (25)	1.5	Twice daily for 28 days	-*	76

* side-effects complicating illness due to rhinovirus

infections. Scott et al (22) conducted further field studies to assess tolerance and prophylactic effect of highly purified IFN-α (Le) at dosages already established as protective against rhinovirus challenge (4.4 MU per day) as well as at lower dosage levels. Thus, they gave IFN-α (Le) at 4.4 MU, 1.52 MU and 0.44 MU or placebo to 68 students who participated in a 28 day study. Interferon at the lower doses did not protect against virus infection and/or colds. However, at the higher dosage level (4.4 MU per day) although it prevented infection and apparently colds, significant unwanted effects, mainly local inflammation, were detected. Furthermore, at this minimal effective dose, when given for a period of more than a week, fewer than half the volunteers were able to continue taking it for a month without significant discomfort. Further confirmation of these findings came from further studies by Hayden et al (23) with IFN-alfa-2b. They found that this IFN, even when given at a low dose of 1.2 MU twice daily, resulted in considerable side effects. In this study medication was stopped after 12 days because of the frequent occurrence of nasal irritation manifested by blood tinged nasal mucus (44% of the IFN group versus 15% placebo.) Furthermore, IFN did not decrease the occurrence of illnesses as compared with placebo although it seems to reduce the incidence of laboratory documented infections.

Douglas et al (24) in Australia conducted further clinical evaluation of IFN-alfa-2b, this time with a dosage level of 1 MU twice daily for 28 days. They found at the conclusion of their study that 11% of the IFN treated group and 2% of the placebo had experienced frequent or constant nasal discomfort and that 20% of IFN treated individuals versus 1% of placebo experienced blood tinged mucus. Although the IFN treated recipients were shown to have significantly fewer rhinovirus-associated episodes than those who received placebo, no clinical benefit was demonstrated. This had led them to conclude that IFN-alfa-2b would have to be given at a dosage level of not less than 2 MU but not greater than 10 MU and preferably only for short term prophylaxis rather than one month.

Monto et al (25) conducted field trials with the same interferon (IFN-alfa-2b) given over a period of four weeks. Four hundred individuals participated in these trials and received either 1.5 MU of IFN or placebo twice daily. They found that, although this interferon prevented rhinovirus infection with an efficacy of 76%, the incidence of side effects, including blood tinged mucus, in 49% of those who received IFN as compared with 16% of those who received placebo, occurred at an unacceptable rate. They suggested, as others had done, that interferon should only be used for short term prophylaxis to protect individuals who might be at particular risk, e.g. those with chronic bronchitis or asthma. Indeed, Herzog et al (26) investigated the prophylactic effect of a low dose (1.5 or 0.3 MU) of intra-nasal recombinant leucocyte interferon (IFN-alfa-2a) in 147 families. Treatment was initiated within two days after the appearance of an index case in the family and medication was continued daily for five days. Although local tolerance was excellent, low doses were not found to have a prophylactic effect and both regimes had no therapeutic effect on an established common cold. However, prophylaxis with 1.5 MU did shorten the duration of the cold (median of two days vs four days in the placebo). Furthermore, medication with 1.5 MU also reduced the median total clinical score when compared with placebo (10.5 vs 30).

In placebo-controlled double-blind studies conducted in both Australia (27) and the USA (28) it has recently been established that IFN-alfa-2b can be used effectively in preventing a rhinovirus cold in the family setting. In these studies, once the index case was identified as having developed the signs and symptoms of a cold, other members of the family who were in contact with the index case received 5 MU per day for seven days. Medication was initiated within forty-eight hours of onset of illness in a family member. The effect was quite dramatic in that both studies, although independently conducted, demonstrated about 88%

efficacy in preventing colds caused by rhinoviruses and 33 - 43% fewer days with nasal symptoms relative to placebo. A summary of these findings is shown in Table 6.

Table 6. Prevention of natural colds in the family setting: summary of data from field trials with interferon alpha-2.

Reference	Dose MU	% fewer* days with nasal symptoms	% fewer episodes of colds*		
			Total	Rhinovirus	Other respiratory viruses
(27)	5 per day for 7 days	33	41	86	Nil
(28)	5 per day for 7 days	43	39	88	Nil

*relative to placebo treated group

Used in this way IFN was well tolerated and only minor side effects were recorded. These family studies, therefore, have now paved the way for further evaluation of these IFNs, this time in the prevention of rhinovirus colds among individuals with chronic obstructive airways diseases, e.g. chronic bronchitis and asthma. Therefore, it will be of particular interest to see the outcome of such trials. If they are successful, interferons will have found a clinical application in the prevention of the common cold among those who may be at particular risk of developing lower respiratory tract complications.

Synthetic antirhinovirus agents

Over the past eight years or so a number of very potent antirhinovirus compounds have been synthesised. These compounds (Fig. 2) are active against rhinoviruses at MICs of <0.25 µg/ml and some at concentrations as low as 0.0078 µg/ml (Table 2). Furthermore, their apparent therapeutic

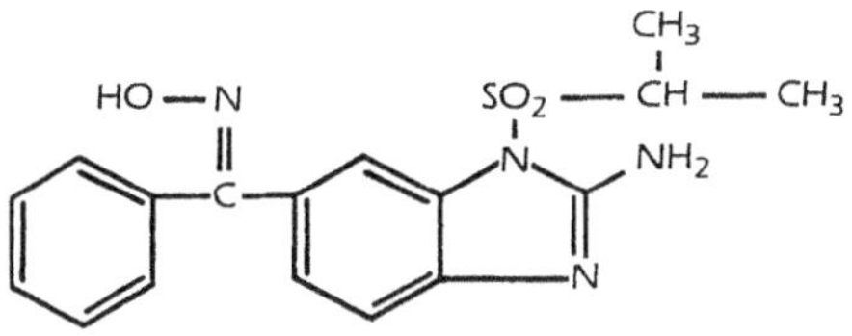

[2-amino-1-(isopropylsulphonyl)-6-benzimidazole phenyl ketone oxime (enviroxime)

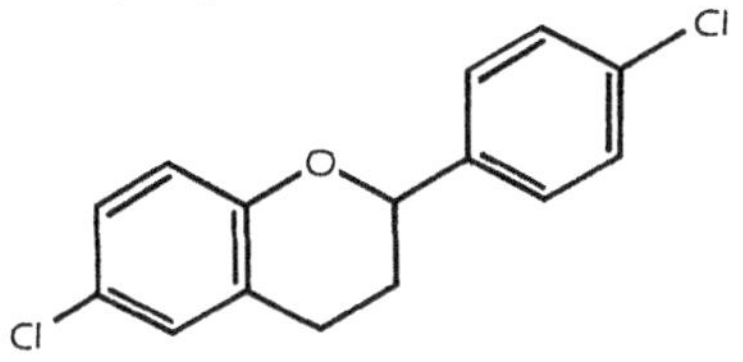

4',6-Dichloroflavan

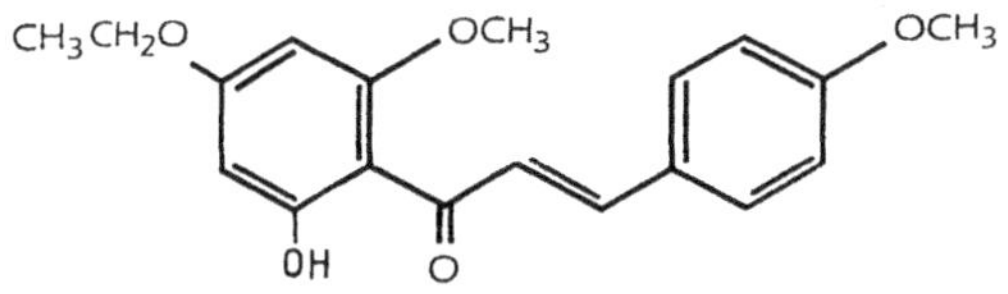

4'-ethoxy-2'-hydroxy-4,6' dimethoxychalcone (Ro-09-0410)

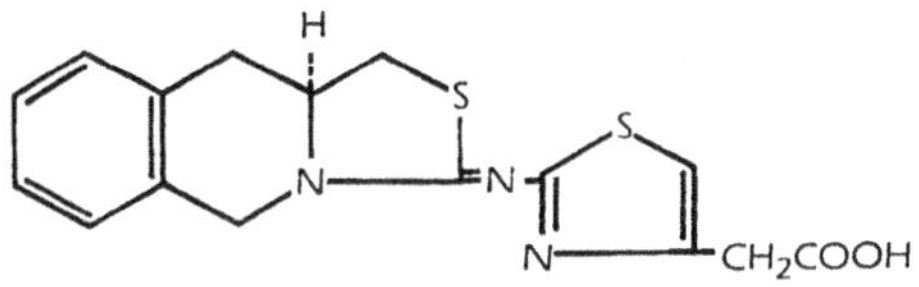

2-[(1,5,10,10a-tetrahydro-3H-thiazolo [3,4b] -isoquinolin-3-ylidene) amino] -4-thiazole acetic acid (S) or 44-081 R.P.

Fig. 2. Antirhinovirus compounds that have undergone clinical evaluation in volunteers.

index is also high (Table 2). In addition, experiments in animals suggested that when given orally they were non-toxic and that suitable formulation produces blood and tissue levels of active drugs in excess of those required to inhibit the replication of rhinovirus in culture. Therefore, clearly these antivirals were ready for clinical evaluation in man. Table 7 summarizes the outcome of the various double-blind placebo-controlled trials conducted in human volunteers to assess the prophylactic efficacy of these compounds in the prevention of experimental rhinovirus infection. As summarized in this Table, none of these synthetic antirhinovirus compounds were found to protect against experimental infection when given intranasally, orally or both.

Table 7. Summary of efficacy data on synthetic antirhinovirus compounds in the prevention of experimental rhinovirus infections.

Compound (Ref.)	% Reduction relative to placebo		
	Significant colds*	Virus shedding	Seroconversion or AB rises
1) Enviroxime (IN and oral) (29)	33.4	33.4	0
2) Enviroxime (IN) (30)	0	0	0
3) Enviroxime (IN) (31)	0	17	0
4) Dichloroflavan (oral) (32)	28.4	0	1.8
5) Ro-09-0415 (oral) (33)	17.4	16.7	0
6) 44-081 R.P. (IN) (35)	0	0	0
7) Dichlorflavan (IN) (34)	15	0	0

* Mild, moderate or severe

Thus, Phillpotts et al (29) at the Common Cold Unit were able to show some 33.4% reduction in the incidence of significant colds and virus shedding when enviroxime [2-amino-1-(isopropylsulphonyl)-6-benzimidazole phenyl ketone oxime] was given both orally (25 mg) and intranasally (284 μg/nostril by spray) four times daily starting one day before and continuing for five days after challenge with a sensitive strain of rhinovirus (HRV-9). Only the reduction in rhinorrhoea relative to placebo reached statistical significance. Furthermore, oral enviroxime was poorly tolerated as about two-thirds of volunteers complained of nausea and abdominal pain. Two groups in the United States, Hayden and Gwaltney (30) and Levandowski et al (31) evaluated enviroxime when administered intranasally (284 μg/nostril, four or five times daily). Again, they were not able to show any significant reduction in the incidence of colds, virus shedding or serological evidence of infection as compared with placebo.

Phillpotts et al (32), in placebo controlled double-blind studies, evaluated 4,6-Dichloroflavan given by the oral route at a dosage regimen of 1 mg/kg three times daily (3 doses before and 13 doses after intranasal challenge with HRV-9) and failed to show any prophylactic effect. Similarly, another antirhinovirus drug, Ro-09-0415, a phosphorylated 'prodrug' of a potent anti-rhinovirus compound, 4'-ethoxy-2'-hydroxy-4,6' dimethoxy-chalcone (Ro-09-0410) has undergone clinical evaluation in volunteer experiments. It was administered orally at a dose of 1200 mg twice daily to volunteers for seven days, starting one day before challenge with HRV-9 (33). Again, no prophylactic effect was demonstrated and medication did not prevent virus infection. Both dichloroflavan and Ro-09-0415 when given orally reached plasma concentrations in excess of those required to inhibit the replication of HRV-9 *in vitro*.

It was concluded from these studies that the drugs may not have reached the tissues of the nasal mucosa in sufficient concentrations to inhibit the replication of rhinovirus 9 in the nasal epithelium and therefore intranasal administration of these two compounds was thought to be desirable. Unfortunately, both compounds were found too irritant to be administered intranasally and therefore further intranasal evaluation had to await the development of a new formulation.

A new formulation of dichloroflavan became available in 1984, as a 5% w/v aqueous wet-milled suspension (40 mg). Initial experiments to study the tolerance and pharmacokinetics suggested that single and repeated dosages in man had shown that intranasal administration was well tolerated and that drug persisted in the nasal cavity at levels in excess of its minimal inhibitory concentration of a sensitive rhinovirus strain in vitro for 3 to 3½ hours after dosing (Gilbert and Freestone, personal communication). Because of this further clinical evaluation of this compound to assess its efficacy when given intranasally was clearly warranted. In double-blind placebo-controlled trials conducted in human volunteers, we gave dichloroflavan five times per day as a 5% w/v aqueous suspension (40 mg) intranasally for five doses before and 21 doses after intranasal challenge with rhinovirus type 9, a virus known to be sensitive to this compound. As shown in Table 7 dichloroflavan did not protect volunteers from illness or infection (34). Indeed, there was some indication that drug may have increased the severity of clinical signs and symptoms. We concluded that the intranasal administration of fine particles of a hydrophobic compound such as dichloroflavan may not provide sustained delivery of adequate concentration of drug to the site of virus replication, the ciliated nasal epithelial cells (34). An intranasal preparation of the chalcone Ro-09-0410 has also recently become available and we are now at the stage of evaluating it clinically.

Another synthetic compound, 2-[(1,5,10,10a-tetrahydro-3*H*-thiazolo[3,4b] isoquinolin-3-ylidene) amino]-4-thiazoleacetic acid(S), 44-081 R.P., has been found to inhibit the replication of a sensitive rhinovirus (HRV-EL) at concentrations of 0.7 μg/ml (35). This compound was thought to have two important properties that could have particular advantage over previously evaluated compounds. First, it is readily soluble in water at neutral pH and this allowed intranasal administration without the need for dissolving or suspending agents. Second, this compound was found to render cells pretreated with it resistant to rhinovirus, and it was postulated that this might circumvent its rapid clearance by the mucociliary apparatus, a situation analogous to the antiviral state induced by IFN. Because of these properties a clinical trial was thought worthwhile. The compound or placebo was therefore given intranasally in a 0.15 ml dosage (300 μg) with a Calmar Albert spray 6 times daily per nostril starting the day before virus challenge and continuing for 6 days. Results showed that medication with this compound like other previously evaluated antirhinovirus substances did not result in significant beneficial effect clinically nor was there any significant effect on virus shedding or serological evidence of infection, despite the presence of 2.5-fold higher concentrations of the drug in nasal washings than the MIC of the drug against the challenge virus (HRV-EL) in cell cultures. It was concluded that even at these concentrations, there was not enough antiviral activity to prevent or inhibit virus replication in the nasal epithelium.

Treatment of the common cold

There have been few attempts to study whether

antirhinovirus agents can modify the course of illness following experimental and naturally occurring rhinovirus infection. Thus, Phillpotts et al (36) in double-blind placebo-controlled trials, studied the therapeutic efficacy of enviroxime. Volunteers were infected with HRV-9 and treated with enviroxime (284 µg/nostril) 44 hours after virus challenge six times a day for 5 days. Although there was a statistically significant reduction in clinical score on the last day of medication, the therapeutic efficacy of enviroxime could not be demonstrated conclusively, since the compound had no effect on the reduction of nasal secretion weight or the quantity of virus shed. Miller et al (37) conducted a very large community based controlled field trial of enviroxime given intranasally to assess its therapeutic effect when tested against naturally occurring common colds. Generally there were no consistent statistically significant differences between treated and untreated groups and it was therefore concluded that no therapeutic effect could be demonstrated.

Hayden and Gwaltney (38) assessed the therapeutic efficacy of recombinant interferon alpha-2 (IFN-alfa-2b) in experimental infections with rhinovirus type 39. Thus adult volunteers were given 9 MU of this IFN or placebo intranasally 3 times a day for 5 days beginning 28 hours after rhinovirus inoculation i.e. before disease was evident. Although they were able to show some beneficial effect, e.g. significant reduction in the duration and quantity of viral shedding and a modest effect on nasal symptom scores, this effect was mainly seen when IFN was given as nasal drops and not spray. However, it was generally concluded that IFN-alfa-2b was unlikely to be therapeutically useful in treating naturally occurring rhinoviral colds.

Synergy between antirhinoviral agents

Combination medication could have an important role

to play either in the prophylaxis or treatment of a rhinovirus infection,especially if one could demonstrate powerful synergistic effects between various antirhino-viral agents. Such synergistic activity may allow the delivery of high antirhinoviral activity to the target tissues and possibly with lower toxicity especially if the drugs in combinations need only be used at lower concentrations. This has led us to investigate initially the *in vitro* interaction between various antirhinoviral agents (39). As Table 8 shows, to our surprise, we were able to demonstrate significant synergistic effects between various interferons, interferons and synthetic antirhinovirus compounds, and between synthetic compounds themselves when in combination. The most powerful combination was in fact between interferons and synthetic compounds particularly IFN-alfa-2b or HuIFN-γ with enviroxime. In such combination the antirhinoviral activity was some 100-fold higher than that of each of these agents when used alone as tested in the appropriate cell-culture system. These studies were later extended and confirmed in experiments using human embryonic tracheal or nasal epithelial cultures which in our view represented a system very close to that of the nasal epithelium *in vivo*. Thus this powerful antirhinoviral activity was not confined to the cell-culture system. These findings therefore have set the stage for evaluating this concept of synergy in human volunteers experiments.

Prospects for rhinovirus chemoprophylaxis and therapy

It is clear from the above review that considerable progress has been achieved in developing potent anti-rhinoviral agents, whether interferons or synthetic substances. Furthermore, studies in human volunteers have now established that a large number of interferons are effective in preventing experimental and naturally

Table 8. Synergy between antirhinoviral drugs against RV9 (100 $TCID_{50}$/ml) using FIC index*

	FIC index of drug combinations					
	HuIFN α-2	HuIFN β	HuIFN β x 401	HuIFN γ	Enviroxime	DCF
HuIFN β	0.20					
HuIFN β x 401	0.10	0.13				
HuIFN γ	0.18	0.12	0.10			
Enviroxime	0.09	0.17	0.18	0.06		
DCF	0.18	0.29	0.18	0.06	0.50	
Ro-09-0410	0.10	0.15	0.13	0.06	0.37	0.50

$$\text{* FIC index} = \frac{(\text{MIC of drug A in comb.})}{(\text{MIC of drug A alone})} + \frac{(\text{MIC of drug B in comb.})}{(\text{MIC of drug B alone})}$$

The interpretation of the indexes as follows:

FIC index < 0.5 Significant synergism
FIC index 0.5 - 0.9 Suggestive of synergism
FIC index 1 Effects are additive
FIC index 1.1 - 1.9 Indifference or partial antagonism
FIC index > 2 Antagonism

occurring rhinovirus colds. Field studies in families suggest that IFN may be usefully applied in the prevention of colds within the family setting by protecting other members of the family who are in contact with the index case. This concept could prove particularly useful in protecting a member of the family who may be at particular risk of developing lower respiratory tract complication following a rhinovirus infection e.g. an individual with chronic obstructive airways disease. Indeed, a number of interferons are at the moment being evaluated as possible prophylaxis against rhinovirus infection in individuals with chronic obstructive airways disease e.g. bronchitis or asthma. However, the above review also highlighted the fact that for IFNs to be effective they need to be given at high concentrations. Because of this they can only be used for short-term prophylaxis e.g. within 7 days since prolonged use, as many of the above studies showed resulted in unacceptable effects, mainly local inflammation, nasal irritation and ulcerations and in a proportion of individuals nasal bleeding. Clearly, IFNs would have a much wider application as a long-term prophylaxis for instance to protect susceptible individuals for 1 - 3 months during the rhinovirus season had it not been for its toxicity. Therefore, it would clearly be desirable to search for or develop new less toxic interferons. A recent encouraging discovery was that recombinant interferon-Beta$_{serine}$ (IFN-β_{ser}) when given at a daily dose of 12 MU or 3 MU by nasal spray for 25 consecutive days was found to be well tolerated and no significant differences in the frequency of respiratory symptoms or nasal complaints were recorded among those who received it as compared with those who received placebo (40). Since HuIFNβ_{ser} has already been found to be effective in the prevention of experimental rhinovirus infection in volunteers (17) this IFN may well offer a clinically acceptable means for seasonal prophylaxis of common colds. If the planned experiments on synergy (interferon-alfa and enviroxime) prove successful in volunteers, it may be possible to use interferons at a lower dosage than that required at present when these molecules are used alone. If this should prove to be the case, it would be possible perhaps to use interferons in combination with synthetic anti-rhinovirus drugs for seasonal prophylaxis at concentrations that are not toxic. Alternatively, molecules may be prepared by recombinant DNA technology, that may have high therapeutic ratio, e.g. would be devoid of or have minimal side-effects yet have potent antiviral activity. Such molecules might then be used either for long-term prophylaxis or possibly even therapeutically since the administration of high concentrations would be possible. Indeed, although the studies of Hayden and Gwaltney (38) did not show very significant therapeutic effect with IFN-alfa-2b when given at a dose of 9 MU, there was a general reduction in both virus shedding and clinical scores, and hence, it is not inconceivable that molecules with better therapeutic ratio may have a more dramatic effect on the course of the common cold illness.

With synthetic antirhinovirus compounds, clearly, there is a problem of delivery since a majority of these compounds are extremely potent - MIC 0.0078 - 0.25 µg/ml. Further, recent unpublished data suggest that even when a new and very potent synthetic antirhinovirus compound (MIC 0.002 µg/ml) was administered to volunteers at concentrations in excess of that required for interferon to establish its prophylactic efficacy intranasally, still no prophylactic effect could be established (Al-Nakib and Tyrrell, unpublished data). One aspect of this lack of prophylactic efficacy is thought to be the rapid removal of these compounds by the mucociliary clearance mechanism in the nose and hence despite frequent dosing (6 times per day), not enough antiviral activitiy is retained in the nasal epithelium to prevent or even limit rhinovirus replication in these tissues. With some compounds, such as dichloroflavan, the drug was still not protective despite the fact that with single and repeated dosages the drug persisted in the nasal cavity at levels in excess of the inhibitory concentration of a sensitive rhinovirus strain for 3 - 3½ hours after dosing (34). In this case, intranasal administration may not have provided sustained delivery of adequate concentration of the drug to the ciliated epithelial cells. These findings therefore highlight the need to design formulation vehicles capable of delivering high concentrations of drug to the nasal mucosa. Alternatively, new molecules may have to be synthesised that when given orally enter the respiratory secretions in adequate antiviral concentrations. Indeed, the anti-influenza compound ICI 130,685 was found to be present at concentrations 4 - 7 higher in secretions than in the blood after oral administration (41).

The success of IFN as an antirhinoviral agent when given intranasally may be partly due to the fact that these molecules attach to receptors in the nasal mucosa. Thus, provided enough can be administered, sufficient amounts of IFN will reach these receptors and establish the antiviral state despite continuous attrition by the mucociliary clearance mechanism. By analogy with IFNs, targeting of molecules may therefore be a useful approach to deliver synthetic antirhinovirus agents. Indeed, recent studies by Hayden et al (42) showed that when monoclonal antibodies to the rhinovirus receptor sites were given prophylactically prior to challenge with HRV-39, a transient but significant protective effect was demonstrated among those who received these monoclonals as compared with those who received placebo. This effect was thought to be due to the monoclonals attaching to the rhinovirus receptor sites, thus blocking the attachment of the virus to these receptors and consequently delayed infection. It is expected that experiments such as these may pave the way for the development of synthetic molecules which could block virus receptor sites and hence prevent infection. However, considerable work will need to be done to establish approximately the number of rhinovirus receptors in the nasal epithelium, the individual variation and the rate of turnover prior to and during the course of infection.

The above review highlights the fact that important progress has been achieved particularly regarding the prevention of the common cold with interferons and the development of more potent synthetic antirhinovirus agents. However, there are still some important issues that need to be addressed, particularly the optimal mode of delivering these compounds to the target tissue. Once problems such as these are surmounted, we feel, perhaps, that we should expect some important progress towards reaching those thresholds that we have all been expecting now for sometime, concerning the prevention and treatment of the common cold.

The possible applications of antirhinovirus drugs

When considering the development of antivirals it is important to consider how they might be used. As indicated above an effective antirhinovirus drug can prevent or treat about half of all the cases of the common cold, but it can be asked whether this is worth doing or practicable and if it is possible what are the relative advantages and disadvantages.

The advantages could include the patient being saved from losing time at work or study. Furthermore, recent work shows that, at least for some demanding tasks, human performance declines during colds (43). Thus in addition to those few who would not be able or allowed to work with a cold, e.g. opera singers, pilots, there are probably many more who would work better as well as feel better if their colds were prevented or treated.

A number of patients with chronic diseases are at increased risk of relapses after colds and similar mild infections e.g. nephritis, juvenile chronic arthritis, and in patients with chronic or relapsing lower respiratory disease, obstructive airways disease such as chronic bronchitis, asthma or "wheezy bronchitis". Infection with rhinoviruses produce exacerbations and indeed the virus may invade the lower respiratory tract. Any effective way of preventing or treating rhinovirus infections would be saving such patients weeks of disabling illness or even hospital admission.

As colds are shortlived and self-limiting it is likely that an effective therapy would only reduce its severity or duration to a limited extent. On the other hand this is probably true of antibiotic treatment of mild bacterial infections, such as streptococcal pharyngitis. Yet it is likely that the use of antibiotics in a group of such cases ensures that the occasional patient who would otherwise have had a severe infection has a mild illness like the rest, and greatly reduces the frequency of various complications, such as purulent otitis media and mastoiditis. Thus it might be decided that therapy might still be worthwhile for normal subjects, to ensure that they all have mild colds.

It is impossible to say at the moment when or whether antirhinovirus treatment could logically become a part of clinical medicine. It is our opinion that short-term IFN prophylaxis for individuals for whom it is particularly important to be free of colds on a foreseeable future date is worthwhile, and data may soon be available to justify it for patients with chronic chest disease exposed to colds in their families. On the other hand continuous prophylaxis for healthy individuals who get only occasional mild colds might not be practicable, because they would not comply with the need for regular medication, or else it would be reckoned to be too costly or, even worse, the long exposure to medication might increase unacceptably the frequency of adverse drug effects.

As indicated above therapy might have a less marked effect than prophylaxis, except in preventing a cold going on to severe disease - on the other hand patients who are sick are much more compliant than those who feel well, and the expense and risk of toxic effects would be much less than those of prophylaxis.

We therefore feel that as it seems to be theoretically possible and as the practical value would be significant, efforts should be directed towards developing an effective chemotherapy of rhinovirus infections.

REFERENCES

1. Beare, A.S. and Reed, S.E. In: Chemoprophylaxis and Viral Infections of the Respiratory Tract. (Ed. J. Oxford), CRC Press, Cleveland, 1977, pp. 27-49.
2. Tisdale, M. and Selway, J.W.T. J. Gen. Virol. 64: 795-803, 1983.
3. Ishitsuka, H., Ninomiya, Y.T., Ohsawa, C., Fujiu, M. and Suhara, Y. Antimicrob. Agents Chemother. 22: 617-621, 1982.
4. Ninomiya, Y., Ohsawa, C., Aoyama, M., Umeda, I., Suhara, Y. and Ishitsuka, H. Virology, 134: 269-276, 1984.
5. Reed, S.E., Craig, J.W. and Tyrrell, D.A.J. J. Infect. Dis. 133: A128-A135, 1976.
6. Merigan, T.C., Reed, S.E., Hall, T.S. and Tyrrell, D.A.J. Lancet i: 563-567, 1973.
7. Acornley, J.E., Bessell, C.J., Bynoe, M.L., Godtfredsen, W.O. and Knoyle, M.J. Br. J. Pharmacol. Chemother. 31: 210-220, 1967.
8. Greenberg, S.B., Harmon, M.W., Couch, R.B., Johnson, P.E., Wilson, S.Z., Dasco, C.C., Bloom, K. and Quarles, J. J. Infect. Dis. 145: 542-546, 1982.
9. Aoki, F.Y. and Crawley, J.C.W. Br. J. Clin. Pharmacol. 3: 869-878, 1976.
10. Scott, G.M., Phillpotts, R.J., Wallace, J., Secher, D.S., Cantell, K. and Tyrrell, D.A.J. Br. Med. J. 284: 1822-1825, 1982.

11. Scott, G.M., Phillpotts, R.J., Wallace, J., Gauci, C.L., Greiner, J. and Tyrrell, D.A.J. Lancet ii: 186-188, 1982.
12. Phillpotts, R.J., Scott, G.M., Higgins, P.G., Wallace, J., Tyrrell, D.A.J. and Gauci, C.L. Antiviral Res. 3: 121-136, 1983.
13. Hayden, F.G. and Gwaltney, J.M.Jr. J. Infect. Dis. 148: 543-550, 1983.
14. Samo, T.C., Greenberg, S.B., Couch, R.B., Quarles, J., Johnson, P.E., Hook, S. and Harmon, M.W. J. Infect. Dis. 148: 535-542, 1983.
15. Samo, T.C., Greenberg, S.B., Palmer, J.M., Couch, R.B., Harmon, M.W. and Johnson, P.E. J. Infect. Dis. 150: 181-188, 1984.
16. Scott, G.M., Reed, S., Cartwright, T. and Tyrrell, D.A.J. Arch. Virol. 65: 135-139, 1980.
17. Higgins, P.G., Al-Nakib, W., Willman, J. and Tyrrell, D.A.J. J. Interferon Res. 6: 153-159, 1986.
18. Phillpotts, R.J., Higgins, P.G., Willman, J.S., Tyrrell, D.A.J., Freestone, D.S. and Shepherd, W.M. J. Interferon Res. 4: 535-541, 1984.
19. Betts, R.F., Erb, S., Roth, F., Reichman, R.C., Treanor, J., Beutner, K. and Dolin, R. 13th Int. Congress Chemother. Vienna, 60/13-5, 1983.
20. Farr, B., Gwaltney, J.M.Jr., Adams, K.F. and Hayden, F.G. Antimicrob. Agents Chemother. 26: 31-34, 1984.
21. Hayden, F.G., Mills, S.E. and Johns, M.E. J. Infect. Dis. 148: 914-921, 1983.
22. Scott, G.M., Onwubalili, J.K., Robinson, J.A., Doré, C., Secher, D.S. and Cantell, K. J. Med. Virol. 17: 99-106, 1985.
23. Hayden, F.G., Gwaltney, J.M.Jr. and Johns, M.E. Antiviral Res. 5: 111-116, 1985.
24. Douglas, R.M., Albrecht, J.K., Miles, H.B., Moore, B.W., Read, R., Worswick, D.A. and Woodward, A.J. J. Infect. Dis. 151: 731-736, 1985.
25. Monto, A.S., Shope, T.C., Schwartz, S.A. and Albrecht, J.K. J. Infect. Dis. 154: 128-133, 1986.
26. Herzog, C., Berger, R., Fernex, M., Friesecke, K., Havas, L., Just, M. and Dubach, U.C. Antiviral Res. 6: 171-176, 1986.
27. Douglas, R.M., Moore, B.W., Miles, H.B., Davies, L.M., Graham, N.M.H., Ryan, P., Worswick, D.A. and Albrecht, J.K. N. Eng. J. Med. 314: 65-70, 1986.
28. Hayden, F.G., Albrecht, J.K., Kaiser, D.L. and Gwaltney, J.M.Jr. N. Eng. J. Med. 314: 71-75, 1986.
29. Phillpotts, R.J., Jones, R.W., Delong, D.C., Reed, S.E., Wallace, J. and Tyrrell, D.A.J. Lancet ii: 1342-1344, 1981.
30. Hayden, F.G. and Gwaltney, J.M.Jr. Antimicrob. Agents Chemother. 21: 892-897, 1982.
31. Levandowski, R.A., Pachucki, C.T., Rubenis, M. and Jackson, G.G. Antimicrob. Agents Chemother. 22: 1004-1007, 1982.

32. Phillpotts, R.J., Wallace, J., Tyrrell, D.A.J., Freestone, D.S. and Shepherd, W.M. Arch. Virol. 75: 115-121, 1983.
33. Phillpotts, R.J., Higgins, P.G., Willman, J.S., Tyrrell, D.A.J. and Lenox-Smith, I. J. Antimicrob. Chemother. 14: 403-409, 1984.
34. Al-Nakib, W., Willman, J., Higgins, P.G., Tyrrell, D.A.J., Shepherd, W.M. and Freestone, D.S. Arch. Virol. in press, 1986.
35. Zerial, A., Werner, G.H., Phillpotts, R.J., Willman, J.S., Higgins, P.G., Tyrrell, D.A.J. Antimicrob. Agents Chemother. 27: 846-850, 1985.
36. Phillpotts, R.J., Wallace, J., Tyrrell, D.A.J. and Tagart, V.B. Antimicrob. Agents Chemother. 23: 671-675, 1983.
37. Miller, F.D., Monto, A.S., Delong, D.C., Exelby, A., Bryan, E.R. and Srivastava, S. Antimicrob. Agents Chemother. 27: 102-106, 1985.
38. Hayden, F.G. and Gwaltney, J.M.Jr. J. Infect. Dis. 150: 174-180, 1984.
39. Ahmad, A.L.M. and Tyrrell, D.A.J. Antiviral Res. 6: 241-252, 1986.
40. Hayden, F.G., Innes, D.J.Jr., Mills, S.E. and Levine, P.A. J. Interferon Res. 6(Suppl. 1): 31, 1986.
41. Al-Nakib, W., Higgins, P.G., Willman, J., Tyrrell, D.A.J., Swallow, D.L., Hurst, B.C. and Rushton, A. J. Antimicrob. Chemother. 18: 119-129, 1986.
42. Hayden, F.G., Gwaltney, J.M.Jr. and Colonno, R.J. Paper presented at the Annual Meeting Amer. Soc. Virol. Santa Barbara, Calif. U.S.A. June 1986.
43. Smith, A.P., Tyrrell, D.A.J., Al-Nakib, W., Coyle, K.B., Donovan, C.B., Higgins, P.G. and Willman, J.S. Br. J. Psychology, in press.

16

RIMANTADINE AND AMANTADINE IN THE PROPHYLAXIS AND THERAPY OF INFLUENZA A

RAPHAEL DOLIN

Infectious Disease Unit, School of Medicine, University of Rochester, Rochester, New York 14642, U.S.A.

Amantadine (1-adamantanamine hydrochloride) and rimantadine (α-methyl-1-adamantanemethylamine hydrochloride) are antiviral compounds with activity primarily against influenza A virus. Although in vitro and in vivo antiviral activities of these compounds have been recognized since the early 1960's, these compounds have received relatively limited use in western countries for prevention and/or treatment of influenza A infections, although rimantadine is used extensively for that purpose in the Soviet Union. Both compounds have been widely studied in experimentally induced and natural occurring infections in humans, primarily in uncomplicated disease in young adults. Recently, several controlled studies of rimantadine in the prophylaxis and therapy of influenza A infection in children and in elderly subjects have also been carried out in the United States, and additional studies of amantadine have been conducted as well. These investigations have stimulated renewed interests in these compounds, as discussed below.

STRUCTURE AND ACTIVITY

Amantadine and rimantadine are primary symmetrical amines with unusual "bird cage-like" structures as noted in Figures 1 and 2. The in vitro activity of both drugs against influenza A virus is similiar, although rimantadine appears to be somewhat more active against certain influenza A viruses strains both in vitro and in animal studies (1). Both compounds possess some in vitro activity against parainfluenza and rubella viruses, but this activity requires concentrations 50-100 times those required for inhibition of influenza A viruses, and these concentrations are considerably higher than those which can be achieved in humans (2).

The precise mechanism of action of amantadine or rimantadine is not understood in molecular terms. These compounds do not inactivate extracellular virus or interfere with attachment of virus to the cell surface. It has been suggested that these compounds interfere with a step early in infection, such as penetration and/or uncoating of the virus, and perhaps with virus assembly as well. Early studies had indicated that resistance to these drugs segregates with the gene which codes for the M-protein of influenza A virus (3). However, other studies have shown that other genes, including those which code for the hemagglutinin, may also contribute to resistance (4). Recent work has demonstated that resistance to amantadine and rimantadine is associated with an alteration in the trans-membrane portion of the M2 protein, and that single amino acid substitutions at a critical site can confer resistance (5,6). These data suggest that the mechanism of action of these compounds is mediated through interaction with the M2 protein, and perhaps also through interactions between the M2 protein and the hemagglutinin.

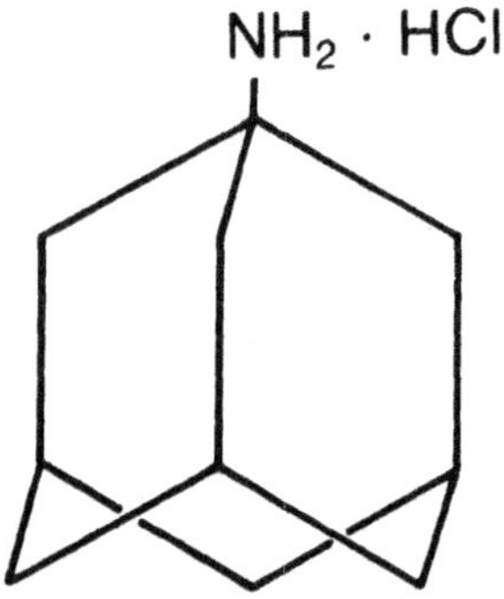

1

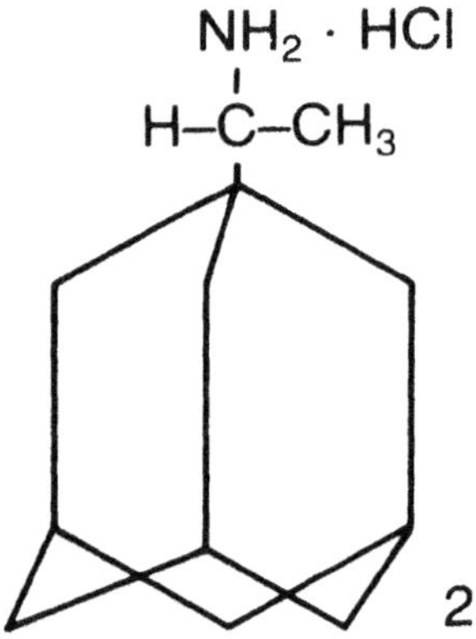

2

Fig. 1. Amantadine (1-adamantanamine hydrochloride).
Fig. 2. Rimantadine (α-methyl-1-adamantanemethylamine hydrochloride). From Ref. 7.

PHARMACOKINETICS

Despite their structural similarities, the pharmacokinetic properties of amantadine and rimantadine differ markedly. Because both drugs can only be administered orally, the pharmacokinetic properties have been determined only after oral administration. Amantadine is well absorbed orally and peak levels of approximately 0.5 ug/ml are reached 2-4 hours after a 200 mg dose. The drug is not metabolized and is excreted largely unchanged by the kidney, with 95% of the dose appearing in the urine. The serum half-life of amantadine is approximately 14 hours, and is markedly prolonged by impaired renal function (7). Amantadine has a large volume of distribution and relatively little is removed by hemodialysis. The drug appears to penetrate pulmonary tissue, and concentrations in respiratory secretions are approximately two-thirds of those in the serum (8).

Rimantadine is also well absorbed orally, but peak levels are approximately half of those of amantadine. Maximum plasma concentrations of approximately 0.3 ug/ml are achieved within 3-8 hours after a 200 mg dose. Rimantadine is extensively metabolized to ortho-, meta-, and para-hydroxylated metabolites, approximately 80% of which are excreted in the urine (9). The half-life of rimantadine is approximately 30 hours, and there is a suggestion that the metabolites may have an even longer half-life. Thus, rimantadine and its metabolites can accumulate in the presence of renal dysfunction. Although the toxic potential of high levels of rimantadine or its metabolites is uncertain, dosage adjustments may be necessary in that setting. Because rimantadine also has a large volume of distribution, little drug is removed by hemodialysis. Rimantadine is concentrated in respiratory secretions to a greater degree than is amantadine, with levels that approach or exceed those found in the serum (8).

PROPHYLAXIS OF INFLUENZA A

The prophylactic efficacies of amantadine and rimantadine in naturally occurring infection have now been established in multiple studies. The majority of these studies have demonstrated significant reductions of influenza-like illness associated with laboratory documented influenza A infection, ranging from 50 to

Table 1 *Influenza-like illness. Laboratory-documented influenza, and infection with influenza A virus among volunteers receiving placebo, rimantadine, or amantadine*

Treatment group (no. of subjects)	No. with laboratory-documented influenza*	No. infected with influenza A virus**
Placebo (132)	27 (21%)	32 (24%)
Rimantadine (133)	4 (3%)***	11 (8%)***
efficacy rate (%)$^+$	85	66
Amantadine (113)	2 (2%)***	7 (6%)***
efficacy rate (%)$^+$	91	74

* Defined as influenza-like illness along with virus isolation or a rise in serum antibody to influenza A virus.
** Defined as influenza A virus isolation or a rise in serum antibody to influenza A virus, irrespective of the presence of illness.
*** $P < 0.001$ as compared with placebo by chi-square analysis.
$^+$ Efficacy rate as calculated by the expression:

$$\frac{\text{rate in placebo recipients} - \text{rate in rimantadine/amantadine recipients}}{\text{rate in placebo recipients}} \times 100$$

100% (10). At the doses which have been most frequently employed (200 mg/day) in young adults, the efficacies of the two drugs appear to be similar, although relatively few studies have been carried out in which the drugs have been compared directly in the same study. Such a study was recently conducted during an outbreak of influenza A/H_1N_1 and H_3N_2, in which the effects of both drugs were compared with that seen in a placebo treated group (11). As noted in Table 1, efficacy rates of 85-91% were observed in reduction of virus specific illness. Both drugs were less efficacious in reduction of infection with influenza A virus as opposed to reduction of illness (Table 1), which is a potentially desirable feature of chemoprophylaxis, since subclinical infection may result in immunity. As was the case in this study, most other studies of prophylaxis with these drugs have been carried out in young adults, and to a lesser extent in children. Recently, Clover and colleagues studied rimantadine administered prophylactically to children in a family setting (12). Children received rimantadine or placebo for 5 weeks after an outbreak of influenza A was recognized in the community. As noted in Table 2, all of the influenza A associated illness occurred in the placebo recipients (efficacy rate of 100%). The rate of influenza A infection and illness among untreated adults also appeared to be reduced (53% efficacy), although this reduction did not reach statistical significance.

Table 2.--Frequency of Infection and Clinical Illness Caused by Influenza A (H_1N_1) Virus in Families in Which Children Received Rimantadine or Placebo

	Children			Adults		
Drug	No.	No. (%) Infected*	No. (%) Ill†	No.	No. (%) Infected‡	No. (%) Ill
Placebo	41	13 (31.7)	7 (17.0)	36	7 (19.0)	1 (2.9)
Rimantadine hydrochloride	35	1 (2.9)	0 (0.0)	34	3 (8.8)	0 (0.0)

*P=.001.
†P=.01.
‡P=.2.
Reprinted with permission from ref. 12.

A placebo controlled trial of the prophylactic effect of rimantadine in the elderly has also recently been completed (13). In this study, 105 elderly nursing home residence received rimantadine 200 mg/day or placebo for six to seven weeks. In this study, rimantadine reduced the rate of influenza-like illness by 63%. The prophylactic effect of rimantadine was observed primarily in individuals who had received influenza A vaccination previously, suggesting that rimantadine may confer additive protection to that afforded by the vaccine. Similar observations regarding an additive effect of chemoprophylaxis and vaccination have been made by investigators in the Soviet Union.

THERAPY OF INFLUENZA A INFECTIONS

Both amantadine and rimantadine have also been demonstrated to be effective in the therapy of influenza A infections (10). Again, controlled studies have been carried out primarily in self-limited disease in young adults and in normal children. These studies have shown a modest, but statistically significant effect of either amantadine or rimantadine when compared to a placebo treated group. In studies in which the therapeutic efficacies of the two drugs have been compared directly, amantadine and rimantadine each reduced the duration of signs and symptoms of illness by approximately 50%; and the efficacies of the two drugs were not significantly different (Fig. 3) (14). A recent study compared the effect of aspirin

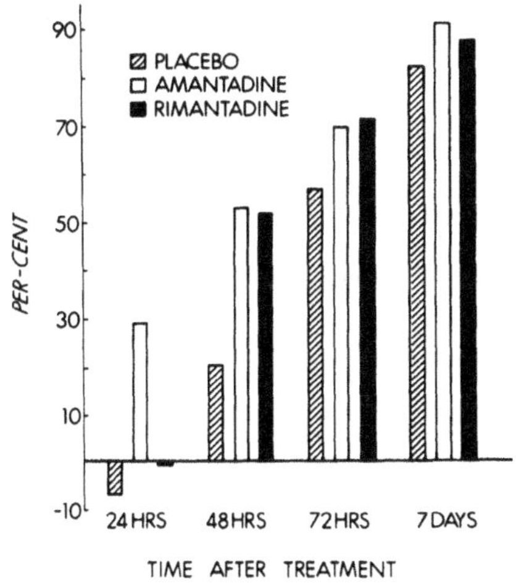

Fig. 3. Percent improvement in subjects with acute influenza A who received 200 mg of amantadine, rimantadine, or placebo for 5 days.
From ref. 14.

with that of amantadine therapy on uncomplicated illness in young adults. This study concluded that amantadine was superior to aspirin in reduction of the serverity and duration of symptoms of influenza, although aspirin was more effective in reduction of fever (15).

Two recent studies examined the therapeutic efficacy of rimantadine with that of acetaminophen in the treatment of uncomplicated influenza A infection in young children. In a study of an A/H_3N_2 outbreak by Hall and colleagues (16), children who received rimantadine showed significantly greater improvement in fever, as well as in severity of illness, compared to those who received acetaminophen. (Fig. 4) In a study during an A/H_1N_1 outbreak by Wright et al.(17), rimantadine and acetaminophen had equivalent clinical effects on illness in children. In this latter study, relatively milder illness was observed, compared to the A/H_3N_2 outbreak discussed earlier, and this may have accounted for the difference in results according to the authors.

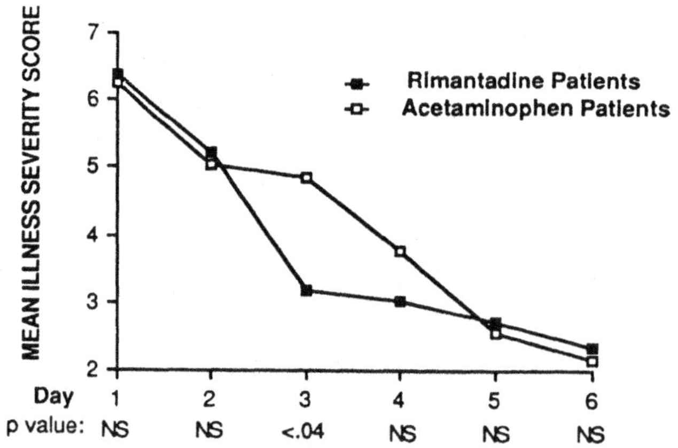

Fig. 4: The mean score for the severity of illness according to day for 37 children treated with rimantadine compared to 32 children treated with acetaminophen.
Reprinted with permission from ref. 16.

There have been few controlled studies to evaluate the efficacy of amantadine or rimantadine in the therapy of influenza in elderly or other high risk subjects. Walters and Paulshock reported a study in which amantadine at 200 mg/day or placebo was administered for 10 days to residents of a home for the elderly during an influenza A/H_2N_2 outbreak (18). Amantadine recipients manifested a more rapid rate of defervescence, but no other beneficial clinical effects were noted among the 32 patients who were studied. Betts and colleagues recently conducted a placebo controlled trial which examined the effect of rimantadine at a dose of 200 mg/day for 5 days in 81 elderly residents of two nursing homes during an influenza A/H_3N_2 outbreak (19). Rimantadine treatment was associated with a lower mean temperature early in illness, and also with a more rapid resolution of clinical illness scores. These effects were seen primarily in subjects who presented with moderate or severe illnesses.

As of this writing, controlled studies of the therapy of complications of influenza A, e.g. pneumonia, have not been reported. An NIH-supported multicenter collaborative study is underway in the United States, which is examining the efficacy of rimantadine versus placebo in that setting.

ADVERSE EFFECTS

Amantadine has been generally associated with a 5-10% rate of central nervous system (CNS) side effects, and occasionally higher rates have been noted. These side effects consist primarily of jitteriness, anxiety, insomnia, difficulty in concentrating, and rarely hallucinations. These side-effects, although troublesome, have been generally mild and reversible upon cessation of administration of the drug. CNS side-effects have been reported more frequently at higher plasma levels of amantadine, although a broad overlap between the levels at which toxic effects were present or absent have been noted (20). On occasion, seizures have been reported with extremely high plasma levels of amantadine. Other side effects have also been reported, including worsening of congestive heart failure, livedo reticularis and arrhythmias, but their causal relationship to amantadine has not been established.

Rimantadine appears to be exceedingly well tolerated, and several studies have reported side effects at no greater frequency than those seen in a placebo-treated group. A large scale study in which amantadine or rimantadine was administered at 200 mg/day for 6 weeks, found that the rates of side effects in the rimantadine and placebo groups were not significantly different, while an excess rate of CNS side effects of 11% was seen in the amantadine treated group (11). (Table 3) Doses of rimantadine of 200 mg/day administered for 6 to 7 weeks have also been well tolerated in elderly subjects (13). Doses of 5 mg/kg/day for children under 10 years of age, and 200 mg/day for children for more than ten years of age have also been very well tolerated when administered for as long as five weeks (12).

Table 3 *Withdrawal rates among recipients of placebo, rimantadine and amantadine*

Treatment/group (No. of subjects)	Withdrawals No. of subjects (%)	Reasons for withdrawal No. of subjects (%)			
		CNS side-effects*	Non-CNS side-effects	Unrelated to side-effects	Unknown
Placebo (148)	16 (11)	6 (4)	1 (0.7)	8 (5)	1 (0.7)
Rimantadine (147)	14 (10)	9 (6)	1 (0.7)	4 (3)	0 (0)
Amantadine (145)	32 (22)**	19 (13)***	4 (3)	3 (2)	6 (4)

* Primarily insomnia, 'jitteriness', and difficulty in concentrating.
** $P < 0.01$ as compared with the placebo, and $P < 0.005$ as compared with rimantadine, by chi-square analysis.
*** $P < 0.01$ as compared with placebo and $P < 0.05$ as compared with rimantadine.

The precise basis for the difference in the rates of side-effects observed between amantadine and rimantadine recipients is unclear at present. A small study in which attempts were made to correlate side-effects with plasma levels of amantadine and rimantadine, concluded that the potential for CNS toxicity was similar for either drug at similar blood levels, although no precise blood level of either drug was predicted for toxicity (20). However, other observers have noted that extraordinarily high blood levels of rimantadine, even in elderly subjects with cerebrovascular dysfunction, have not been associated with CNS toxicity (13). Whether the differences in side effects can be explained on the bases of differences in pharmacokinetics, or whether they reflect other properties of the two drugs remains an unresolved question at present.

The suggestion that side effects may be dose related has stimulated investigators to examine the efficacy of lower doses of amantadine. Payler and colleagues administered amantadine at a dose of 100 mg/day to a highly vaccinated group of English schoolboys, and demonstrated significant protection (21). A recent study of experimentally induced influenza A virus infection in normal volunteers indicated that 100 mg/day was also effective in the prevention of illness, although a direct comparison with the 200 mg/day dose was not carried out (22). Investigators in the Soviet Union have reported high rates of efficacy with doses of rimantadine considerably less than 200 mg/day, although such regimens have not been studied extensively elsewhere.

Because of the concerns for CNS toxicity, the US Public Health Service has recommended that the dose of amantadine be lowered to 100 mg/day in elderly subjects, although the efficacy of this regimen has not been rigorously established in that patient population. From the above discussion, it is apparent that neither the minimal effective dose, nor the optimal regimen of administration has been established for either amantadine or rimantadine. Additional studies to provide information regarding these questions would be important to formulate strategies for optimal utilization of these drugs.

ANTIVIRAL EFFECT AND THE DEVELOPMENT OF RESISTANCE

The majority of clinical trials in which appropriate laboratory studies were conducted, have demonstrated decreased shedding of influenza A virus associated with administration of amantadine or rimantadine. Although naturally occurring influenza A viruses which are resistant to amantadine and rimantadine have on occasion been detected (23), relatively few studies have examined the potential for development of resistance associated with the use of these compounds in human infections. In the previously discussed study carried out by Hall and colleagues, resistant viruses were detected in 27% of rimantadine recipients, compared to 6% of acetaminophen recipients (16). After rimantadine therapy was stopped, the proportion of children who shed virus and the quantity of virus shed were significantly greater in the rimantadine treated group. Despite this, rimantadine therapy was associated with overall beneficial clinical effects, and the clinical significance of either the development of resistance or of the somewhat prolonged virus shedding remains unclear. In the similarly designed study conducted by Wright and colleagues, virus resistant to rimantadine was noted in both rimantadine and acetaminophen treated groups (17). Recently conducted studies of the therapy of influenza A infections in young adults and in the elderly did not show prolonged virus shedding in rimantadine recipients nor high rates of resistant virus (24). The reasons for the differences in these observations are uncertain, but may be related to the higher levels of virus replication which occur in children compared to adults.

Additional studies are needed to assess the potential epidemiologic and clinical significance of the emergence of viruses resistant to amantadine and rimantadine.

SUMMARY

Amantadine and rimantadine are antiviral compounds with antiviral activity primarily against influenza A viruses. The mechanisms of action of these two compounds are similar, and appear to involve interaction with the transmembrane portion of the M2 protein.

Amantadine and rimantadine have been demonstrated to be effective in the prophylaxis and therapy of naturally occurring influenza A infections in multiple clinical studies. The majority of these studies have been carried out in young adults with uncomplicated disease. Recent studies have also indicated that rimantadine is effective in the prophylaxis of influenza A in young children, and is as effective or more effective than acetaminophen in the therapy of influenza in young children as well. Limited studies of rimantadine in the elderly have suggested that rimantadine is also effective in the prophylaxis and therapy of influenza A in elderly subjects. Controlled studies to examine the effect of rimantadine on the therapy of complicated influenza A are underway in the United States.

At the doses most commonly employed, rimantadine and amantadine appear to be similarly efficacious, but amantadine is associated with higher rates of central nervous system side effects. The precise basis for the difference in reactogenicity is unclear at present. Lower doses of amantadine are being evaluated in an attempt to reduce the rates of side effects.

REFERENCES

1. Hayden, F.G., Cote, K.M, Douglas, Jr., R.G. Antimicrob. Agents Chemother., 17:865-870, 1980.

2. Hoffmann, C.E. In: Selective Inhibitors of Viral Function, (Ed. W.A. Carter) CRC Press, Cleveland, Ohio, pp. 199-211, 1973.

3. Lubeck, M.D., Schulman, J.L, Palese, P. J. Virol., 28:710-716, 1978.

4. Scholtissek, C., Faulkner, G.P. J. Gen. Virol., 44:807-815, 1979.

5. Hay, A.J., Wolstenholme, A.J., Skehel, J.J., Smith, M.H. Embo J., 4:3021-3024, 1985.

6. Belshe, R. B., Hay, A.J., Skehel, J.J., Hall, C.B., Betts, R. Abstract, Proceedings of the 7th International Congress of Virology, August, 1987.

7. Dolin, R. In: Antimicrobial Agents Annual 1 (Eds. P.K. Peterson and J. Verhoef), Elsevier Science Publishers BV, 1986.

8. Hayden, F.G., Hoffman, H.E., Antimicrob. Agents Chemother., 28:216-221, 1987.

9. Wills, R.J., J. Resp. Dis. (In Press) 1987.

10. LaMontagne, J.R., Galasso, G.J., J. Infect. Dis. 138:928-931, 1978.

11. Dolin, R., Reichman, R.C., Madore, H.P. et al. N. Engl. J. Med., 307:580-584, 1982.

12. Clover, R.D., Crawford, S.A., Abell, T.D., Ramsey, Jr., C.N., Glezen, W. P., Couch, R.B. Am. J. Dis. Child. 140:706-709, 1986.

13. Dolin, R., Betts, R.F., Treanor, J.J., et al. In: Program and Abstracts, 23rd Interscience Conference on Antimicrobial Agents and Chemotherapy, Abstract No. 691. American Society for Microbiology, Washington, D.C., 1983.

14. Van Voris, L.P., Betts, R.F., Hayden, F.G. et al. J. Am. Med. Assoc., 245:1128-1131, 1981.

15. Younkin, S.W., Betts, R.F., Roth, F.K., Douglas, Jr., R. G. Antimicrob. Agents Chemother., 23:577-582, 1983.

16. Hall, C.B., Dolin, R., Gala, C.L. et al. Pediatrics (In Press), 1987.

17. Thompson, J., Fleet, W., Lawrence, E. et al. J. Med. Virol., 21:249-255, 1987.

18. Walters, H.E., Paulshock, M. Med. 67:176-179, 1970.

19. Graman, P.S., Treanor, J.J., Domurat, F., Dolin, R., Betts, R.F. In: Program and Abstracts, 26th Interscience Conference on Antimicrobial Agents and Chemotherapy, Abstract No. 90. American Society of Microbiology, New Orleans, LA, 1986.

20. Hayden, F.G., Hoffman, H.E., Spyker, D.A. Antimicrob. Agents Chemother., 23:458-464, 1983.

21. Payler, D.K., Purdham, P.A. Lancet, 1:502-504, March 3, 1984.

22. Sears, S.D., Clements, M.L. Antimicrob. Agents Chemother. Submitted for Publication, 1987.

23. Heider, H., Adamczyk, B., Presber, H.W., et al. Acta Virol. 25:395-400, 1980.

24. Dolin, R., Betts, R.F. Unpublished data 1987.

17

RIBAVIRIN AEROSOL TREATMENT OF INFLUENZA, RESPIRATORY SYNCYTIAL AND PARAINFLUENZA VIRUS INFECTIONS OF MAN

V. KNIGHT AND B. GILBERT

Department of Microbiology and Immunology, Baylor College of Medicine, One Baylor Plaza, Houston, Texas 77030

ABSTRACT

Ribavirin aerosol treatment of respiratory syncytial virus infection in infants has recently been approved in the United States and Canada. Much of the clinical data which supported the approval is summarized in this chapter. We believe that compelling evidence is also presented for the activity of ribavirin aerosol in the treatment of influenza A and B infections. Limited data suggest that parainfluenza virus infections might also respond to the treatment. The chapter also describes the aerobiologic principles as well as practical information which underlie aerosol treatment. We shall follow with great interest the future experience with this new antiviral, ribavirin, and its administration in small particle aerosol.

INTRODUCTION

Inhalation of small particle aerosol as a means of administration of antiviral therapy in man was first performed by us in 1979 (1) when, following extensive studies in animals by U.S. Army scientists (2-6), normal volunteers were exposed to amantadine aerosol for several hours. The methodology and the equipment used was described in a separate report (7). Subsequent studies disclosed favorable results in the treatment of college students with naturally acquired influenza with a small particle aerosol of ribavirin (8). Since that time a number of studies have been made which show a beneficial effect in the treatment of influenza A and B and respiratory syncytial virus infections. Limited information is also available suggesting favorable effect of ribavirin aerosol in the treatment of parainfluenza virus infection. The following report will review those studies to define the present status of ribavirin aerosol treatment of these viral infections.

Respiratory tract deposition of aerosols of hygroscopic particles.

The site of deposition of inhaled particles in the respiratory tract is a function of particle size. The problem is complicated, however, by the fact that most aerosols used for treatment are hygroscopic (particles of water or aqueous solutions of drug) and change in size due to loss or accretion of water when the ambient relative humidity decreases or increases. Table 1 shows an estimate of the fractional deposition of hygroscopic aerosol particles at various levels of the respiratory tract (9).

Table 1. Deposition of 1.5 micron hygroscopic particles within the respiratory tract.

	% Deposition Hygroscopic Particles	% Deposition Nonhygroscopic Particles
Nose	36*	25
Pharynx to secondary bronchi	1**	0
Tertiary bronchi to respiratory bronchioles	25**	10
Alveolar ducts	21**	13
Total	83	48

*Initially upon inhalation, 24% of 1.5 μm diameter particles, which are assumed to increase to 2 μm in diameter due to accretion of water, are deposited. Upon exhalation, particles have increased to 4μm in diameter due to the further addition of water, so that 12% of the total of inhaled particles will be deposited in the nose at exhalation, or a total deposition of 36%.

**Retention as 4 μm diameter particles.

In this table particles at ambient humidity have an approximate mean diameter of 1.5 microns. When inhaled, they enlarge in the nose to 2 microns in diameter due to the increasing relative humidity at this site. As they pass into the highly saturated air of the trachea and below they increase to a diameter of about 4 microns. Table 1 shows that 24 percent of particles of 2 microns in diameter on inhalation will deposit in the nose, and that a total of 46 percent of the four micron particles will deposit in the lung. Upon exhalation an additional 12 percent of the inhaled particles will deposit in the nose for a total deposition of 83 percent of inhaled particles. The foregoing illustration was chosen because the Collison aerosol generator (Figure 1) produces particles in this range of size.

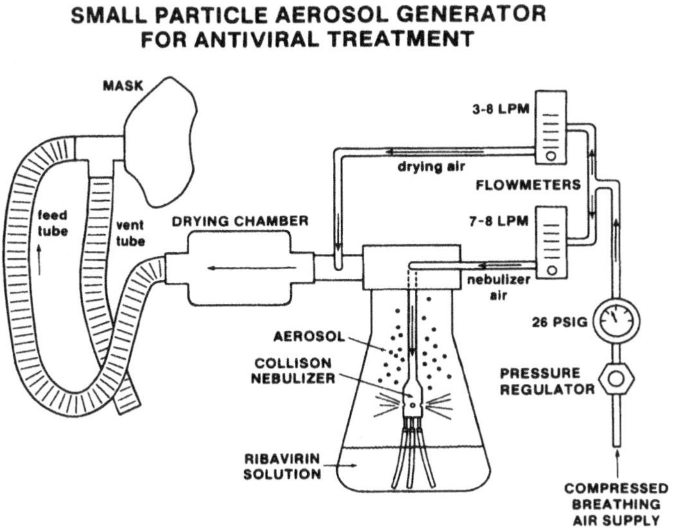

Fig. 1.

Particles are generated from liquid using compressed air passing through an orifice of critical size (0.013 inches). The small particles making up the aerosol are carried upward in the stream of air to the drying chamber. The drying chamber receives additional dry air from the wall source in the hospital or from a portable compressor. The dry air ($\leq$ 10 percent relative humidity) reduces the relative humidity in the drying chamber to about 70 percent with a consequent reduction of particle size to 1.5 microns diameter. This aerosol is inhaled by the patient.

Characterization of the particle size of a Collison generator of the type used for treatment is shown in Figure 2.

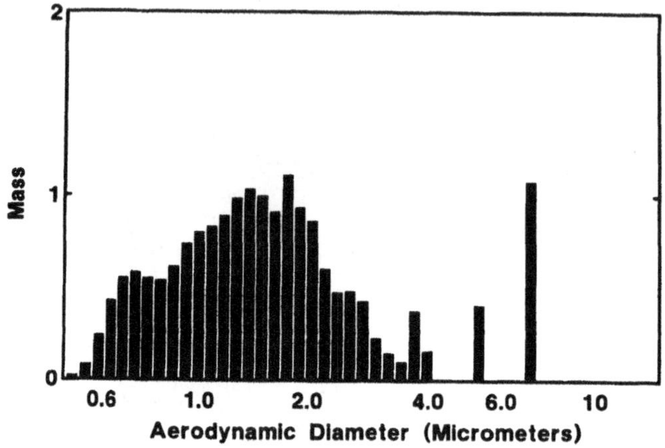

Fig. 2.

The aerodynamic mass median diameter of particles generated from distilled water is 1.43 microns and few particles are larger than 6.0 microns in diameter.

Experiments in this laboratory showed a total deposition of aerosol from the Collison generator of about 63 percent with nose in, mouth out breathing. This is in substantial agreement with the foregoing data when correction is made for mouth out breathing instead of nose out breathing as described in Table 1. Based on the assumption that patients with influenza or other viral disease associated with nasal obstruction would breath appreciably through the mouth a respiratory tract deposition fraction of 70 percent was selected for estimation of retained dosage. Non-hygroscopic particles such as latex beads, carbon particles, mineral particles, etc., do not change size in different relative humidities and will have the deposition pattern shown in the third column of Table 1.

Deposition of particles administered through a respirator.

Since the nose is bypassed by treatment through a respirator, 4 micron particles would be an efficient size to administer because of the greater efficiency of deposition of such particles in the lung. As indicated earlier, particles are 4 microns in diameter when produced by the Collison generator and they can be passed directly into the respiratory tubing without mixing with drying air, which is the system we have adapted in such cases (Figure 3).

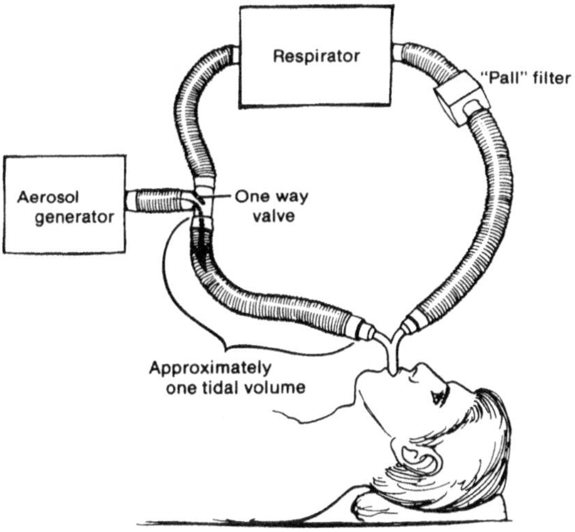

Fig. 3.

Without direct measurement we have estimated a deposition fraction of 60 percent for particles administered into the lung by respirator. It seems probable that the regular, deep respiration controlled by the respirator will deposit particles more efficiently than might occur with spontaneous respiration.

Estimation of aerosol dosage.

A characteristic of the Collison generator is that water vapor is lost in excess of particles containing drug. This will lead to a gradual concentration of drug in the aerosol generator flask of about 17 percent in 12 hours. This will lead to a proportionate increase in drug concentration in aerosol. The mean concentration of ribavirin in aerosol during a 12 hour run is about 200 μg/L. We have used this value for estimation of dosage. Since aerosol deposition is determined by aerosol concentration times minute volume (tidal volume times respirations per minute) times fractional deposition, each patients dose will depend on his values for these variables. However, since tidal volume (and minute volume) is related to body weight and metabolism, the dosage can be estimated from these values. We have estimated that the average deposited dose is 0.82 mg/Kg of body weight per hour for both children and adults inhaling the aerosol through the nose. The dose is slightly smaller when aerosol is administered through a respirator. If the aerosol generator is continuously operated for periods longer than 12 to 15 hours,the concentration of drug increases rapidly and may reach 500-600 μg/L in the aerosol. Such concentrations have been used repeatedly in the treatment of infants and adults with severe pulmonary involvement associated with respiratory syncytial virus or influenza virus infection without untoward consequences.

Operation of aerosol generator with a respirator.

Many patients with influenza or respiratory syncytial virus infection involving the lower respiratory tract who required assisted ventilation have received successful treatment with ribavirin aerosol (Figure 3). The output line of the aerosol generator is attached to the inspiratory line from the respirator to the patient at a distance from the patient such that the internal volume of the respirator tubing about equals the tidal volume of the patient. During the time the patient is exhaling the tubing fills with ribavirin aerosol; when the patient inhales, the aerosol will be carried into the respiratory tract of the patient. To prevent possible loss of pressure in the respirator, a one-way valve is inserted between the respirator and the aerosol generator to prevent back flow into the aerosol generator and loss of pressure in the respirator tubing. The integrity of this system needs to be checked periodically. In respirators in which exhaled air is recycled a filter should be inserted into the expiratory line from the patient to the respirator to prevent aerosol from entering the respirator and clogging valves. The PALL filter No. BB-50T has served this purpose effectively. The filter should be changed every two or three hours, whenever deposits of drug build up on the filter. An aerosol generator can function for 18-20 hours without difficulty if the filters are changed regularly.

Ribavirin aerosol treatment of influenza.

Based on the favorable results of experimental influenza virus infections of mice (11), ribavirin aerosol treatments were evaluated in a series of studies of natural influenza A and B virus infections in students at Texas A and M University (8,10,12,13). The essentials of the studies were that they were made in the same college location in the years 1981 through 1984. The student population each year was nearly identical in epidemiological characteristics. The format of the investigation changed only in minor details from year to year. The viral etiology of the natural influenza varied and included principally influenza A(H1N1) virus infections, a lesser number with influenza B and a small number with influenza A(H3N2) virus infections.

In 1981 and 1982 patients were randomly selected for ribavirin or placebo treatment but the clinical observers were aware of which patients were treated and which were controls. In 1983 and 1984 examining physicians did not know which patients were treated or controlled. Patients were unaware of their treatment status in any year. Qualitative and quantitative virus cultures, antibody measurements and other laboratory work were performed by technicians unaware of the treatment status of the patients. Oral temperatures were recorded every four hours during the study in the student dispensary where the study was conducted.

Patients were admitted to the study if their temperature was 101°F or greater and they had had systemic symptoms compatible with influenza less than 24 hours duration. Following admission examination and testing, treatment status was randomly determined. Extensive clinical laboratory studies including blood counts, liver function tests and electrolyte measurements were made at admission, at three days, at eight days after admission and at 30 days after admission. Quantitative cultures for influenza virus in nasal wash fluids were made once or twice daily during the stay in the hospital.

All patients meeting the clinical criteria described above were entered into the study, but the analysis of results was limited to patients from whom an influenza virus was recovered. In some years, more than one kind of influenza virus was isolated from different individuals, but the viral etiology was not known until after completion of clinical studies. About 80 percent of patients admitted to the study shed influenza virus. The non-influenza patients were usually negative on culture for viral or bacterial pathogens.

The 1981 study of influenza in college students.

An example of one of the several studies is the treatment of students infected with influenza A(H1N1) virus in 1981 (8). The fourteen students who were treated and the 17 students in the control group shed virus. The mean maximum daily temperatures were elevated similarly at admission, 39.4°C treated, and 39.2°C controls. In both groups of patients mean maximum daily temperatures were the same the day of and the day after admission, but on the third day the mean maximum daily temperature was 36.9°C in treated patients and 37.4°C in control patients ($P = 0.003$). The duration of fever from admission to a sustained temperature of less than 37.4°C was 22.8 hours in treated patients and 38.1 hours in control patients ($P = 0.008$). Symptom scores based on a detailed bedside examination in a standard format were 2.6 for systemic illness in both treated and control patients at admission but about 18 hours later treated patients scored 1.5 while control patients scored 2.2 ($P = 0.004$), and this trend of difference between treated control patients continued thereafter. Minor respiratory symptoms showed similar trends in favor of an effect of therapy. Virus shedding correlated with duration of fever. By 18 hours after start of treatment the mean titer of virus per 0.1 mL of nasal wash fluid in treated patients had dropped from about 300 $TCID_{50}$ to 1 $TCID_{50}$. In control patients, after 18 hours, the titer of virus in secretions had not reduced from the 400 $TCID_{50}$ at admission. At 30 hours, titers in both groups of patients decreased, associated with recovery from illness, but the difference between treated and control remained highly significant through 68 hours of observation. Despite treatment, antibody response was excellent in both treated and control patients, and in fact, was slightly higher in treated patients.

The fourteen treated patients received ribavirin aerosol for 23 hours during the first 36 hours in the hospital. The average total dosage was 1.1 grams (50mg/hour retained in the respiratory tract). There was no local or systemic toxicity and no laboratory evidence of toxicity of ribavirin. In summary, the study revealed uniform and significant differences in height and duration of fever, systemic symptoms and virus shedding favoring therapeutic activity of ribavirin aerosol.

Summary of six influenza outbreaks in college students treated with ribavirin aerosol.

Using the methodology just described, ribavirin aerosol was tested at the same location in 1981, 1982, 1983 and 1984. As described in the preceding section it was found that there was a close correlation in the intensity of virus shedding, height of fever and symptomatology with the duration of fever. Table 2 shows the duration of fever in treated and control patients in the period 1981 to 1984 according to the strain of virus causing illness in a particular year.

Table 2. Mean hours from start of treatment to afebrile (<100°F).

Virus	Year	Patients		P-Value
		Control	Treated	
H3N2	83	55.5(3)	25.1(2)	0.017
B	82	55.0(7)	36.1(9)	0.047
H1N1	84	48.6(20)	29.9(18)	0.004
	83	33.5(14)	30.3(13)	0.3
	82	42.7(11)	31.4(8)	0.066
	81	35.5(17)	21.1(14)	0.015

Values represent time in hours. Value in parentheses are number of patients studied. P-values for control and treated patients were obtained from analysis of data by Student's t-test, two-tailed. Statistical analysis of A(H1N1) data by Student's t-test, two-tailed; Control: 1982 v. 1981 or 1983, 1981 or 1983 v. 1984, not significant. Treated: 1981 v. 1982, P = 0.016; 1981 v. 1983, P = 0.032; 1981 v. 1984 and 1982 v. 1983, not significant.

There was a total of 64 treated and 72 control patients studied in these six outbreaks. Ribavirin aerosol treatment was associated with a significant shortening of febrile illness in four outbreaks, borderline significance in one and no significant difference in another. Treated patients had fever ranging from 21.1 to 36.1 hours in the six studies while fever persisted in control patients from 33.5 to 55.5 hours. Fever in controls in the 1983 influenza A(H1N1) outbreak lasted only 33.5 hours in 14 control patients, barely longer than the 30.3 hours in 13 treated patients. It was our opinion that the failure to demonstrate therapeutic activity was related to the mildness of the illness in these patients. It seems probable that once initiated fever and systemic illness might require a certain minimum time to resolve. With the numbers of patients in these studies it was necessary to have about 12 hours or more difference in the duration of fever between treated and control patients to establish a significant effect.

The dosage of ribavirin could have affected the outcome in different years, but dosage did not correlate with the therapeutic effect. The total average dose given in the first 36-40 hours in the hospital increased from 1.15 grams in 1981, to 1.93 grams in 1982, to 3.12 grams in 1983, but decreased to 2.4 grams in 1984. The least favorable response occurred in 1983 when the total dose was highest. These dose differences were largely based on increasing the time of inhalation of aerosol. (In later years there was some increased efficiency of the aerosol equipment that made a minor contribution to the increased dosage.)

More recently, the effect of dosage has been investigated by Wyde et al. (14). The purpose was to discover whether or not a major increase in the concentration of ribavirin in the aerosol would increase the effectiveness of treatment. Those workers found that increasing the concentration of ribavirin from 200 µg/L to 600 µg/L in the aerosol resulted in complete protection of mice from experimental influenza virus infection when treatment was given only 4 hours per day for four days. A dose of 200 µg/L of aerosol required 12 hours of daily treatment for four days to produce a similar level of protection. The latter dose is the one currently in use for human patients. There would be many advantages to the high dose, especially a shorter treatment period in man if similar results can be obtained in the treatment of human illness. Since we had previously noted that aerosol concentrations increase greatly due to excess evaporation of water when the generator has been operated for more than 12-15 hours, the favorable response of some very ill patients, infants and adults may have represented an effect of this larger dose. It is also noted that we have not detected toxicity or intolerance to aerosol treatment in patients receiving prolonged treatment with ribavirin aerosol.

Treatment of influenzal pneumonia.

Experience with influenzal pneumonia is summarized in Table 3 (15).

Table 3. Influenzal pneumonia treated with ribavirin aerosol.

Date of Admission Viral etiology	Age/Sex	Underlying Illness	Secondary bacterial Infection	Respirator Required	Hours of Treatment*	Course	Reference
12-30-80 A/Bangkok/79(H3N2)	61/M	Acute myocardial infarction	No	Yes	60.5	Recovery, no sequelae	Knight et al. 1981
3-16-84 Hemadsorbing agent, probable A/Philippines/2/82(H3N2)	32/F	None	Ps. aeruginosa	Yes	101.0	Recovery, no sequelae	Knight et al. 1986a
2-4-84 B/Singapore/82	33/M	Pulmonary fibrosis from chemical exposure	No	No	51.0	Recovery, no sequelae	Gilbert et al. 1986
1-1-86 B/Singapore/82	36/M	Diabetes, alcoholism	Streptococcus B	Yes	63.0	Recovery, 16 days, no followup	Present report

*Estimated dosage to the respiratory tract, 50 mg/hr.

Two patients with influenza A(H3N2) virus and two patients with influenza B virus infection as the cause of pneumonia received substantial treatment with ribavirin aerosol. Three patients had underlying diseases. Three of the patients also had adult respiratory distress syndrome at admission (all except the patient admitted on 2/4/84 with influenza B virus pneumonia). These three critically ill patients were intubated and required assisted ventilation. All four patients improved promptly and were discharged from the hospital in two to three weeks after start of treatment.

Ribavirin aerosol treatment of respiratory syncytial virus infection.

Following the demonstration of the activity of ribavirin aerosol in the treatment of influenza in 1981 (8), studies using this methodology were tested by Hall et al. (16) against experimental respiratory syncytial virus (RSV) infection of college students. They found that six subjects who shed RSV and were treated with ribavirin had significantly fewer symptoms, less fever, and a higher percentage were no longer shedding RSV on day 7 and 8 after inoculation than 7 control patients. The treatment was well tolerated and there was no significant effect of ribavirin treatment on pulmonary function measured by spirometry performed at the start of treatment, at five days and two months after treatment. Subsequently, Taber et al. (17) reported that ribavirin aerosol significantly reduced illness in patients with uncomplicated bronchiolitis associated with RSV infection. Virus shedding was not reduced in treated patients in the four day period of observation. Hall et al. (18) showed more substantial therapeutic effect in a study in babies with RSV pneumonia. These patients were treated for an average of 20 hours per day for 3 to 6 days, substantially longer than the regimens of 12 hours per day for five days followed by Taber et al. (17). In addition to more rapid clinical improvement in treated patients, Hall et al. (18) found a significant reduction in the titer of RSV in secretions at the end of treatment in ribavirin treated patients compared to control patients, and the number of days of shedding virus was lower in treated patients. Another finding of importance confirmed in Hall's et al. studies, was more rapid return toward normal of arterial oxygen values in treated patients compared to controls. Untreated patients typically exhibit hypoxemia for weeks or months after recovery from acute illness, probably indicating slow healing of pulmonary lesions of RSV infection. It is speculated that chronic lung disease might originate from persistent pulmonary changes of RSV infection. Long term followup will be required to evaluate this potential benefit of treatment.

Hall et al. (19) later reported on the treatment of infants with underlying cardiopulmonary disease (congenital heart diseases or bronchopulmonary dysplasia) treated with ribavirin aerosol. Among 13 such patients (6 treated, 7 controls) the rate and degree of improvement was greater in treated than in control patients. Virus shedding was reduced and arterial oxygen values improved more at the end of treatment in treated patients. Treatment was given almost continuously for a minimum of three days. At the end of treatment, treated patients had improved clinically 45 percent from their status at admission compared to only 21 percent for control patients ($P<0.01$). Improvement in treated patients was appreciable during the first 24 hours of treatment and continued thereafter; control patients actually became worse during the first day of treatment.

In contrast to studies with influenza, two series of patients with RSV infection treated with ribavirin have shown a lesser antibody response than control patients. In the first study (17) antibody responses were lower in treated than in control patients ($P = 0.045$) and in the second study Rosner et al. (20) measured antibody in nasopharyngeal secretions and found that only 43.8% of treated patients developed RSV-IgE response in contrast to 75% in controls ($P = 0.037$). IgA-RSV responses were also less in nasopharyngeal specimens from treated patients, but the differences were less marked than the difference in IgE-RSV responses. Rosner et al. (20) suggested that the reduced IgE-RSV response at the time of the primary infection due to treatment might result in less wheezing at the time of subsequent infections.

Recently, Barry et al. (21) have shown a beneficial effect of ribavirin aerosol in the treatment of bronchiolitis due to RSV infection with results similar to those reported by Taber et al. (17).

Conrad et al. (22) studied ribavirin aerosol treatment of RSV disease proved by virus isolation in infants. They compared the response to treatment of 33 high risk, seriously ill infants infected with RSV to the course of 97 infants with uncomplicated bronchiolitis. They felt that they could not justifiably withhold treatment of seriously ill patients for the purpose of drug evaluation. The more seriously ill patients treated with ribavirin showed prompter resolution of illness than did the untreated patients. The greatest improvement occurred in treated patients between the first and second day of treatment. By use of a therapeutic intervention score they found that recovery in treated patients was significantly more rapid than in the control patients despite their initially milder illness.

Ribavirin aerosol treatment of RSV and parainfluenza virus type 3 infections in immunodeficient infants.

Gelfand et al. (23) treated an 8 month old infant with severe combined immune deficiency disease (SCID) and parainfluenza virus type 3 pneumonia with ribavirin aerosol. At start of treatment the infant was critically ill and exhibited impaired respiratory function with a PaO_2 of 55 mm Hg, fever, radiographic changes, dyspnoea and great debilitation. Numerous viral particles were detected in respiratory secretions by electron microscopy. He was given four, five day courses of ribavirin aerosol and later two, ten day courses of treatment. After each treatment virus disappeared from secretions and major clinical improvement occurred. However, when treatment was stopped the patient relapsed. He received a marrow transplant from his father after the last treatment following which immune function returned.

A second patient, a 5 1/2 month old boy with RSV pneumonia and SCID, responded well to a single course of ribavirin aerosol and did not relapse during a one-week follow-up when he was successfully transplanted with bone marrow from his mother. McIntosh et al. (24) treated another immunodeficient 5 month old female who had pneumonia associated with RSV and parainfluenza virus type 3. She was critically ill, hypoxic and required endotracheal intubation. She received two five day and two 10 day courses of ribavirin aerosol treatment. During the last two 10 day treatments the drug concentration in the reservoir was increased from 20 to 25 mg/mL. This will produce an aerosol concentration of about 250 micrograms per L during the first 12 hours of treatment; thereafter it would increase two or three fold from the starting concentration. The patient received treatment 20 hours per day. During the treatment period there was a slow, steady improvement leading to virtual clearing of pneumonia by the eight week. Parainfluenza virus type 3 disappeared from secretions by the 12th day although RSV was detected 64 days after start of the treatment program following which it disappeared. There was no intolerance or toxicity from the treatment. Marrow transplant was performed on day 100 without success, although the virus infection had cleared; at day 255 a second marrow transplant was successful. In summary, despite the presence of RSV specific IgA, IgG and IgM antibody in secretions, recovery from viral infection did not occur until after prolonged treatment with ribavirin.

The foregoing reports have described a definite and substantial effect of ribavirin against RSV infection in infants. This includes all of the common manifestations of the disease, uncomplicated bronchiolitis, RSV pneumonia and pulmonary infection with RSV in infants with underlying congenital heart disease, bronchopulmonary dysplasia, prematurity, leukemia and immune deficiency. Two patients with severe combined immune deficiency disease with parainfluenza virus type 3 pneumonia improved with ribavirin treatment. In both cases the virus disappeared promptly from secretions, suggesting that this virus is more susceptible to ribavirin aerosol than is RSV.

DISCUSSION

This chapter has presented a description of the theoretical basis and the methodology of aerosol treatment with ribavirin. Since minute volume and metabolic need for gas exchange can be approximately correlated with body weight, deposited dosage of ribavirin aerosol was calculated to be about 0.82 mg/Kg of body weight per hour. Twelve or more hours daily of treatment were empirically selected and were found to be associated with more rapid recovery. Experiments in animals suggest that higher doses in a shorter period of time might also be successful.

The clinical studies showed a substantial therapeutic effect in the treatment of RSV infection in all of its manifestations in infants. The parainfluenza virus infection occurred in children with severe combined immune deficiency disease and for that reason recovery seemed especially significant.

College students with uncomplicated influenza caused by both A and B viruses recovered more quickly with the treatment. A small study of influenzal pneumonia caused by both A and B viruses revealed uniform, prompt recovery in four patients critically ill with this disease.

We noted that illness and virus shedding in RSV infected infants respond less rapidly to ribavirin aerosol than influenza in college students. The interpretation of these findings probably should not be that influenza is more responsive to treatment than RSV infection, since the relative effect of the drug in the two infections in terms of the untreated disease may be about equivalent. This question emphasizes that criteria for what constitutes optimum antiviral effect, are as yet, inadequately defined. Such criteria will become available when more drugs with a greater range of activity and modes of action are studied. For the present, however, we can with confidence report that antiviral therapy for influenza A and B virus and RSV is available, and for the latter two diseases, no other treatment is effective. We believe that parainfluenza virus infection will prove to be responsive to the treatment as well.

REFERENCES

1. Knight, V., Bloom, K., Wilson, S.Z., and Wilson, R.K. Antimicrob. Ag. Chemother. 16:572-578, 1979.
2. Young, H.W., Dominik, J.W., Walker, J.S., and Larson, E.W. J. Clin. Microb. 5:131-136, 1977.
3. Walker, J.S., Stephen, E.L., and Spertzel, R.O. J. Infect. Dis. 133: A140-144, 1976.
4. Stephen, E.L., Dominik, J.W., Moe, J.B., and Walker, J.S. Antimicrob. Ag. Chemother. 10:549-554, 1976.
5. Arensman, J.B., Dominik, J.W., and Hilmas, D.E. Antimicrob. Ag. Chemother. 12:40-46, 1977.
6. Berendt, R.W., Walker, J.S., Dominik, J.W., and Stephen, E.L. Antimicrob. Ag. Chemother. 11:1069-1070, 1977.
7. Wilson, S.Z., Knight, V., Moore, R., and Larson, E.W. Proc. Soc. Exp. Biol. Med. 161:350-354, 1979.
8. Knight, V., McClung, H.W., Wilson, S.Z., Waters, B.K., Quarles, J.M., Cameron, R.W., Greggs, S.E., Zerwas, J., and Couch, R.B. Lancet 2: 945-949, 1981.
9. Knight, V. Lea and Febiger, Philadelphia, PA, p.3, 1973.
10. McClung, H.W., Knight, V., Gilbert, B.E., Wilson, S.Z., Quarles, J.M., and Divine, G. J. Amer. Med. Assoc. 249:2671-2674, 1983.
11. Wilson, S.Z., Knight, V., Wyde, P.R., Drake, S., and Couch, R.B. Antimicrob. Ag. Chemother. 17:642-648, 1980.
12. Wilson, S.Z., Gilbert, B.E., Quarles, J.M., Knight, V., McClung, H.W., Moore, R.V. and Couch, R.B. Antimicrob. Ag. Chemother. 26:200-203, 1984.
13. Gilbert, B.E., Wilson, S.Z., Knight, V., Couch, R.B., Quarles, J.M., Dure, L., Hayes, N., and Willis, G. Antimicrob. Ag. Chemother. 27:309-313, 1985.
14. Wyde, P.R., Wilson, S.Z., Gilbert, B.E. Antimicrob. Ag. Chemother. submitted
15. Knight, V., Gilbert, B.E.,Wilson, S.Z. J. Royal Soc. Med., in press.
16. Hall, C.B., Walsh, E.E., Hruska, J.F., Betts, R.F., and Hall, W.J. JAMA 249:2666-2670, 1983.

17. Taber, L.W., Knight, V., Gilbert, B.E., McClung, H.W., Wilson, S.Z., Norton, J., Thurston, W., Gordon, W.H., Atmar, R.L., and Schlaudt, W.R. Pediatrics 72:613-618, 1983.
18. Hall, C.B., McBride, J.T., Walsh, E.E., Bell, D.M., Gala, C.L., Hildreth, S., Ten Eyck, L.G., and Hall, W.J. N. Eng. J. Med. 308:1443-1447, 1983.
19. Hall, C.B., McBride, J.R., Gala, C.L. Hildreth, S., Schnabel, K.C. JAMA 254:3047-3051, 1985.
20. Rosner, I.K., Wellwer, R.C., Edelson, P.J., Geraci-Ciardullo, K., and Sun, M. J. Infect. Dis., in press.
21. Barry, W., Cochburn, F., Cornall, R., Price, J.F., Sutherland, G., and Vardag, A. Arch. Dis. Childhood 61:593-597, 1986.
22. Conrad, D.A., Chistenson, J.C., Wauer, J.L., and Marks, M.I. Pediatric Infect. Dis. (in press).
23. Gelfand, E.W., McCurdy, D., Rao, C.P. and Middleton, P.J. Lancet 2:732-733, 1983.
24. McIntosh, K., Kwachek, S., Cairus, L.M., Burns, J., and Goodspeed, B. Amer. J. Dis. Child. 138:305-308, 1984.

18

ANTIVIRAL THERAPY OF HIGHLY PATHOGENIC VIRAL DISEASES

J.B. McCORMICK and S.P. FISHER-HOCH

Special Pathogens Branch, Division of Viral Diseases, Center for Infectious Diseases, Centers for Disease Control, Atlanta, Georgia 30333, U.S.A.

INTRODUCTION

There are many severe diseases caused by infection with enveloped negative-stranded RNA viruses which could benefit significantly from antiviral therapy. These diseases include rabies caused by a rhabdovirus, Lassa fever, Argentinian and Bolivian hemorrhagic fevers caused by arenaviruses, Ebola and Marburg hemorrhagic fevers caused by Filoviridae, other hemorrhagic fevers associated with bunyavirus infections [Crimean-Congo hemorrhagic fever (CCHF), Rift valley fever and hemorrhagic fever with renal syndrome (HFRS)], as well as many of the togavirus diseases such as encephalitis, Yellow fever and Dengue fever. The status of antiviral therapy differs for each of these infections, but with few exceptions little clinical trial data exist. The purpose of this review will be to summarize the relevant data for antiviral therapy of these diseases with special emphasis on those human RNA virus infections already known to respond to antiviral therapy in clinical trials.

Many of the infections mentioned above cause hundreds of thousands of cases and thousands of deaths each year and all are associated with a high case fatality ratio (1-4). For example, the case fatality ratio of rabies is nearly 100 %; it is 90 % for Ebola hemorrhagic fever, and 15-20 % for Lassa fever. However, since most of them are endemic in underdeveloped countries, the resources available for seeking effective therapy have been very limited compared to those devoted to diseases more common in affluent countries.

THERAPY OF VIRAL (RNA) HEMORRHAGIC INFECTIONS

In vitro studies

Past approaches to therapy of these diseases have included the highly toxic interferon inducers, more recently interferons themselves and various antiviral compounds. The more successful antiviral compounds include nucleoside analogues such as ribavirin, 3-deazaguanine (3-DG), 3-deazauridine (3-DU), and (S)-9-(2,3-dihydroxypropyl)adenine (S-DHPA) (5). More recent research has identified some newer nucleoside analogues such as tiazofurin and selenazole (6). Rimantadine and amantadine have had limited success (7). The high toxicity level of many of these drugs has limited their use in clinical trials. The few antiviral drugs which have been shown to have acceptable toxicity and clinical efficacy in the treatment of negative-strand virus infections include rimantadine for the treatment of influenza (8), and ribavirin for the treatment of respiratory syncytial virus infections in children and for the treatment of Lassa fever (9,10).

Arenaviruses

One of the most dramatic clinical successes to date has been the therapy of arenavirus infection. There are four arenaviruses which are known to cause disease in humans : lymphocytic choriomenengitis (LCM), Argentinian and Bolivian hemorrhagic fevers (AHF and BHF), and Lassa fever (LF). LCM is the only one of the three diseases which does not have a high case fatality, the others have case fatalities of 15-20 % when untreated.

The earliest studies were done with interferons and interferon inducers, both of which appeared to accelerate the disease process in monkeys infected with Machupo virus (11). Studies of LCM in mice show that high levels of interferon are associated with more severe disease and death (12). Furthermore, interferon levels in patients with AHF have been shown to be quite high, especially in those with severe disease. Interferons may be induced during the course of the disease, and may play an important role in the pathologic process (13). These observations suggest that treatment of human arenavirus infections with interferons may in fact be contraindicated.

A more successful approach has been the effective treatment of Junin virus infection in animals and humans with convalescent-phase plasma. A randomized trial of 188 patients with Argentinian hemorrhagic fever demonstrated that convalescent-phase plasma reduced the mortality from 16 % to 1 % if the patients were treated the first 8 days of illness (14). The efficacy of the plasma therapy appears directly related to the concentration of neutralizing antibodies in the plasma administered (15). Similar success was obtained in treating monkeys infected with Machupo virus (16). However, in both instances a late neurologic syndrome often develops, which may be related to the antibody neutralization titer in the convalescent-phase plasma (14-16). The use of plasma for the treatment of Lassa fever has not been promising in humans (10), despite the efficacy of plasma therapy in some animal studies (see further). Plasma therapy of Ebola and CCHF have been advocated but efficacy has not been demonstrated.

Although a number of the nucleoside drugs mentioned above appear to inhibit arenavirus replication in vitro, only ribavirin has been tested in animal models as well as in human infections (10,17,18). Therefore, a detailed description will be made of ribavirin therapy of these diseases.

Ribavirin. Ribavirin inhibits the replication of several RNA viruses and selected DNA viruses in vitro and in vivo (19). For RNA viruses the activity appears to be confined to the negative-stranded viruses such as myxoviruses, rhabdoviruses, bunyaviruses and arenaviruses, as well as the retroviruses (18,20). It does not appear to be active against double-stranded RNA viruses such as rotavirus; nor is it active against the highly letal negative-stranded RNA viruses Ebola and Marburg (Elliott LH, McCormick JB, unpublished data). Ribavirin acts as an inosine or guanosine analog in vitro, and can deplete the guanosine component of nucleotide pools in cells grown in the presence of ribavirin (20). Several experiments have suggested that the mechanism of inhibition of replication is through the blocking of 5'capping of messenger RNA (20, 21). This observation is consistent with the guanosine analogy of ribavirin since the final step in 5'capping is a guanylylation of the terminal ribonucleic acid at the 5'end of viral messenger RNA (20). Other activities of ribavirin have been suggested, such as inhibition of viral RNA polymerase (20), but these are less well established. Ribavirin is stable at room temperature, and is highly water soluble. It has been formulated for oral and intravenous use as well as for administration by respiratory insufflation, though it is currently licensed in the United States only for respiratory therapy.

- Ribavirin pharmacology. The distribution of radiolabeled ribavirin in monkeys has been demonstrated in all organ systems, such as the liver, where many of the arenaviruses replicate to a high level (22,23). Intravenous drug is absorbed more rapidly than intramuscular drug, but its half-life in plasma following i.v. administration is twice that by i.m. inoculation (24).By eight hours following a single dose of ribavirin, 50 % of the drug has been excreted in the urine. However, the intracellular level of drug in red cells increases for 24 hours before decreasing (25). The concentrations of drug in the plasma vary much more widely in single compared to multiple daily regimens. Doses above 30 mg/kg/day are required to reach arenavirus therapeutic levels in tissues of primates (25). The highest levels of drug are found in the red blood cells, and apparently reflect an accumulation of the drug. The half life of a radiolabeled marker on ribavirin in red blood cells is about 7 days (26). The highest tissue levels seen following a single dose were in the liver, kidney and adrenal, and the lowest in the brain (25).

The drug is phosphorylated intracellularly and the triphosphate is believed to be the active form of the drug (20). The drug is excreted primarily through the urinary tract and is principally excreted in a monophosphate form (20).

- Side effects of ribavirin. The principal side effect of ribavirin is the reduction in hemoglobin. The mechanism of this reduction appears to be hemolysis of older red blood cells through interference with the cell homeostasis. The red cell reduction is reversible and does not appear to result from suppression of the bone marrow (27). This is a cumulative effect, and the decrease in hemoglobin is not seen until the second week of therapy. The drug has no measurable effect on the number of white blood cells, and actually causes an increase in the number of platelets, although it has no effect in vitro on platelet function. Some teratogenic effects were shown when the drug was given at 10 mg/kg/day early in pregnancy in rats and rabbits but none was observed in baboons at doses of 120 mg/kg/day given at critical points during pregnancy (28).

- In vitro activity of ribavirin against arenaviruses. Ribavirin was shown to inhibit in vitro both Machupo and Lassa viruses at a level of 32 μg/ml and to completely inhibit growth at 100 μg/ml. No studies have been done on the specific mechanism of the inhibition of arenavirus replication, though it is presumed likely to be through inhibition of 5'capping of messenger RNA, as observed with other negative-strand RNA viruses (20). Only ribavirin triacetate has been found to inhibit arenavirus replication to nearly the extent seen with ribavirin. All other substitutions or alterations in the ribose or triazole compounds have markedly reduced or eliminated the activity of the drug (20).

- Antiviral activity of ribavirin in primates. Rhesus monkeys which have been infected with a lethal dose of Lassa virus survived when treated with ribavirin (50 mg/kg loading dose, and 30 mg/kg/day for 14 days) within 5 days of challenge. While monkeys treated on day 5 had more severe illness than those treated before, all had limited viremia compared to untreated, fatally infected monkeys (17). Monkeys treated with the same dose of ribavirin on day 7 after inoculation did not survive. Monkeys treated with the same dose of ribavirin as well as immune plasma did survive, suggesting a synergistic effect of ribavirin and immune plasma (28). More recently, primate studies have shown Lassa fever convalescent-phase plasma with a log neutralization index of > 10^2

pfu/ml is required to successfully prevent lethal infection in monkeys, but the plasma must be strain-specific and must be given very early in the course of infection (29).

Therapy of human arenavirus infection

Lassa fever. Human infection by Lassa virus is widespread in West Africa with an antibody prevalance of up to 50 % in some villages (1). The virus is transmitted to humans through contact with infected urine or secretions from persistently infected *Mastomys natalensis* rodents as well as from human to human. The disease Lassa fever ranges in severity from mild or asymptomatic infection to a severe fatal illness (30) and affects all ages and sexes of people (1). Lassa fever produces a number of severe complications including high maternal mortality (especially in the third trimester), fetal wastage, and deafness in many persons who recover from the acute illness (30). In some geographic areas Lassa fever is a major cause of mortality in the adult population, and as such is a prime candidate for effective antiviral therapy (30). Fatality in hospitalized cases of Lassa fever is 15-20 %, and in one study Lassa fever was responsible for 15-30 % of adult medical hospital admissions and 30 % of hospital medical deaths (30).

Two studies of the effectiveness of ribavirin against Lassa fever have been conducted (10). The first was a trial of orally administered ribavirin compared to convalescent-phase plasma. Patients with a clinical diagnosis of Lassa fever were randomly assigned to a drug or plasma treatment group. The patients treated with plasma were given a single unit (250-300 ml) of convalescent-phase plasma with an immunofluorescent antibody (IFA) titer of > 128. The second group was given a 2 gram loading dose of ribavirin followed by 1 gram per day in divided doses every 8 hours for 10 days.

A second trial of ribavirin in patients with Lassa fever included two randomly assigned groups of clinically diagnosed patients who had, in addition to the diagnosis of Lassa fever, a hospital admission value of serum aspartate aminotransferase (AST) of > 150 IU/l. The patients in the first group received intravenous ribavirin, while the second group received intravenous ribavirin and a unit of convalescent-phase plasma at the time of admission. The ribavirin was given as a 2 gram loading dose followed by 1 gram every 6 hours for 4 days. The dose was then reduced to 0.5 grams every 8 hours for 6 more days, a total of ten days of therapy.

The data from both studies were ultimately evaluated for a fatal or nonfatal outcome. In addition, patients were grouped by two risk factors on hospital admission, the level of AST and the level of viremia (31). Survivor analysis of these two variables had shown that levels of AST of 110 IU/1 or above were significantly associated with increased case fatality, and that a level of 150 IU/1 was associated with a case fatality of 55 % (10). Similarly, an admission viremia of $10^{3.6}$ $TCID_{50}$/ml and above was associated with a case fatality of 76 % (31). Patients in the risk groups treated with ribavirin at any time in illness had significantly lower case fatality than untreated patients with the same risk factors (Table 1) (10). (Untreated patients were those studied 3 or 4 years earlier at the same hospitals prior to the availability of ribavirin). The day of illness treatment was begun had a significant effect on outcome. Thus, patients with AST and viremia risk factors who were treated within the first 6 days of illness experienced a 5-9 % case fatality. Those with the same risk factors in whom treatment was initiated more than 6 days after the onset of illness the case fatality was 20-47 % (Table 2). The case fatality was significantly less for ribavirin-treated patients in all categories than for either non-treated patients or for plasma-treated controls. Furthermore, patients treated with ribavirin had significant declines in viremia regardless of outcome, whereas patients who were untreated or treated with plasma and who died showed no decrease in viremia, consistent with the observation that outcome is closely related to the inhibition of virus replication (10). These data show that ribavirin can prevent death in Lassa fever when given at any point in the illness, but that it is more effective when given early. It would appear that ribavirin has an effect on the viremia levels both early and late in illness. It seems likely, therefore that the pathogenesis of

Table 1. Outcome of Lassa fever patients by treatment and risk factor.

	AST ≥ 150			Viremia ≥ 3.6 logs/ml		
	Lived	Died	% Fatal	Lived	Died	% Fatal
No therapy	27	33	55	11	35	76
Ribavirin :						
i.v.[a]	51	12	19	21	10	32
oral[a]	12	2	14	7	3	30
Plasma[b]	14	14	50	9	12	57

[a]Significantly better than no therapy and than plasma therapy.
[b]Not significantly different from untreated.

Table 2. Outcome of Lassa fever patients treated 7 days or later after onset of illness, by treatment and risk factor.

	AST ≥ 150			Viremia ≥ 3.6 logs/ml		
	Lived	Died	% Fatal	Lived	Died	% Fatal
No therapy	20	22	52	6	21	78
Ribavirin :						
i.v.[a]	32	11	26	10	9	47
oral[a]	8	1	11	3	2	40
Plasma[b]	4	8	66	5	7	58

[a]Significantly better than no therapy and than plasma therapy.
[b]Not significantly better than no therapy.

the infection is less reversible later in illness. Therefore, just a reduction in viremia is not sufficient for recovery in all patients. Patients coming late in disease will require further assessment to see if other drugs may be used to stabilize the shock state sufficiently long to allow the viremia to be reduced.

Other arenaviruses. Two other arenaviruses, Junin and Machupo viruses, also cause the severe human diseases of Argentinian hemorrhagic fever (AHF) and Bolivian hemorrhagic fever (BHF) respectively. As in Lassa fever these infections are the result of contact with infected urine or other secretions of persistently infected rodents. There is less evidence of significant human to human infection from these infections, and the diseases are much more limited in the populations affected than is Lassa fever. Nevertheless, they are serious infections in humans with a mortality of 15-20 % when untreated (13), and are also candidates for antiviral therapy. The use of convalescent-phase plasma for the treatment of AHF has in fact been highly successful (14). It does, however, depend on collection of plasma from persons known to have had the disease, checking the plasma (or screening the donor) for antibodies to hepatitis, and proper storage until its use. No effort has yet been made to produce an immune globulin against AHF. The expense of collecting, testing and storing the plasma continues to be highly compensated for by the efficacy of the treatment. The system works well also because the disease is well localized to a relatively small rural area of Argentina. Such a system might work less well if the affected area were much larger. In addition, the advent of Acquired Immunodeficiency Syndrome (AIDS) and the other retroviral diseases transmissable by blood products may increase the risk of using plasma in the future.

Other viral hemorrhagic fevers

Other viral hemorrhagic fevers are also candidates for antiviral therapy, and some are in various stages of study. Hantaan virus, the cause of HFRS, has been shown sensitive to ribavirin in vitro and in a very limited animal study using the natural host Apodemus agrarius (5). There are no satisfactory primate models of HFRS, however, a clinical trial of ribavirin in humans with HFRS is currently being conducted. Hantaan virus is insensitive to all of the interferons in vitro (5). Tiazofurin and its selenium analogue are active in vitro against Hantaan virus, but no in vivo data are available (6). Rift valley fever has been shown to be sensitive to a number of compounds in vitro and in vivo including ribavirin, interferon and selenazole and 2-amino-1,3,4-selenazole (5). Treatment of Rift valley fever virus infections of hamsters by ribavirin at 20 mg/kg b.i.d. has been successful in reducing fatality by 80 %. Ribavirin therapy of monkeys with a loading dose of 50 mg/kg and 30 mg/kg/day resulted in protection from a fatal dose of Rift valley fever virus (5). CCHF virus has been shown sensitive to ribavirin at concentrations of 15-20 μg/ml in vitro; however, no animal therapeutic studies have yet investigated the efficacy of ribavirin against CCHF virus in vivo. This disease is widespread throughout Africa, Asia and the Middle East, causing severe human disease, including several nosocomial outbreaks. In addition, it causes epizootics in domestic animals, thus it is a prime candidate for antiviral therapy.

OTHER APPROACHES TO TREATMENT OF VIRAL HEMORRHAGIC FEVERS

More recent understanding of the immune and physiologic responses to acute infections has suggested new approaches to the treatment of many acute viral diseases. Such an approach would be directed at suppressing or otherwise altering specific host responses to the infection without interfering with the immune response, a sort of "time buying" approach until the virus is cleared by the host immune response. A recent example of such an approach stems from studies of the pathophysiology of Lassa fever. The cause of death in this disease appears to be associated primarily with shock (32,33). Recent studies have excluded disseminated intravascular coagulation as a significant component of the disease process (32). However, studies of endothelial and platelet function suggest that Lassa virus infection produces important dysfunctions of both of these cell systems (32), even though platelets are to be found in adequate numbers and endothelial cells appear normal by histological examination. Studies initiated in primates and then in humans show that in the face of a platelet count of 2 x $10^6/mm^3$ aggregation was abnormal by day 6 after infection (days 6-9 of disease for humans) becoming virtually absent just prior to death. In patients who improve, the platelet aggregation rapidly returns to normal. The same studies in monkeys have shown that endothelium is grossly dysfunctional, as judged by the extensive extravasation of fluid into extravascular spaces, and by failure of prostacyclin production. The view that these lesions are central to the shock associated with Lassa fever, and the potential of prostacyclin treatment in stabilizing patients with severe Lassa fever through the shock phase of the illness is presently being assessed (34).

SUMMARY

There are many severe diseases caused by arboviruses and hemorrhagic fever viruses. These include many of the severe encephalitides, Yellow fever, Dengue fever as well as Lassa fever, Ebola and Marburg diseases, and Crimean/Congo hemorrhagic fever. Although all of these are severe, often lethal, infections, only Lassa fever and Argentinian hemorrhagic fever have been shown amenable to intervention by antiviral therapy by ribavirin and convalescent-phase plasma, respectively. Therapeutic studies of hemorrhagic fever with renal syndrome and Crimean/Congo hemorrhagic fever with ribavirin are currently underway. However, no candidate antiviral agents have yet been produced against the viruses causing Yellow fever, Dengue fever, Ebola-Marburg virus infections, nor against the encephalitides caused by the togaviridae. The frequency of human infection, severity of disease and sequelae, and the high case fatality of many of these latter diseases justify an increased effort in searching for new, effective antiviral drugs against these relatively neglected infections.

REFERENCES

1. McCormick, J.B., Webb, P.A., Krebs, J.W., Johnson, K.M. and Smith, E.S. J. Infect. Dis. 155 : 437-444, 1987.
2. Chen Hua Xin, Qiu Fu Xi, Dong, Bi Jin et al. J. Infect. Dis. 155 : in press, 1987.
3. Report of an International Commission. Bull. WHO 56 : 271-293, 1978.
4. Shope, R.F. In: Tropical and Geographic Medicine (Eds. K.S. Warren and A.F. Mahmoud), McGraw Hill, New York, 1984, pp. 612-620.
5. Canonico, P.G., Kende, M., Luscri, B.J. and Huggins, J.W. J. Antimicrob. Chemother. 14 (Suppl. A) : 27-41, 1984.
6. Huggins, J.W., Robins, R.K. and Canonico, P.G. Antimicrob. Agents Chemother. 26 : 476-480, 1984.

7. Koff, W.C., Elm, J.L. Jr. and Halstead, S.B. Am. J. Trop. Med. Hyg. 30 : 180-189, 1981.
8. Van Voris, L.P., Betts, R.F., Hayden, F.G., Christmas, W.A. and Douglas, R.G. Jr. J. Amer. Med. Assoc. 245 : 1128-1131, 1981.
9. Hall, C.B., McBride, J.T., Walsh, E.E., Bell, D.M., Gala, C.L., Hildreth, S., Ten Eyck, L.G. and Hall, W.J. New Engl. J. Med. 308 : 1443-1447, 1983.
10. McCormick, J.B., King, I.J., Webb, P.A., Scribner, C.L., Craven, R.B., Johnson, K.M., Elliott, L.H. and Belmont-Williams, R. New Engl. J. Med. 314 : 20-26, 1986.
11. Stephen, E.L., Scott, S.K., Eddy, G.A. and Levy, H.B. Texas Reports Biol. Med. 35 : 449-454, 1977.
12. Pfau, C.J., Gresser, I. and Hunt, K.D. J. Gen. Virol. 64 : 1827-1830, 1983.
13. Levis, S.C., Saavedra, M.C., Ceccoli, C., Falcoff, E., Feuillade, M.R., Enria, D.A.M., Maiztegui, J.I. and Falcoff, R. J. Infect. Dis. 149 : 428-433, 1984.
14. Maiztegui, J.I., Fernandez, N.J. and de Damilano, A.J. Lancet ii : 1216-1217, 1979.
15. Enria, D.A., Briggiler, A.M., Fernandez, N.J., Levis, S.C. and Maiztegui, J.I. Lancet ii : 255-256, 1984.
16. Eddy, G.A., Wagner, F.S. Scott, S.K. and Mahlandt, B.J. Bull. WHO 52 : 723-727, 1975.
17. Jahrling, P.B., Hesse, R.A., Eddy, G.A., Johnson, K.M., Callis, R.T. and Stephen, E.L. J. Infect. Dis. 141 : 580-589, 1980.
18. Stephen, E.L., Jones, D.E., Peters, C.J., Eddy, G.A., Loizeaux, P.S. and Jahrling, P.B. In: Ribavirin. A broad spectrum antiviral agent (Eds. R.A. Smith and W. Kirkpatrick), Academic Press, New York, 1980, pp. 169-183.
19. Witkowski, J.T., Robins, R.K., Sidwell, R.W. and Simon, L.N. J. Med. Chem. 15 : 1150-1154, 1972.
20. Smith, R.A. In: Clinical Applications of Ribavirin (Eds. R.A. Smith, V. Knight and J.A.D. Smith), Academic Press, New York, 1984, pp. 1-18.
21. Goswami, B.B., Borek, E., Sharma, O.K., Fujitaki, J. and Smith, R.A. Biochem. Biophys. Res. Commun. 89 : 830-836, 1979.
22. Walker, D.H., McCormick, J.B., Johnson, K.M., Webb, P.A., Komba-Kono, G., Elliott, L.H. and Gardner, J.J. Am. J. Pathol. 107 : 349-356, 1982.
23. McCormick, J.B., Walker, D.H., King, I.J., Webb, P.A., Elliott, L.H., Whitfield, S.G. and Johnson, K.M. Am. J. Trop. Med. Hyg. 35 : 401-407, 1986.
24. Connor, J.D., Hintz, M., Van Dyke, R., McCormick, J.B. and McIntosh, K. In: Clinical Applications of Ribavirin (Eds. R.A. Smith, V. Knight and J.A.D. Smith), Academic Press, New York, 1984, pp. 107-123.
25. Catlin, D.H., Smith, R.A. and Samuels, A.I. In: Ribavirin. A Broad Spectrum Antiviral Agent. Academic Press, New York, 1980, pp. 83-98.
26. Canonico, P.G., Kende, M. and Huggins, J.W. In: Clinical Applications of Ribavirin (Eds. R.A. Smith, V. Knight and J.A.D. Smith), Academic Press, New York, 1984, pp.65-77.
27. Shulman, N.R. In: Clinical Applications of Ribavirin (Eds. R.A. Smith, V. Knight and J.A.D. Smith), Academic Press, New York, 1984, pp. 79-92.
28. Jahrling, P.B., Peters, C.J. and Stephen, E.L. J. Infect. Dis. 149 : 420-427, 1984.
29. Jahrling, P.B. and Peters, C.J. Infect. Immun. 44 : 528-533, 1984.
30. McCormick, J.B., King, I.J., Webb, P.A., Johnson, K.M., O'Sullivan, R., Smith, E.S., Trippel, S. and Tong, T.C. J. Infect. Dis. 155 : 445-455, 1987.

31. Johnson, K.M., McCormick, J.B., Webb, P.A., Smith, E.S., Elliott, L.H. and King, I.J. J. Infect. Dis. 155 : 456-464, 1987.
32. Fisher-Hoch, S.P., Mitchell, S.W., Sasso, D.R., Lange, J.V., Ramsey, R. and McCormick, J.B. J. Infect. Dis. 155 : 465-474, 1987.
33. Fisher-Hoch, S.P., Platt, G.S., Neild, G.H., Southee, T., Baskerville, A., Raymond, R.T., Lloyd, G. and Simpson, D.I.H. J. Infect. Dis. 152 : 887-894, 1985.
34. Fisher-Hoch, S.P., Price, M.E., Craven, R.B., Price, F.M., Forthall, D.N., Sasso, D.R., Scott, S.M. and McCormick, J.B. Lancet ii : 1227-1229, 1985.

19

PERSPECTIVES IN THE USE OF ANTIVIRAL AGENTS FOR PREVENTION AND TREATMENT OF RESPIRATORY VIRUS INFECTIONS

S.J. SPERBER, F.G. HAYDEN

Departments of Internal Medicine and Pathology, University of Virginia School of Medicine, Charlottesville, Virginia

THE NEED FOR ANTIVIRAL CHEMOPROPHYLAXIS AND THERAPY

Impact of Respiratory Viral Infections

Respiratory viral infections are the most common infectious diseases of man and a major cause of morbidity and mortality globally. In developing countries respiratory viruses remain an important cause of childhood mortality (1). Acute respiratory diseases, which are frequently caused by viral infections, account for 20% of childhood mortality and in some parts of the world are the commonest cause of death in children (1). It is estimated that 2.2 million deaths occur throughout the world annually as a result of acute respiratory diseases. These diseases account for about 13% of all deaths in Africa, Central America and the developing countries of Asia, compared to 3% in North America (1). In developed countries certain viral infections continue to cause significant mortality in high-risk groups, particularly infants and the elderly. In the United States, influenza virus epidemics have been estimated to have been associated with over 10,000 excess deaths during 18 of the past 28 years (2).

Viral respiratory tract infections of adults and children are important targets for antiviral chemoprophylaxis and therapy because of their frequency and cummulative morbidity. According to 1981 estimates, the average individual in the United States suffers 1.05 nontrivial upper respiratory illnesses or bouts of influenza per year (3). These episodes account for 36% of days lost from work and 54% of days lost from school due to acute conditions (3). The common cold has been estimated to cause about 250 million days of restricted activity and about 30 million days lost each from work and school (4). Other studies have found that the average adult experiences 2.3 (5) to 5.6 (6) common cold episodes each year. Rhinoviruses alone cause nearly one infection per adult per year (7). Annual expenditures in 1985 for over-the-counter cold symptom treatments were estimated to be $556 million (8), and others have estimated that sales of proprietary cold remedies exceed one billion dollars per year (4). In the United States the economic costs of an influenza pandemic has been placed at 5-6 billion dollars (9), and in epidemic years the costs related to excess hospitalizations are measured in hundreds of million of dollars (10).

Respiratory viruses, including those like rhinovirus that typically produce upper respiratory symptoms, have been implicated in exacerbations of asthma and chronic obstructive airways disease (11-13). Respiratory syncytial virus infections commonly cause lower respiratory tract involvement in the form of bronchiolitis and pneumonia and have been associated with attacks of childhood asthma and prolonged pulmonary function abnormalities (13). Influenza virus infections are well-documented causes of acute deteriorations in asthma and chronic bronchitis patients and have been associated with prolonged small airways functional abnormalitities and airway hyperreactivity in previously healthy adults. The available evidence suggests that the successful use of antiviral agents for prevention and possibly therapy of respiratory viral infections could offer substantial clinical benefit to those with preexisting airways disorders. The role of respiratory viral infections in the pathogenesis of chronic pulmonary disease in adults (14) is unsettled but effective interventions could potentially modify the development of lower respiratory tract problems in certain groups.

Certain respiratory viruses, especially influenza (2) and RSV (15), are nosocomially transmitted agents which infect both hospital personnel and patients. Infected patients often have serious complications and prolonged hospitalizations following these infections. Such groups comprise additional target populations for the use of antiviral agents to prevent the consequences of respiratory viral infections.

Etiologic Agents and Related Clinical Syndromes

The respiratory viral pathogens accounting for the majority of illnesses are listed in Table 1. These viruses share the respiratory tract mucosa as the initial and principal sites of replication and disease expression. A number of other viruses can cause infections with respiratory tract manifestations, often as part of systemic infection, such as Epstein-Barr virus mononucleosis or cytomegalovirus pneumonitis, but these infections are considered in other chapters. The respiratory viruses differ from each other in fundamental biochemical characteristics and the nature of their host cell interactions, as well as in antigenic diversity, epidemiologic patterns, and clinical manifestations of infection. Differences in virus-host cell interactions will necessitate the development of virus-specific agents for most infections. The differing epidemiologic patterns of infection need to be considered in identifying effective strategies for the use of antiviral agents and designing field studies of drug efficacy.

Table 1. Characteristics of the Respiratory Viruses

Virus (Family)	Nucleic Acid	Lipid envelope	Virion diameter (nm)	#antigenic serotypes
Rhino (picorna)	ssRNA	-	20-30	100
Influenza (orthomyxo)	ssRNA (segmented)	+	80-120	3 types-A,B,C, (3 A subtypes, many A & B strains)
Parainfluenza (paramyxo)	ssRNA	+	150-300	types 1-4
Respiratory syncytial (paramyxo)	ssRNA	+	150-300	1 (2 or 3 subtypes recognized)
Adeno (adeno)	dsDNA	-	60-90	41
Corona (corona)	ssRNA	+	75-160	$\geq$4

Abbreviations: ss= single-stranded, ds = double-stranded

This large group of viruses causes a relatively small number of overlapping clinical syndromes (Table 2), the manifestations of which depend on the site and agent of infection. The common cold, pharyngitis, laryngitis, tracheobronchitis, and pneumonia are recognized as typical clinical expressions of infection at different levels of the respiratory tract. In children, croup (laryngotracheobronchitis) and bronchiolitis are additional age-related syndromes. Many unrelated viral pathogens, and some nonviral respiratory pathogens such as mycoplasma, are capable of causing similar clinical syndromes. Not only are there potentially multiple etiologies for each syndrome, but each type of virus is capable of causing different clinical syndromes. These observations suggest that antiviral chemoprophylaxis against all or most respiratory viral illnesses would necessitate continuous use of a safe broad spectrum agent or combination of agents. This possibility appears remote at the present time.

Table 2. Common Respiratory Syndromes and Associated Viral Etiologies

Virus	Common Cold	Pharyngitis	Tracheo-bronchitis	Croup	Bronchiolitis	Pneumonia
Rhino	+++	+	$\pm$			
Influenza	+	+	+++	++[a]	+[a]	++
Para-influenza	+	+	++	+++[a]	++[a]	++[a]
RSV	+	+	++	++[a]	+++[a]	+++[a]
Adeno	++	+++[b]	+		+	++[b]
Corona	++	$\pm$				

Abbreviations: +++ = most commonly associated virus, ++ = commonly associated, + = associated, $\pm$ = possibly associated
a = infants and young children, b = military personnel

Rapid Viral Diagnosis

Rapid viral diagnosis is an integral part of the successful application of virus-specific antiviral agents for management of respiratory tract infections. Factors which are helpful in making a specific diagnosis include knowledge of the age and general condition of the host, time of year, and epidemiologic characteristics of the viruses under consideration. Certain clinical presentations, such as RSV bronchiolitis in young children or febrile respiratory illness in an adult during a known community outbreak of influenza, are sufficiently characteristic to allow a high likelihood of accurate etiologic diagnosis.

However, clinical and epidemiologic evaluation will usually not provide sufficient information to identify a specific viral pathogen, and laboratory documentation of infection is required. Serologic studies are helpful in retrospect but too slow to be beneficial for the therapeutic decisions in an individual patient. Viral isolation may also not be rapid enough to be of value in making clinical decisions for individual patients. Between two and ten days are required to isolate most respiratory viruses (16). Specimen quality and method of collection are important related variables. For example, nasopharyngeal washes or aspirates have a higher yield of virus recovery than do throat swabs for some respiratory pathogens including RSV (17) and rhinovirus, whereas throat swabs are better for isolation of adenovirus (18). At the community level, the virology laboratory can provide surveillance data, helpful in implementing antiviral prophylaxis or treatment.

The rapid detection of viral antigens by immunofluoresence and enzyme-linked immunoassays is gaining wider availability. Several types of sensitive diagnostic test systems are commercially available for RSV. The development of rapid accurate diagnostic methods is needed for other respiratory viruses. In the future, effective intervention will probably depend on the use of narrow spectrum agents in conjunction with rapid etiologic diagnosis.

Limitations of Immunization

Active immunization would in theory be the optimal means to prevent respiratory viral disease. One major problem with immunization is the vast antigenic diversity of the respiratory viruses (Table 1). Rhinoviruses, which account for about 40% of common colds, currently number 100 different serotypes. Even for infections like influenza where effective vaccines exist, the vaccine may lose efficacy because of antigenic changes in the epidemic strains and because of waning host immunity. These limitations necessitate reformulation of the influenza vaccine and annual administration to high risk individuals. The latter requirement has probably contributed to the relatively low utilization rates of inactivated influenza vaccines in the United States. Considerable investigation has been directed at developing attenuated, intranasally-administered vaccines which would be expected to stimulate local immune responses and perhaps provide more solid and longer-lasting protection (19,20). The development of new vaccines may enable implementation of more effective immunization programs. Use of the oral, attenuated adenovirus vaccine by military recruits has proven effective in controlling acute respiratory disese caused by adenovirus types 4 and 7 in this specific population (21). For other viral pathogens, like respiratory syncytial virus, the development of a safe and effective vaccine for infants and young children has not been successful. An inactivated vaccine led to enhanced illness when natural infection was acquired (22), and an attenuated vaccine was not protective (23). According to the author of one infectious diseases monograph "Human respiratory viruses are among the most sucessful animal viruses in the world. Many show regular antigenic variation, and ... are perhaps entering their golden age, with an almost unlimited supply of susceptible hosts..." (24).

Present Status of Antiviral Agents

Both the development and acceptance of effective antivirals for respiratory viral infections has been slow. In the United States effective chemotherapy is currently limited to two licensed drugs, oral amantadine for influenza A virus infections and aerosolized ribavirin for RSV bronchiolitis and pneumonia. This situation contrasts with the extensive and growing number of effective antimicrobic agents used for treating nonviral respiratory tract infections. Advances in the field will depend in part on identifying other potent, selective antivirals.

Paradoxically, confusion also exists concerning the appropriate use of the few available antiviral drugs. For example, in 1968 amantadine became the first systemic antiviral agent licensed for use in the United States. Despite its documented efficacy in the treatment and especially prevention of influenza A virus-induced illness, amantadine has only recently began to gain acceptance by the medical community.

APPROACHES TO THE CONTROL OF RESPIRATORY VIRAL ILLNESS

Prophylaxis of Infection

Methods for controlling respiratory viral illness can be divided into those which prevent transmission of infection and those used for treatment of an established infection. Strategies aimed at protecting uninfected contacts require knowledge of the epidemiology and modes of transmission of the particular virus. In general, transmission requires exposure of a susceptible respiratory tract mucosa to infectious virus. The infective doses, initial sites of infection (nasal passages, conjunctiva, pharynx, or lower respiratory tract) and the associated mechanisms of transmission (direct contact with secretions, exposure to large droplets or inhalation of small particle aerosols) have not been defined for naturally occurring infections but appear to differ for the different respiratory viruses. These differences may have especially important implications for the use of topical antiviral agents. For example, intranasal application of an antiviral may not provide protection against viruses like influenza which can initiate infection in the lower respiratory tract.

Environmental or barrier control measures. For rhinovirus, exposure of the hands to a virus-ladden fomite or hand with subsequent self-inoculation of the conjunctiva or nasal mucosa is an efficient mechanism for spreading infection under experimental conditions (7). Thus, measures aimed at preventing rhinovirus infections might involve mechanical barriers, such as paper handkerchiefs, to prevent contamination of hands and the environment; barriers to interrupt the process of self-inoculation; and practices to remove infectious virus from contaminated hands, such as efficient hand washing. In an experimental setting, the use of paper handkerchiefs reduced rhinovirus contamination of the hands of ill subjects and decreased the frequency of transmitting infection to volunteers exposed by hand-contact with ill subjects (25).

Barrier measures have been investigated to reduce the risk of nosocomial RSV transmission to other infants and hospital staff. The routine use of masks and gowns has been shown to be ineffective in reducing transmission of RSV to hospital personnel caring for infants with the disease (26) or to other infants (27). However, the use of disposable eye-nose goggles by hospital staff has been found to reduce the risk of nosocomial RSV infection in both infants and hospital personnel (28).

Considerable investigation is currently directed at the use of compounds that cause direct or contact inactivation of infectious virus. Such virucidal agents could supplement the effects of mechanical barriers and have been formulated in various ways to determine the feasibility of interrupting transmission. Paper handkerchiefs impregnated with virucidal materials such as citric acid and malic acid appear to be effective in blocking transmission of rhinoviruses under experimental conditions (29,30). Also, under experimental conditions, disinfection of contaminated fomites (31) or the application of virucidal substances (iodine) on the fingertips have been shown to reduce the risk of rhinovirus transmission (32). Virucidal hand lotions may have a similar effect in preventing hand contamination with subsequent self-innoculation or transmission to others. Although of considerable potential value, these approaches have not yet been proven to be effective in preventing natural rhinovirus infections.

Attention to the circulation of air from hospital rooms of persons infected with viruses known to be transmitted by aerosol such as influenza may serve to decrease nosocomial transmission. Attempts to interrupt spread of respiratory infections by air purification with ultraviolet light was found to reduce the risk of measles but not common cold transmission in schools (33).

Immunization. Passive immunization is a theoretical possibility where the agent of infection and the timing of exposure are known. Systemic administration of specific antiviral antibodies has been shown to be effective for certain viral infections like measles and varicella, which have a respiratory tract portal of entry followed by viremic dissemination. Whether systemic or topical administration of antibody would be useful in prophylaxis of infections limited to the respiratory tract remains to be determined. Such an approach would not be feasible for the common cold but might have potential value in high-risk patients after known exposure to viruses of known antigenic type, such as RSV or parainfluenza virus. Passive immunization with intravenous globulins rich in antibody to RSV for prophylaxis and treatment has proven promising in the cotton rat model and is currently under investigation for treatment in a primate model of this disease (34).

Antiviral chemoprophylaxis. Effective chemoprophylaxis generally requires administration of the agent for the duration of the exposure. This period depends on the prophylaxis strategy (seasonal, postexposure, institution-based), the viral pathogen, and in the case of postexposure use, the duration of illness and viral shedding in the index patient.

In formulating prophylactic antiviral strategies for syndromes such as the common cold, it would be desirable to employ an agent active against a broad range of potential pathogens. Interferons have in vitro activity against most respiratory viruses but intranasal interferon alpha-2 has been ineffective in field studies for preventing infections due to viruses other than rhinovirus (35,36). Reduction in the severity of symptoms among those infected with parainfluenza was observed during one prophylaxis study with interferon alpha-2 (36), but this study and others (35,37) have not found reductions in parainfluenza infections. The success of intranasal alpha interferons for prophylaxis of influenza A and B has also been limited (38-40).

One concern regarding antiviral chemoprophylaxis is the successful prevention of infection, such that the protected patient remains suseptible to infection upon subsequent exposure. Optimally a chemoprophylactic agent would allow subclinical infection and associated natural immunization to occur without the burden of illness. In rhinovirus infections, high dose intranasal interferons can prevent both infection and illness after experimental challenge, whereas lower doses allow infection but protect against illness (41). Oral amantadine and rimantadine are more protective against influenza A virus-induced illness than against laboratory-documented infection (42) but still leave a substantial proportion of recipients susceptible to later infection.

Since most respiratory viral infections are self-limited illnesses, potential antiviral agents must have very high therapeutic indices. In particular, with drugs used for prophylaxis in healthy children or adults, only a fraction of whom will be expected to contract the infection, the antiviral must be free from significant toxicity. Because of the frequency of respiratory viral infections, antiviral agents can be anticipated to have repeated use and must also be free of significant cummulative side effects.

Treatment of Established Infection

Objectives in the treatment of an established respiratory viral infection are to reduce the severity of symptoms and decrease the risk of serious morbidity or secondary complications. Treatment of such infections may be directed principally at providing symptomatic relief with no effect on viral replication or at inhibiting viral replication. These approaches are not exclusive, and combinations of antiviral and symptomatic drugs may ultimately provide the greatest benefit.

Preventing transmission of infections is also an important goal of treatment.

This could be accomplished by use of antiviral drugs which limit viral replication or perhaps by reducing illness manifestations (mucus production, cough, sneezing) that are important in transmitting infection. For example, in a mouse model of influenza, rimantadine treatment was effective in reducing pulmonary virus titers and the risk of transmitting infection to untreated contact animals (43). Drugs which reduce symptoms without affecting viral growth could theoretically reduce the risk of transmission but this has not been tested directly.

Symptomatic regimens. Symptomatic treatments for respiratory viral disease have a wide acceptance and utilization throughout the world. In Western nations oral antihistamines and oral or intranasal decongestants are major symptomatic therapies for the common cold and have been shown to provide some reduction in cold symptoms (44). Careful study of the pathogenesis of illness production in different infections may identify host responses that could be specific targets for symptomatic interventions. For example, release of histamine does not appear to play a significant role during rhinovirus colds, whereas high concentrations of kinins have been found in the nasal secretions of symptomatic subjects (45). Drugs which modify the production or biologic effects of kinins might be candidates for symptom control.

Adverse effects of such commonly used treatment regimens need to be considered in the cost-benefit analysis of their value. Placebo-controlled trials have found mild central nervous system side effects with oral antihistamines used alone (46) or in combination with amantadine (47). In experimental rhinovirus colds (48), aspirin therapy was associated with increased viral shedding compared to placebo. If such an effect occurred in naturally occuring rhinovirus infection, it is possible that aspirin treatment could increase the risk of spread of infection to others. The association of Reye's syndrome and salicylate use during certain viral infections including influenza (49) demonstrates another undesirable unexpected adverse consequence of symptomatic therapy.

Certain traditional folk remedies which are culturally accepted as providing symptoms benefit may have biologically relevant activities that could explain their usefulness. Ingestion of hot chicken soup has been advocated by many generations as a remedy for the common cold. This common remedy may provide symptomatic relief because of its ability to increase nasal mucus velocity (50). In China, traditional herbal remedies have been used for many generations to treat upper respiratory tract disease or mild disease of the lower respiratory tract. For severe illness the herbal therapy is combined with Western therapies (1). A flavone isolated from the Chinese medicinal herb Agastache Folium has been shown to have potent in vitro anti-picornavirus activity, perhaps by inhibition of viral replication (51).

Additionally, it has been suggested that radix astragalus, used in traditional Chinese medicine, stimulates IgA secretion and induces interferon production when given orally or by aerosol (1). As such this remedy may have or be capable of inducing specific antiviral activity.

Antiviral chemotherapy. Antiviral therapy would be expected to provide benefit, if ongoing viral replication was central to the pathogenesis of symptom development. Since most respiratory infections are self-limited and have relatively short periods of viral replication, early initiation of treatment is needed to exert a beneficial effect. If the severity or duration of clinical illness is not related to the degree of viral replication, then specific antiviral chemotherapy would probably fail to substantially affect the clinical course. Host responses, such as the release of inflammatory mediators or immune-mediated injury, may play greater roles than on-going viral replication in causing disease in certain infections. Assessments of the duration and degree of viral replication, reflected in the concentrations of virus shed in respiratory secretions, and their relationship to the degree of illness and possibly long-term sequelae are important in this regard.

Another concern in the use of antivirals for treatment of respiratory virus infections, as in prophylaxis, is the possible inhibition of normal host immune responses to infection. This could theoretically leave the host susceptible to reinfection with the same virus serotype. No differences in seroconversion rates have been observed with the use of amantadine or rimantadine for treatment of uncomplicated influenza compared with controls (52, 53). A related concern is the possibility of a rebound in viral replication if treatment ceases before host responses are sufficient to control the infection. Increased prevalence of viral shedding without associated clinical deterioration has been described after low dose, short-term intranasal interferon prophylaxis of experimental rhinovirus infections (41) and after oral rimantadine therapy for natural influenza A virus infections in children (54). The clinical significance of these observations remains to be determined, but suggests the need for longer duration of treatment. Prolonged or recrudescent infections are of particular concern in children with primary or acquired immunodeficiency syndromes (55).

With widespread use, the development of resistance to antiviral agents is an increasing concern. Drug resistant influenza A viruses are readily recovered under laboratory conditions and have been reportedly isolated from non-drug exposed patients (56). During the course of experimental avian influenza (57) amantadine and rimantadine-resistant viruses were readily isolated from drug treated birds. These viruses were also shown to infect and cause illness in contact birds receiving amantadine prophylaxis. Because large amounts of replicating virus are exposed to drug during therapeutic use, it may increase the likelihood of selecting drug-resistant virus relative to prophylaxis. Monitoring of the drug susceptibilities of respiratory viruses will be an important part of future studies.

Topical Administration of Respiratory Antivirals

Although differing in a number of characteristics, the respiratory viruses share a common affinity for the human respiratory tract. For the use of antivirals, this means that effective drug concentrations must be achieved at the site of infection within cells of the respiratory epithelium. Certain antivirals like interferons may exert activity through interaction with cellular receptors and induction of intracellular mediators, but adequate delivery of agents to susceptible areas of respiratory mucosa remains a major practical problem. The use of orally or parenterally administered antivirals is often limited by systemic toxicity. Alternatively, the topical application of antivirals to the respiratory mucosa may achieve high regional antiviral activity and reduce the risk of systemic side effects. This approach has been used successfully with aerosolized ribavirin, in contrast to oral administration (58), and with intranasal interferon.

The nature of the drug and its delivery system are important variables in topical application of antivirals to the respiratory tract. Optimal characteristics of solubility and vehicle composition have not been well defined, but water solubility appears to be advantageous. The convoluted and extensive surface area of the nasal mucous membranes and mucociliary clearance mechanisms have been obstacles to the delivery of intranasal medications. Limited studies of intranasally administered interferons have suggested that they are cleared rapidly like particulate materials. Additionally, beta interferon may be directly inactivated by nasal secretions (59). Interventions to maintain higher local concentrations, such as the use of oral antihistamines to decrease clearance or of saturated cotton pledgets, reduce the amount of interferon needed to achieve an antiviral effect but are probably not of practical value (60). Studies using radiolabelled albumin solutions have found that the intranasal distribution of coarse nasal sprays administered to volunteers is much poorer than that of nasal drops (61), especially if the subjects are upright rather than supine (62). Depending on the volume, velocity, and particle size of the spray, materials are frequently deposited in the vestibule or anterior nasal passages and do not distribute well to the nasopharynx. In volunteer studies of experimental rhinovirus colds, administration of interferon alpha-2 by nasal drops appeared to be associated with greater antiviral effects and clinical benefit than when given by nasal spray (63). Another concern with topical administration is the development of local toxic reactions. Intranasal interferon use avoids the dose-related toxicities of systemic administration but may be associated with the frequent occurance of local irritation and mucosal histopathologic changes (64).

The type of delivery system is also critical when administering drugs by aerosol to the lower respiratory tract. The development of efficient and reliable small particle aerosol generator units has been central to the successful application of aerosolized ribavirin. Currently available delivery systems are limited to use in institutionalized patients because of the requirement for prolonged and often continuous exposure periods and because of the need to monitor the apparatus. In situations where oral administration may be unreliable and where no parenteral form exists, aerosol may provide an alternative means of drug delivery. Aerosolized rimantadine was as effective as oral rimantadine in experimental influenza illness (65) and one study suggested a beneficial effect of intermittent aersolized amantadine in uncomplicated influenza (66). On the other hand, aerosol therapy may cause local irritation or exacerbate preexisting lung disease. In this regard aerosolized amantadine was associated with reversible abnormalities (67), whereas aerosolized ribavirin has been remarkably well tolerated in infants with RSV bronchiolitis and pneumonia.

Strategies For Use

Because of the frequency of respiratory viral infections, clinical use of antiviral agents depends on developing strategies that are effective and safe during long-term or repeated drug administration. In addition, these strategies must be economically feasible and accepted by both physicians and the public. Medication taken on a seasonal basis to prevent common colds would not be successful in the market place if its cost were excessive. Acceptability will be influenced by the complexity of the dose regimen, route of administration, particular target population, and demonstration of cost-effectiveness. Acceptability will also depend on accessibility. If the availability of antivirals for respiratory tract infections is limited to physician-based prescribing, then use will be restricted principally to therapeutic administration and perhaps to prophylaxis in high-risk groups.

These points raise several important questions. If antiviral agents become available to prevent or treat common colds, should they be available without prescription, as are most symptomatic cold remedies? What is the potential for misuse by consumers and what are the risks of selecting drug-resistant viruses? What is the potential effect that effective respiratory antivirals would have on the current widespread use of antibacterials for upper respiratory illnesses? And could these potential changes in antibiotic prescribing have an impact on bacterial resistance patterns?

Seasonal prophylaxis. Viruses which occur predictably in a given locale are potential targets for seasonal prophylaxis. This approach can be most effectively used against viral infections that occur in epidemics of short duration. This pattern is classically shown by the influenza viruses which cause annual outbreaks, lasting 6-8 weeks in a particular region. Daily oral administration of amantadine or rimantadine during community outbreaks of influenza has been proven to be protective (70-90% efficacy) against influenza A virus-induced illness in placebo-controlled studies (42). However, premature discontinuation of prophylaxis can result in an increased risk of infection (68). Prophylaxis might serve not only to protect the subject receiving medication, but also to confer protection on untreated contacts by reducing the number of infectious sources in the family unit or other closed population. For example, recent studies have found that rimantadine prophylaxis of school children during influenza A outbreaks reduces the risk of illness in untreated family contacts (69).

Many other viruses, including respiratory syncytial virus, parainfluenza virus types 1 and 2, rhinovirus, and coronavirus have unique seasonal patterns, but the longer duration of their periods of activity, the variable age-associated illness rates, and the occurrence of overlapping periods of activity are important variables in considering seasonal prophylaxis with antiviral agents which are virus-specific. The broad *in vitro* antiviral spectrum of interferons and the encouraging results in volunteer models of experimental infection engendered hopes that long-term or seasonal use would protect against a wide range of respiratory viruses. As discussed above, intranasal interferon alpha-2 has been proven to prevent only natural rhinovirus infections in studies to date, and its prolonged use (weeks) has been associated with frequent nasal side effects (64). These shortcomings appear to preclude its long-term use in healthy adults. Studies of intranasal interferon alpha-2 in asthmatic children have found some evidence of protection against respiratory illness during intermittent use over a period of three months (70), but further studies are needed in high-risk groups. A recent tolerance trial with interferon beta-serine$_{17}$ indicates that it may have lower potential for side effects but its efficacy remains to be established (71).

Postexposure prophylaxis. Prophylaxis after exposure to a person with respiratory illness, such as commonly occurs in the family setting, is an effective approach where there are high rates of secondary transmission. The prinicipal reservoir of many viruses appears to be the upper respiratory tract of school children (72). The common pattern of transmission involves spread within the school system and subsequently within the household. Epidemiologic studies have shown this to be the case for rhinovirus colds, as well as for influenza and respiratory syncytial virus infections, where school-aged children are frequently implicated as introducing the virus into the household. Short-term (7 days) prophylatic use of intranasal interferon has been recently shown to be an effective means of preventing transmission of rhinovirus colds in families (73,74). One study of oral amantadine (10 days) found marked reductions in influenza illness occurence in amantadine-treated household contacts compared to placebo (75), but a subsequent study by the same investigators failed to confirm their initial observations (76). Further studies are needed to determine whether this approach is an effective and practical one.

Institution-based prophylaxis. Work sites as a place of viral transmission have received increasing study. In contrast to the household setting, rhinovirus infections do not appear to spread very efficiently among persons at work. Other viruses, particularly influenza and RSV, are well documented nosocomial pathogens (2,15). Use of antivirals to prevent transmission of these agents from patient to staff and from staff to patient would be one strategy for controlling this problem. One placebo-controlled study found oral amantadine administration to hospitalized patients was efficacious in preventing nosocomial influenza during a documented community outbreak (77). An uncontrolled study suggested that amantadine administration to patients and staff may prevent the further spread of an established nosocomial outbreak of influenza (78).

Prophylaxis of other instutitional populations, such as those in nursing homes, boarding schools, or day care centers is appropriate where there is a documented risk of outbreak occurence. Amantadine and rimantadine have proven prophylactic efficacy in such settings (79).

Treatment. Therapeutic strategies for the use of an antiviral agent depends on the frequency of infection and its associated morbidity including both short and long-term sequelae. Since use is limited to those individuals who are symptomatic, some drug side effects may be acceptable, particularly in individuals who are at high risk for serious complications. Important targets for therapy include influenza virus infections in children and adults, and respiratory syncytial virus, parainfluenza virus, and adenovirus infections in children. Oral amantadine and rimantadine have significant therapeutic effects in uncomplicated influenza A virus infections of adults. Amantadine has been shown to speed the resolution of small airways functional abnormalities (80). However, it remains undetermined whether either drug can prevent the complications of influenza in high-risk patients or accelerate the resolution of established viral pneumonia (81). Similarly, aerosolized ribavirin has significant antiviral and clinical effects in RSV bronchiolitis and pneumonia of hospitalized children (82), but it is unclear whether its use can decrease the need for ventilatory support during short-term management or reduce the risk of long-term complications.

Other viruses, such as rhinovirus and coronavirus, are such common causes of infection that they have a high cumulative burden of morbidity and economic losses. They may be associated with important complications, including exacerbations of airways disease in those with chronic lung disease and bacterial infections of the ear and sinuses in previously normal patients. For these reasons, these infections are also appropriate targets for treatment. However, recent studies using intranasal interferon for treatment of natural colds have found no evidence of symptomatic benefit compared to placebo (83). An earlier study of intranasal enviroxime also found that it was ineffective in natural colds (84).

DEVELOPMENTAL TESTING OF RESPIRATORY ANTIVIRAL AGENTS

Types of Antiviral Agents

The development of selective antiviral agents depends on identifying drugs which inhibit virus-specific events with no or minimal effects on host cell function. Neutralizing antibodies can interact with virus particles to prevent attachment. Agents which competitively bind to specific host receptor sites could also prevent virus attachment. For example, human rhinoviruses can be divided into a major group representing nearly 90% of serotypes that share a single type of host cell receptor (85) and a minor group which utilizes a different receptor. Monoclonal antibodies directed against the major human cell receptor sites are potent inhibitors of rhinovirus replication *in vitro* (86). Intranasal administration of rhinovirus receptor monoclonal antibody has recently been shown to delay the onset of symptoms and virus shedding following experimental rhinovirus infection in man (87).

Preventing penetration of virus into host cells and viral uncoating are also sites for selective inhibition of viral replication. Certain types of chemical compounds (chalcones, WIN 51711) appear to directly bind to the rhinovirus protein-shell and inhibit replication by essentially trapping the genome inside the virion particle (88,89). To be effective such antivirals must have pharmacological properties that allow direct interaction with the virus particle. The adamantane compounds (amantadine, rimantadine) may exert antiviral activity at the stage of uncoating of the viral genome although their exact mechanism of action is undetermined (90).

With the identification of specific viral enzymes it may be possible to develop antiviral agents which act preferentially on the enzyme or substrate (91). Synthetic peptides are being developed to competitively inhibit the function of critical viral enzymes. For example, the amino acid sequence of the fusion activity region of the influenza HA has been identified and small peptides of the region have been synthesized with the hope of competitively inhibiting its activity. Similar fusion sequences have been synthesized from the parainfluenza F proteins (92). Glycoprotein biosynthesis can be inhibited by interfering with intracellular transport, proteolytic cleavage, or glycosylation of the protein (91). Cleavage of influenza HA is required for virus infectivity and is performed by host cell proteolytic enzymes. Certain protease inhibitors have been shown in animal models to reduce influenza virus titer and NA activity (93).

Other potential sites for antiviral action include inhibition of viral nucleic acid transcription or translation. For example, ribavirin monophosphate inhibits inosine monophosphate dehydrogenase, an enzyme responsible for the synthesis of guanine necleotides (94). After conversion to the triphosphate, ribavirin inhibits steps in the capping and elongation of mRNA. The assembly and release of viruses from infected cells are additional possible targets for antiviral agents.

An approach to more effective antiviral chemotherapy might be the use of combination therapy. By using two agents with different modes of action, an enhanced or synergistic antiviral effect can occasionally be achieved. Synergistic anti-rhinovirus activity has been demonstrated in vitro when various interferons are combined with other anti-rhinoviral agents (95). Interferons also have a synergistic effect in vitro in combination with rimantadine or ribavirin against influenza viruses (96). In vitro and in vivo animal model data indicate that combinations of ribavirin and amantadine or rimantadine also offer promise (97,98), but clinical trials have not yet been conducted.

Preclinical Testing

The obvious goal of research in antiviral chemotherapy is the development of agents which are both effective and safe when used in man. As for other antivirals, efficacy and toxicity testing are performed initially in appropriate cell culture and animal model test systems. The relationships between inhibitory concentrations under in vitro conditions and achievable drug levels in blood or respiratory secretions have not been established for respiratory antivirals. This relates to both a lack of standardized in vitro test methods and an incomplete understanding of relevant pharmacokinetic parameters in man. Such correlations would be very useful in helping to select optimal dosing regimens. For influenza A viruses, a plaque inhibition assay (99) found inhibitory concentrations of amantadine and rimantadine (0.2-0.4 μg/ml) that could be readily achieved in blood and respiratory secretions in man and appears to predict the therapeutic efficacy of these agents. For other compounds, like ribavirin which undergoes intracellular phosphorylation to its active form, correlations between concentrations of the parent drug active in vitro and achievable drug levels in vivo are even less certain.

The identification of a compound which has potent in vitro activity and little or no cell toxicity unfortunately does not guarantee similar effects in vivo. For example, enviroxime which has significant anti-rhinovirus activity in cell and organ culture had little effect on rhinovirus replication when administered intranasally (84), and oral administration was associated with unacceptable gastrointestinal side effects. Additionally, different serotypes of a certain virus may have differing susceptibilities to certain agents. Rhinovirus sensitivity to various interferons, for instance, has been shown to vary with serotype in certain cell culture systems (100).

Adequate small animal models that reflect the virologic and clinicopathologic events seen in man do not exist for some human respiratory virus infections such as rhinovirus or coronavirus. A recently described mouse model of rhinovirus infection remains to be validated. Certain nonhuman primates have been used for antiviral testing (101), but maintaining these animals and wide-scale testing are very costly. The testing of certain antivirals, like interferons, is also limited by species-specific activity and toxicity. Furthermore, considerable differences may exist in the pharmacology of drugs between different species of test animals and man. For example, amantadine metabolism and excretion differ markedly in rodents, other small animals, and man (102). Such observations highlight the difficulty in extrapolating the results of in vitro or animal model testing to practical applications of antivirals in man.

Volunteer Testing

Because of the limitations of available animal models and the uncomplicated course of most respiratory viral infections in adults, human volunteers experimentally challenged with one of the respiratory viruses are often used to determine drug efficacy under carefully controlled conditions. Subjects, whose susceptibility to infection is based on serum antibody titers, are inoculated with a quantity of virus known to cause infection and usually an associated illness. The study drug is evaluated in a double-blind, randomized comparison with an appropriate placebo given by the same route. By altering the time of initiating drug administration in relation to virus challenge, it is possible to assess prophylactic and/or therapeutic activity. These models have also been used to address the questions of dose regimen and mode of drug administration. Human models of experimental rhinovirus, coronavirus, influenza A and B viruses, respiratory syncytial virus, and parainfluenza virus infections have been used in antiviral studies.

Studies in these models of induced infection have correctly predicted the prophylactic efficacies of oral amantadine and rimantadine for influenza A virus and of intranasal interferons for rhinovirus and the theraputic activity of aerosolized ribavirin in RSV infections. In other circumstances the results observed in experimentally induced infections have not corresponded to observations from field studies of naturally occuring infections. Such discrepancies could relate to a number of factors, including the variable virologic and clinical course of both experimental and natural respiratory viral infections, differences in the pathogenesis of infection between the induced models and natural illness, and the relatively small sample sizes studied in volunteer trials. For example, these models usually employ intranasal administration of the virus, sometimes in high concentrations as for influenza viruses, whereas the site of acquisition, the infectious inoculum and duration of exposure may be different in natural conditions.

Studies of drug toxicities and pharamacokinetics in uninfected volunteers, particularly members of target populations, can provide important data for selecting drug regimens. For example, a placebo-controlled trial of of the structurally related drugs, amantadine and rimantadine, found significant differences in toxicity that related to differences in pharmacokinetics (103). There is a need for long-term studies to evaluate cumulative toxicity and the effect of repeated use on efficacy for those agents which may be subject to frequent and repetitive usage. For interferons, the question of immunogenicity is important, as the production of secretory neutralizing antibody might reduce efficacy.

Less controlled conditions occur when such studies are conducted in naturally occurring illness. One important new variable is determining the specific viral etiology of the subject's clinical syndrome. Other variables include the increased difficulty encountered in assessing compliance with the treatment regimen and the potential for unblinding of the study, either of which could bias study results. One study with vitamin C was flawed by the ability of study subjects to determine their treatment status (104,105). Ascorbic acid (vitamin C), which has been espoused for the prevention and treatment of the common cold, was found to have negligible effectiveness in controlled clinical trials (106-108). Similarly, a study claiming effectiveness for oral zinc lozenges in treating common cold symptoms (109) may have been compromised by the lack of an appropriately blinded placebo (110).

Pharmacokinetics and Drug Delivery

The effective application of antiviral agents depends on the site of virus replication in the host. Knowing that rhinovirus replication occurs in the nasal mucosa, for example, enables targeting of anti-rhinoviral agents to this site. On the other hand, anti-influenza agents need to be distributed to the lower respiratory tract. Assessing the delivery of systemically administered antivirals to the respiratory tract mucosa has not been standardized. Direct measure of active drug concentrations in respiratory mucosal biopsies or scrapings may appear to offer the best assessment. However, *in vitro* studies with amantadine suggest that much intracellular drug, which is concentrated in lysosomes, is not biologically active and that concentrations in the extracellular medium are more predictive of antiviral activity (111). Consequently, measurements of drug that penetrates into respiratory secretions after systemic administration may be more predictive of activity in some instances. Such measurements are complicated by the difficulty of obtaining appropriate respiratory secretions for analysis. Nasal washings have been used but introduce a variable dilution effect. A method for collection of induced nasal secretions that provides small volume samples of much higher concentration, as assessed by IgA measurements has been described (112). This technique was employed to demonstrate that after oral administration of equivalent dosages, rimantadine nasal secretion levels are similar to those of amantadine and higher than those present in the plasma (113), observations which may in part explain its clinical effectiveness despite lower plasma levels than amantadine (103).

In the case of topically applied antiviral agents, such as aerosolized ribavirin or intranasal interferon, the value of measuring nasal mucus or lower respiratory tract secretion levels is also uncertain. Collection of nasal secretions after intranasal application of a drug is confounded by the fact that drug is probably recovered from sites where it is not exerting an antiviral effect. Careful studies of the respiratory tract distribution, systemic absorption, and possible systemic effects of topically applied antivirals are important in the development of such agents. A practical problem is possible interference with virus isolation, when residual drug is present in respiratory specimens. Although this has been reported not to be a problem associated with ribavirin (114), the addition of anti-interferon antibody to collection broth is necessary to increase the yield of rhinovirus in samples containing interferon (115).

SUMMARY

Respiratory viruses continue to be major causes of morbidity and mortality. Several antiviral agents with clinical usefulness in respiratory viral infections are currently available, but more effective agents for treatment and prophylaxis are needed. As newer agents become available and indications expand for available antivirals, physician and patient education will be needed to foster their optimal utilization. Increased knowledge of the epidemiology of respiratory viruses, their mechanisms of transmission and disease production, and rapid diagnosis should allow development of effective strategies for their application. Barrier-type methods of prevention, perhaps incorporating virucidal compounds, may prove useful in decreasing transmission of some respiratory viruses. Combinations of specific antiviral agents with drugs that provide symptomatic relief might provide the best means of treatment. Until effective methods for control are available for most respiratory viruses, medical considerations and economic incentives will continue to stimulate research on antiviral therapy and chemoprophylaxis.

REFERENCES

1. WHO Scientific Group on Viral Respiratory Diseases. WHO Technical Report Series 642: 7-49, 1980.
2. Centers for Disease Control. Morbidity and Mortality Weekly Report 35:317-331, 1986.
3. National Center for Health Statistics. Vital and Health Statistics, Series 10 141: 11-18, 1982.
4. Couch, R.B. J. Infect. Dis. 150:167-173, 1984.
5. Gwaltney, J.M. Jr., Hendley, J.O., Simon, G. and Jordan. W.S. N. Engl. J. Med. 275:1261-1268, 1966.
6. Dingle, J.H., Badger, G.F. and Jordan, W.S. Jr. Illness in the Home. Press of Western Reserve University, Cleveland, 1964, pp. 19-96.
7. Gwaltney, J.M., Jr. Viral Infections of Humans (Ed. A.S. Evans), Plenum Press, New York, 1982, pp.491-517.
8. Douglas, R.G., Jr. N. Engl. J. Med. 314:114-115, 1986.
9. Choppin, P.W. In: Virology in Medicine (Eds. H. Rothschild and J.C. Cohen) Oxford University Press, New York, 1986, pp. 3-45.
10. Barker, W.H. In: Options for the Control of Influenza (Eds. A.P. Kendal and P.A. Patriarca), Alan R. Liss, Inc., New York, 1986, pp. 75-87.
11. Horn, M.E., Reed, S.E. and Taylor, P. Arch. Dis. Child. 54:587-592, 1979.
12. Isaacs, D., Flowers D., Clarke, J.R., Valman, H.B. and Macnaughton, M.R. Arch. Dis. Child. 58:500-503, 1983.
13. Carlsen, K.H., Orstavik, I., Leegaard, J. and Hoeg, H. Arch. Dis. Child. 59:310-315, 1984.
14. Monto, A.S. and Ross, J.W. Am. J. Epidemiol. 107:57-64, 1978.
15. Hall, C.B., Powell, K.R., MacDonald, N.E., Gala, C.L., Menegus, M.E., Suffin, S.C. and Cohen, H.J. N. Engl. J. Med. 315:77-81, 1986.
16. Herrman, E.C., Jr. and Herrman, J.A. In: Viral Chemotherapy (Ed. D. Shugar), Permagon Press, New York, 1984, pp.53-54.

17. Masters, H., Lauer, B., Webber, K., Groothuis, J. and Wren, C. In: Program and Abstracts of the Twenty-sixth Interscience Conference on Antimicrobial Agents and Chemotherapy. Am. Soc. Microbiol. 1986, abst. #82, p. 111.
18. Menegus, M.A. and Douglas, R.G., Jr. In: Principles and Practices of Infectious Diseases, 2nd Edition, (Eds. G.L. Mandell, R.G. Douglas, Jr., and J.E. Bennett), John Wiley and Sons, New York, 1985, pp. 138-149.
19. Couch, R.B., Kasel, J.A., Glezen, W.P., Cate, T.R., Six, H.R., Taber, L.H., Frank, A.L., Greenberg, S.B., Zahradnik, J.M. and Keitel, W.A. J. Infect. Dis. 153:431-440, 1986.
20. Johnson, P.R., Feldman, S., Thompson, J.M., Mahoney, J.D. and Wright, P.F. J. Infect. Dis. 154:121-127, 1986.
21. Top, F.H., Jr. Yale J. Biology and Med. 48:185-195, 1975.
22. Kapikian, A.Z., Mitchell, R.H., Chanock, R.M., Shvedoff, R.A. and Stewart, C.E., Am. J. Epidemiol. 89:405-421, 1969.
23. McIntosh, K., Arbeter, A.M., Stahl, M.K., Orr. J.A., Hodes, D. S. and Ellis, E.F. Pediatr. Res. 8:689-696, 1974.
24. Mims, C.A. The Pathogenesis of Infectious Diseases, Academic Press, New York, 1977, p. 129.
25. Hayden, G.F., Hendley, J.O. and Gwaltney, J.M., Jr. J. Infect. Dis. 152:403-407, 1985.
26. Murphy, D., Todd, J.K., Chao, R.K., Orr, I. and McIntosh, K. J. Pediatrics 99:746-750, 1981.
27. Hall, C.B. and Douglas, R.G., Jr. Am. J. Dis. Child. 135:512-515, 1981.
28. Gala, C.L., Hall, C.B., Schnabel, K.C., Pincus, P.H.. Blossom, P., Hildreth, S.W., Betts, R.F. and Douglas, R.G.,Jr. JAMA 256:2706-2708, 1986.
29. Hayden, G.F., Gwaltney, J.M., Jr., Thacker, D.F. and Hendley, J.O. Antiviral Res. 5:103-109, 1985.
30. Dick, E.C., Hossain, S.U., Mink, K.A., Meschievitz, C.K., Schultz, S.B., Raynor, W.J. and Inhorn, S.L. J. Infect. Dis. 153:352-356, 1986.
31. Gwaltney, J.M., Jr. and Hendley, J.O. Am. J. Epidemiol. 116:828-833, 1982.
32. Gwaltney, J.M., Jr., Moskalski, P.B. and Hendley, J.O. J. Infect. Dis. 142:811-815, 1980.
33. Andrewes, C., The Common Cold. W.W. Norton, New York, 1965, pp. 156-157.
34. Hemming, V.G., Prince, G.A., Horswood, R.L., London, W.T., Murphy, B.R., Walsh, E.E., Fischer, G.W., Weisman, L.E., Baron, P.A. and Chanock, R.M. J. Infect. Dis. 152:1083-1087, 1985.
35. Douglas, R.M., Albrecht, J.K., Miles, H.B., Moore, B.W., Read, R., Worswick, D.A. and Woodward, A.J. J. Infect. Dis. 151:731-736, 1985.
36. Monto, A.S., Shope, T.C., Schwartz, S.A. and Albrecht, J.K. J. Infect. Dis. 154:128-133, 1986.
37. Hayden, F.G., Gwaltney, J.M., Jr. and Johns, M.E. Antiviral Res. 5:111-116, 1985.
38. Solov'ev, V.D. Bull WHO 41:683-688, 1969.
39. Merigan, T.C., Hall, T.S., Reed, S.E. and Tyrrell, D.A.J. Lancet 1:563-567, 1973.
40. Hayden, F.G. In: Antiviral Chemotherapy, New Directions for Clinical Application and Research (Eds. J. Mills and L. Corey), Elsevier, New York, 1986, pp. 28-39.
41. Samo, T.C., Greenberg, S.B., Palmer, J.M., Couch, R.B., Harmon, M.W. and Johnson, P.E. J. Infect. Dis. 150:2,181-188, 1984.
42. Dolin, R., Reichman, R.C., Madore, H.P., Maynard, R., Linton, P.N. and Webber-Jones, J. N. Engl. J. Med. 307:580-584, 1982.
43. Schulman, J.L. Proc. Soc. Exp. Biol. Med. 128:1173-1178, 1968.

44. Crutcher, J.E. and Kantner, T.R. J. Clin. Pharmacol. 21:9-15, 1981.
45. Naclerio, R.M. Gwaltney, J.M., Jr., Hendley, J.O., Eggleston, P., Baumgarten, C.R., Lichtenstein, L.M. and Proud, D. Clin. Res. 33: 613A, 1985.
46. Howard, J.C., Jr., Kantner, T.R., Lilienfield, Princiotto, J.V., Krum, R.E., Cruther, J.E., Belman, M.A. and Danzig, M.R. JAMA 242:2414-2417, 1979.
47. Millet, V.M., Dreisbach, M. and Bryson, Y.J. Antimicrob. Agents Chemother. 21:1-4, 1982.
48. Stanley, E.D., Jackson, G.G., Panusarn, C., Rubenis, M. and Dirda, V. JAMA 231:1248-1251, 1975.
49. Waldman, R.J., Hall, W.N., McGee, H. and Van Amburg, G. JAMA 247:3089-3094, 1982.
50. Saketkhoo, K., Januszkiewicz, A. and Sackner, M.A. Chest 74:408-410, 1978.
51. Ishitsuka, H., Ohsawa, C., Ohiwa, T., Umeda, I. and Suhara Y. Antimicrob. Agents Chemother. 22:611-616, 1982.
52. Wingfield, W.L., Pollack, D. and Grunert, R.R. N. Engl. J. Med. 281:579-581,1969.
53. Van Voris, L.P., Betts, R.F., Hayden, F.G., Christmas, W.A. and Douglas, R.G., Jr. JAMA 245:1128-1131, 1981.
54. Hall, C.B., McBride, J.T., Gala, C.L., Dolin, R. and Markovitz, D. In: Options for the Control of Influenza, (Eds. A. P. Kendal and P. A. Patriarca), Alan R. Liss, Inc., New York, 1986, pp. 331-341.
55. Fishaut, M., Tubergen, D. and McIntosh, K. J. Pediatr. 96:179-186, 1980.
56. Heider, H., Adamczyk, B. Presber, H.W., Schroeder, C., Feldblum R. and Indulen, M.K. Acta Virol. 25: 395-401, 1981.
57. Webster, R.G., Kawaokay, Bean, W.J., Beard, C.W. and Brugh, M. J. Virol. 55:173-176, 1985.
58. Smith, C.B., Charette, R.P., Fox, J.P., Cooney, M.K. and Hall, C.E. J. Infect. Dis. 141:548-554, 1980.
59. Harmon, M.W., Greenberg, S.B. and, Couch, R.B. Proc. Soc. Exp. Bio. Med. 152:598-602, 1976.
60. Greenberg, S., Harmon, M.W., Johnson, P.E. and Couch, R.B. Antimicrob. Agent. Chemoth. 14:596-600, 1978.
61. Hardy, J.G., Lee, S.W. and Wilson, C.G. J. Pharm. Pharmacol. 37:294-297, 1985.
62. Aoki, F.Y. and Crawley, J.C.W. Br. J. Clin. Pharmac. 3:869-878, 1976.
63. Hayden, F.G. and Gwaltney, J.M., Jr. J. Infect. Dis. 150:174-180, 1984.
64. Hayden, F.G., Mills, S.E. and Johns, M.E. J. Infect. Dis. 148:914-921, 1983.
65. Hayden, F.G., Zylidnikov, D.M., Iljenko, V.I. and Padolka, Y.V. Antiviral Res. 2:147-153, 1982.
66. Hayden, F.G., Hall, W.J. and Douglas R.G., Jr. J. Infect. Dis. 141:535-542, 1980.
67. Knight, V., Bloom, K., Wilson S.Z. and Wilson, R.K. Antimicrob. Agents Chemother. 16: 572-578, 1979.
68. Muldoon, R.L., Stanley, E.D. and Jackson, G.G. Am. Rev. Resp. Dis. 113:487-491, 1976.
69. Clover, R., Ramsey C. Jr., Crawford S. and Abell, T. J. Cellular Biochemistry 9C Supplement: 286, 1985.
70. McIntosh, K. - personal communication.
71. Sundmacher, R., Mattes, A., Neumann-Haefelin, D., Papendick, U. and Springmeoer, U. J. Interferon Res. 6: 30-38, 1986.
72. Gwaltney, J.M., Jr. In: Viral Infections in Oral Medicine (Eds. J. Hooks and G. Jordan), Eslevier, New York, 1982, pp. 243-253.
73. Douglas, R.M., Moore, B.W., Miles, H.B., Davies, L.M., Graham, N.M.H., Ryan, P., Worswick, D.A. and Albrecht, J.K. N. Engl. J. Med. 314:66-70, 1986.

74. Hayden, F.G., Albrecht, J.K., Daiser, D.L. and Gwaltney, J.M. N. Engl. J. Med. 314:71-75, 1986.
75. Galbraith, A.W., Oxford, J.S., Schild, G.C. and Watson, G.I. Lancet 2:1026-1028, 1969.
76. Galbraith, A.W., Oxford, J.S., Schild, G.C. and Watson, G.I. Bull WHO 41: 677-682, 1969.
77. O'Donoghue, J.M., Ray, C.G., Terry, D.W., Jr. and Beaty, H.N. Am. J. Epidemiol. 97:276-282, 1973.
78. Atkinson, W.L., Arden, N.H., Patriarca, P.A., Leslie, N., Lui, K-J. and Gohd, R. Arch. Intern. Med. 146:1751-1756, 1986.
79. Payler, D.K. and Purdham, P.A. Lancet 1:502-504, 1984.
80. Little, J.W., Hall, W.J., Douglas, R.G., Jr., Mudholkar, G.S., Speers, D.M. and Patel, K. Am. Rev. Resp. Dis. 118:295-303, 1978.
81. Couch, R.B. and Jackson, G.G. J. Infect. Dis. 134:516-527, 1976.
82. Hall, C.B., McBride, J.T., Walsh, E.E., Bell, D.M., Gala, C.L., Hildreth, S., Eyck, L.G.T. and Hall, W.J. N. Engl. J. Med. 308:1443-1447, 1983.
83. Just, M., Berger, R., Ruuskanen, O., Ludin, M. and Linder, S. J. Interferon Res. 6:32,1968.
84. Hayden, F.G. and Gwaltney, J.M., Jr. Antimicrob. Agents Chemother. 21:892-897, 1982.
85. Abraham, G. and Colonno, R.J. J. Virol. 51:340-345, 1984.
86. Colonno, R.J., Callahan, P.L. and Long, W.J. J. Virol. 57:7-12, 1986.
87. Hayden, F.G., Colonno, R.J. and Gwaltney, J.M., Jr. - unpublished observations.
88. Ninomiya, Y., Ohsawa, D., Aoyama, M., Umeda, I., Suhara, Y. and Ishitsuka, H. Virology 134:269-276, 1984.
89. Smith, T.J., Kremer, M.J., Luo, M., Vriend, G., Arnold, E., Kamer, G., Rossmann, G., McKinlay, M.A., Diana, G.D. and Otto, M.J. Science 233:1286-1293, 1986.
90. Couch, R.B. and Six, H.R. In: Antiviral Chemotherapy, New Directions for Clinical Application and Research (Eds. J. Mills and L.Corey), Elsevier, New York, 1986, pp. 50-57.
91. Galasso, G.J. J. Antimicrob. Chemother. 14 Suppl A:127-136, 1984.
92. Richardson, C.D. and Choppin, P.W. Virology 131: 518-532, 1983.
93. Oxford, J.S. Br. Med. Bulletin 41:396-400, 1985.
94. Prusoff, W.H., Lin, T-S. and Zucker, M. Antiviral Res. 6:311-328, 1986.
95. Ahmad, A.L.M. and Tyrrell, D.A.J. Antiviral Res. 6:241-252, 1986.
96. Hayden, F.G., Schlepushkin, A.N. and Pushkarskaya, N.L. Antimicrob. Agents Chemother. 25:53-57, 1984.
97. Galegov, G.A., Pushkarskaya, N.L., Obrosova-Serova, N.P. and Zhdanov, V.M. Experimentia 33:905-906, 1977.
98. Wilson, S.Z., Knight, V., Wyde, P.R., Drake, S. and Couch, R.B. Antimicrob. Agents Chemother. 17:642-648, 1980.
99. Hayden, F.G., Cote, K.M. and Douglas, R.G., Jr. Antimicrob. Agents Chemother. 17:865-870, 1980.
100. Came, P.E., Schafer, T.W. and Silver, G.H. J. Infect. Dis. 133 Supplement: A136-139, 1976.
101. Grunert, R.R. Ann. Rev. Microbiol. 33:335-353, 1979.
102. Bleidner, W.E., Harmon, J.B., Hewes, W.E., Lynes, T.E. and Hermann, E.C. J. Pharmacol. Exp. Ther. 150:484-490, 1965.
103. Hayden, F.G., Hoffman, H.E. and Spyker, D.A. Antimicrob. Agents Chemother. 23:458-464, 1983.

104. Karlowski, T.R., Chalmers, T.C., Frenkel, L.D., Kapikian, A.Z., Lewis, T.L. and Lynch, J.M. JAMA 231:1038-1042, 1975.
105. Chalmers, T.C. Amer. J. Med. 58:532-536, 1975.
106. Coulehna, J.L., Eberhard, S., Kapner, L., Taylor, F., Rogers, K. and Garry, P. N. Engl. J. Med. 295:973-977, 1976.
107. Pitt, H.A. and Costrini, A.M. JAMA 241:908-911, 1979.
108. Carr, A.B., Einstein, R., Lai, L.Y.C., Martin, N.G. and Starmer, G.A. Med. J. Aust. 2:411-412, 1981.
109. Eby, G.A., Davis, D.R. and Halcomb, W.H. Antimicrob. Agents Chemother. 25:20-24, 1984.
110. Farr, B.M. and Gwaltney, J.M., Jr. J. Chronic Diseases, in press.
111. Richman, D.D., Yazaki, P. and Hostetler, K.Y. Virology 112:81-90, 1981.
112. Powell, K.R., Shorr, R., Cherry, J.D. and Hendley, J.O. J. Infect. Dis. 136:109-111,1977.
113. Hayden, F.G., Minocha, A., Spyker, D.A. and Hoffman, H.E. Antimicrob. Agents Chemother. 28:216-221, 1985.
114. Knight, V., Wilson, S.Z., Quarles, J.M., Greggs, S.E., McClung, H.W., Waters, B.K., Cameron, R.W. Zerwas, J.M. and Couch, R.B. Lancet 2:945-949, 1981.
115. Hayden, F.G. and Gwaltney, J.M., Jr. Antiviral Res. 3:67-71, 1983.

20

PERSPECTIVES FOR THE TREATMENT OF GASTROINTESTINAL TRACT VIRUS INFECTIONS

L.A. BABIUK, M.I. SABARA and P. FRENCHICK

Veterinary Infectious Disease Organization, 124 Veterinary Road, Saskatoon, Saskatchewan, Canada, S7N OWO

INTRODUCTION

Gastointestinal infections are one of the leading causes of morbidity and mortality in infants and young animals, both in developing and in developed countries. It has been estimated that between 3-10 billion cases of diarrhea occur annually in humans, resulting either directly or indirectly in approximately 10 million deaths (1,2). The largest number of deaths occur in the age group of 6-12 months, although diarrhea is a major factor in the health of children up to 5 years of age. The causes of acute diarrhea are often multifactorial being determined by nutritional and environmental conditions. For example, significant malnutrition may lead to an increased severity and duration of diarrhea. In addition to the nutritional component, the level of contamination of the environment by specific organisms has a significant effect on both incidence and mortality rate of the resulting diarrhea (3). It is estimated that malnurished children in unhygienic situations have a fatality rate of 6 deaths per 1000 cases of diarrhea presented for treatment (1).

In the present review, we will briefly discuss the specific agents involved, the mechanisms of pathogenesis of enteric virus infections and the potential application of vacines and antiviral agents for the control of enteric viral infections. Attempts will be made to correlate the mechanisms of pathogenesis with the potential use of a specific antiviral agent under specific disease conditions.

In addition to viral causes of acute gastroenteritis, other organisms such as bacteria and parasites can cause severe gastrointestinal infections. The incidence of diarrhea due to bacterial or parasitic gastrointestinal infections demonstrates geographic variability. For example, the incidence of enterotoxigenic E. coli (ETEC) is very high in developing countries as opposed to developed countries (4). In this review we will not discuss agents other than viruses nor other noninfectious causes of diarrhea.

CAUSATIVE AGENTS

During the past two decades, improvements in electron microscropic identification of microorganisms and diagnostic tests have greatly increased the efficiency with which viral agents can be identified in fecal samples of animals and humans (5). Although many episodes of nonbacterial diarrhea still remain undiagnosed with respect to the etiologic agent, a wide variety of different viruses have been incriminated as potent causative agents of nonbacterial gastroenteritis. A summary of the various viral agents that have been isolated and associated with acute viral gastroenteritis are listed in Table 1.

The viruses which cause gastrointestinal infections can generally be divided into three groups. In the first group, replication is restricted to the gastrointestinal tract and the viruses induce disease as a direct result of their infection of intestinal cells. The agents in this group generally enter the host directly, via the oral cavity, into the gastrointestinal tract. The disease caused by these agents can be characterized by rapid onset (1-2 days) of acute diarrhea, abdominal cramps, nausea, vomiting and fever. It is this group of viruses that will be addressed predominantly in this review.

The second group of viruses can enter the host via the oral cavity and replicate in the gastrointestinal tract, but do not remain localized, and generally do not cause acute gastrointestinal infections even though they are shed in the feces. These viruses often spread to other target organs such as lymphoid tissues or even to the central nervous system and cause systemic infections (6,7,8). Examples of this group of viruses include hepatitis A, coxsackie virus, polio, etc. The third group of viruses which infect the gastrointestinal tract do so indirectly. Their routes of entry are generally not oral and they reach the gastrointestinal tract via systemic spread, most often via the hematogenous route. Therefore, the gastrointestinal tract serves as a secondary target. In these instances gastrointestinal symptoms may occur but they are not the main clinical features. Examples of such infections are hepatitis B and cytomegalovirus (9,10).

At least nine different virus families can cause infections of the gastrointestinal tract and induce some degree of intestinal damage and diarrhea under appropriate conditions. As is evidenced from Table 1, these viruses fall into a wide variety of different families including RNA and DNA viruses. Until recently it was felt that the viruses that cause direct infection of the gastrointestinal needed to be acid-stable and

TABLE 1 - SUMMARY OF SOME VIRAL AGENTS CAPABLE OF CAUSING GASTROINTESTINAL INFECTIONS IN HUMANS AND ANIMALS[1]

Family	Biochemical/Biophysical Properties	Human/Animal[2]
Picornaviridae - entero	27-32nm Icosahedral (naked) ss. RNA	+/+
Reoviridae - reo - rota	80nm Icosahedral (naked) ds RNA	+/+
Astroviridae	28nm Icosahedral (naked) ss RNA	+/+
Caliciviridae	30nm Icosahedral (naked) ss RNA	+/+
Coronaviridae	75-150nm Helical (enveloped) ss RNA	+/+
Toroviridae - Breda - Berne	110-140nm Helical (enveloped) ss RNA	?/+
Norwalk	23-34nm Icosahedral (naked) ss RNA	+/-
Adenoviridae	70nm Icosahedral (naked) ss DNA	+/?
Parvoviridae	30nm Icosahedral (naked) ds DNA	?/+

1 All viruses listed here enter via the oral route and cause local infections of the gastrointestinal tract. Virus is shed in feces and acts as a source of environmental contamination.
2 + indicates definite role in acute gastrointestinal infections; ? limited information prevents definite inclusion or exclusion as true gastrointestinal agents; - no involvement known at present.

non-enveloped so as to reach the intestine after an encounter with harsh gastric acids and duodenal bile salts. However, the recent identification of coronaviruses and Breda/Berne viruses which are enveloped clearly indicates that some enveloped viruses do have the capacity to withstand the environmental conditions in route to the gastrointestinal tract where infection will occur (11,12,13,14,15).

The diversity of biochemical and biophysical properties of enteric viruses and different mechanims of viral replication makes the identification of a single antiviral agent to control the majority of gastrointestinal viral infections remote. Thus, it appears that a concerted effort will have to be made to develop a large number of chemotherapeutic agents, each directed towards a specific virus. Furthermore, since enteric viral infections are localized, the incubation period is short and the infection is almost always self limiting, it becomes imperative that antiviral drugs be used in a prophylactic rather than in a therapeutic mode.

Another problem associated with antiviral chemotherapy is the inability to always predict the time when a specific infection will occur. Although epidemiological surveys have aided in elucidating this problem, more studies are needed before accurate predictions can be made. For example, in the case of rotavirus infections, there is generally a peak of activity in winter months in temperate climates (16). However, in tropical countries, the pattern of rotaviral infections may vary depending on the country and various climatic conditions (17). In other instances, such as with Norwalk virus there does not appear to be any seasonal variation in disease incidence (18).

PATHOPHYSIOLOGY OF ENTERIC VIRUS INFECTIONS

In most virus infections of the gastrointestinal tract, regardless of whether the virus has a predelection for the epithelial cells at the tip of the villi or in the crypts, there is severe shortening and occasionally fusion of adjacent villi, resulting in reduced adsorptive surface of the intestine (19,20). Following infection of the epithelial cells at the tips of the villi, the mature adsorptive cells are replaced with immature squamous to cubiodal epithelial cells. Until these cells mature, their absorptive capacity and enzymatic activity is greatly reduced. Since these immature cells also appear to be relatively resistant to virus infection, the disease is often self-limiting if dehydration is not so significant as to cause hospitalization or death. In virus infections where the crypt cells are not damaged, the rate of recovery is generally rapid. However, in those viral infections where the crypt cells are infected, there is also shortening of villi, but since there are a limited number of new cells available to migrate up the villi, recovery generally takes longer. As more knowledge is gained regarding the virulence and pathogenesis of various gastrointestinal viruses, it is becoming evident that the virulence of the virus will determine the extent of replication within the gastrointestinal tract. Thus, the less virulent viruses may still kill and cause shortening of villi in localized areas, but the viruses are generally restricted to a very small portion of the gastrointestinal tract. Those viruses which are much more virulent appear to have a greater capability of infecting a larger number of cells throughout the gastrointestinal tract, i.e. they are not localized. Thus, an avirulent or mild virus may only infect a certain portion of the jejunum, whereas a more virulent strain of the same virus may infect the jejunum, illeum and even cells of the colon. Since glucose and sodium adsorption are highest in the proximal and middle part of the jejunum, damage here will cause most severe diarrhea. Therefore, the extent of diarrhea will be correlated with the site at which the virus infection occurs. Table 2 illustrates the sites of replication of some of the enteric viruses. Coinfection with virus that replicate in different areas can occur and these mixed infections are often more severe than single infections due to increased intestinal damage. Mixed infections and more common in animals than they are in humans (21).

In most animal species, viral diarrhea is characterized by profuse watery stools containing increased concentrations of sodium, potassium and chloride. The direct destruction of adsorptive epithelial cells and alteration of microvilli leads to diminished glucose, sodium carrier and Na+, K+-ATPase activity. This results in loss of sodium, potassium, chloride, bicarbonate and water. As the virus kills absorptive cells, there is also a loss of enzymes responsible for digestion of disaccharides. These disaccharides, especially lactose, are osmoticaly active and cause an influx of fluid into the gut lumen. The loss of bicarbonate leads to development of acidosis. Acidosis can further create a K^+-H^+ ion exchange across a cell membrane and inhibits cellular functions required for maintaining normal potassium concentrations with a net loss of potassium from cells. As a result, hypoglycemia occurs, glucose adsorption is impaired which further results in drastic reduction of Na+ adsorption since it is glucose dependent. This series of complex pathophysiological changes, if not promptly corrected, may result in the death of the individual. Fortunately, in many cases, the extent of these pathophysiological changes is not sufficient to require hospitalization.

TABLE 2 - SITE OF REPLICATION OF SOME ENTERITIS VIRUSES

Virus	Horizontal	Longitudinal
Rota	Enterocytes, villus tip	Small intestine
Corona	Enterocytes, top half	Small and large intestine, colon
Breda/Berne	Mid villus, crypts	Small intestine, colon
Astro	?	?
Calici	?	?
Parvo	Crypts, lymphoid	Small and large intestine
Adeno	Enterocytes	Small
Reo	?	?
Entero	?	?

? Insufficient data available for definitive statements to be made.

Effective management of diarrhea requires prompt action to prevent continued loss of fluids and electrolyes. In animals, this is most economically achieved by removal of milk from the diet. This reduces the amount of undigested lactose in the lumen and, therefore, reduces fluid loss and acidosis. Therapy should include administration of balanced electrolyte solutions either orally or by the intravenous route. The use of intravenous fluid replacement and careful monitoring of animals could save a large percentage of severely affected animals; however, the costs are generally too high to recommend this as a standard procedure. In humans, the generalized introduction of oral rehydration therapy has greatly reduced mortality in infants. Antidiarrheal drugs, which reduce gut motility should not be used in viral gastroenteritis.

CONTROL OF GASTROINTESTINAL INFECTIONS

Numerous studies have demonstrated that protection must be directed at preventing the initial infection of the intestinal epithelial cells (22,23). Thus, regardless of whether antiviral agents or vaccines are used to induce protection from gastrointestinal viruses, protection must be directed at the intestinal epithelial cells. As a result, systemic application of drugs or vaccines may not always be efficacious. In this section, we will attempt to address control by vaccination, as well as, the potential for using antiviral drugs to limit viral infection of the intestine.

Vaccination.

The presence of antibody in the lumen of the small intestine, which can neutralize the virus before infection is initiated is at present the most effective means of preventing diarrhea (24,25). Unlike humans, most animals do not transfer antibodies across the placenta. Since animals generally suffer the most severe diarrhea in the first few weeks of life, the most effective method of preventing diarrhea is to hyperimmunize the mothers so that they secrete high levels of antibody in their milk during the neonate's susceptible period. The first convincing evidence that such an approach could work was demonstrated by the practice of feeding Transmissible gastroenteritis virus (TGEV) infected intestines to sows approximately one month prior to farrowing (26). The sows experience a mild subclinical infection and secreted sufficient levels of secretory IgA antibody in their milk to protect their piglets. Since there is a common immune system between the gut associated lymphoid tissue and the mammary gland, such intestinal immunization is very effective in pigs. However, this approach maintains virulent viruses in the environment, so recent work has employed attenuated TGEV given either intranasally or orally. Such vaccines have been only moderately successful since they do not replicate to sufficient levels within the sows to induce high levels of immunity and protection. Similar approaches have been used in other species including cattle to induce high levels of protection in milk (24).

The requirement for local immunity has created an interest in generating active immunity in the neonate or newborn before infection occurs. This is especially attractive in humans where the incidence of most cases of diarrhea does not occur until six months of age. It is also pertinent in rotavirus infections where it appears that although infection of neonates can occur, a very high percentage of these neonates are asymptomatic shedders (27). As individuals increase in age, the percentage of asymptomatic carriers decreases and the clinical form of the disease is manifested. Although it is not known why children become more susceptible to rotavirus infections as they increase in age between 6 to 24 months of age versus neonates, it has been postulated that there may be age-related changes in the ability of the epithelial cells of the intestine to bind virus. This is in direct contrast to that seen in most other species where the cells appear to be extremely susceptible in the very young, and after a few weeks or months, animals become more resistant to infections (28).

The need for providing local immunity dictates that virus vaccines be administered orally and induce a local immune response. This can be achieved by using 1) live attenuated or naturally avirulent strains, 2) live attenuated heterologous strains, 3) attenuated genetically altered strains, or hyrid viruses, 4) live bacterial vectors capable of colonizing the intestine and contain the appropriate viral antigens, 5) purified viral antigens or synthetic peptides coated for appropriate delivery to the intestine. At present all of these various approaches are being tried with varied success.

One of the greatest impediments to using live virus vaccines for gastrointestinal diseases is that it is very difficult to culture most of these agents in vitro to titers that will provide sufficient quantities of antigen for economical vaccine production. Although there is a tremendous amount of interest in developing effective vaccines against gastrointestinal virus infections, the trials are only now beginning and it is too early to confirm their potential efficacy. However, we are optimistic that with the potential tools at our disposal the problems will be overcome and it will be possible to produce a variety of effective vaccines against the majority of important gastrointestinal viral infections.

Anti-viral chemotherapy.

At present a large number of compounds have been identified with potential anti-viral activity in vitro. Although few of these compounds have either been tested or proven to be effective in vivo against viruses which cause gastrointestinal infections, the important advances made in this area over the last decade warrant serious consideration for the potential application of some of these drugs to the control of infections of the gastrointestinal tract. In this section we will discuss the potential sites of activity of a selected number of the drugs and speculate as to their applications in gastrointestinal tract infections. In addition, the limitations of using antiviral drugs in these infections will be addressed. The problems of delivery and maintenance of drugs at the site of infection so as to prevent initiation of infection (prophylaxis) rather than therapy will be emphasized.

In gastrointestinal tract infections as in other viral infections, the targets of antiviral drugs must be at sites that are unique to the virus and not have any physiological effects on the specific cells of the gastrointestinal tract. As is the case in other virus infections the antiviral drugs must be directed at various stages of virus replication cycle such as virus attachment, penetration, uncoating, macromolecular synthesis, viral maturation and assembly. Regardless to which stage of virus replication the antiviral drug is targeted, the major problem with using antiviral drugs in controlling viral induced gastrointestinal infections is the rapidity with which diarrhea occurs following the initial infection. Secondly, in many cases, diarrhea is often self limiting; thereby dictating that antiviral drugs must be used in a prophylactic rather than a therapeutic mode. Thus, prevention of viral activity at the early stages of infection appears to be the most promising but also creates the most problems regarding delivery and maintenance of the antiviral drug within the gastrointestinal tract for considerable periods of time.

Inhibitors of viral attachment are not frequently considered to be the major targets of viral replication, yet in many cases this is the mechanism whereby antibody neutralizes virus. If initial attachment could be prevented this could potentially be one of the best approaches for limiting gastrointestinal infections. This is especially important where onset of disease is extremely rapid. Recently there has been a considerable amount of activity in the area of identification of specific viral receptors on host cells (29). Once these specific receptors are identified it may be possible to either develop compounds or specific synthetic peptides which are capable of interfering with viral cellular interactions at the receptor level. The advent of anti-idiotypic antibodies to identify receptors for specific viruses will be extremely useful in this regard (30). However one difficulty with developing a specific anti-receptor approach is that there are many different viruses which can cause gastrointestinal infections (Table I). As a result, at present it appears that one would require a specific receptor blocking compound or peptide for each different virus. Whether this would be economical to produce remains to be determined. Nevertheless it appears useful to continue the search for such compounds which may have broad antiviral activity. Some compounds have recently been identified with such activity (31).

Recently, purified bovine lecithin was shown to inhibit *in vitro* replication of a number of viruses including rotavirus (32). Although there was speculation as to the mechanisms of action, no definitive proof was given. The results do however suggest that inhibition is at the level of viral receptor cell interactions. Since this is a natural product, and is non-toxic, it provides a possible useful approach to antiviral chemotherapy at the early stages of infections.

The step which occurs either simultaneously or immediately after attachment of viruses is viral penetration. Once again, neutralizing antibody plays a major role in preventing viral infectivity and replication at this stage. Thus this stage of virus replication is also attractive as a site for targeting antiviral drugs. Examples of such compounds include protease inhibitors which prevent cleavage of important glycoproteins which are required for viral penetration and virus-cell fusion (33,34). Although at present these examples are mainly limited to respiratory virus infections such as paramyxovirus and respiratory syncytial virus, similar approaches may be possible for some of the viruses which cause gastrointestinal infection. For example, it is well known that proteolytic cleavage of the 84K protein of rotavirus is essential for infectivity (35). Although it is thought that cleavage is not required for viral attachment, it is required for efficient penetration and uncoating (36). Whether additional protease inhibitors present in the intestine of animals and human beings would have deleterious effects on the physiological functions of the gastrointestinal tract remains to be determined.

An extension of the observation that proteolytic cleavage is required to provide active sites for initiation of infectivity, resulted in the synthesis of active oligopeptides with antiviral activity for measles and influenza virus (37). The fact that these oligopeptides are not toxic to cells provides an exciting approach to antiviral therapy in the gastrointestinal tract. However, studies in animals are required to evaluate the usefulness of such an approach to chemotherapy of viral infections of the gastrointestinal tract. Although the oligopeptides are non-toxic, in many cases they may act on the target cell membrane and consequently alter the physiological functions of the cell disrupting their activity and therefore themselves lead to diarrhea. The fact that the epitilial cells of the intestine turn over rapidly dictates that all of the newly susceptible cells continue to be exposed to the oligopeptide to remain refractory to virus infection. Because these oligopeptides are cleaved by the same enzymes that act upon the viruses, high doses are required to prevent virus cleavage. However, the synthesis of non-cleavable analogs of these peptides may reduce the required dose. Whether this will be economically feasible in diseases where it is not possible to predict the exact timing of infections remains to be determined.

Recently aromatic mono- and diamidines with antiviral activity have been identified (38). Althought the mechanism of activity is unknown they appear to act by inhibiting arginine-specific esterproteases and thereby preventing virus envelope-cell membrance fusion and entry (39). Monensin, a carboxylic ionophore, in addition to inhibiting viral glycoprotein processing, also inhibits penetration of viruses into cells (40). The fact that this compound is already used in cattle as a feed additive and an antiparasitic agent suggests that oral administration of it or similar derivatives may also have antiviral activity.

Other compounds that interfere with the uncoating of both envelope and naked viruses have been identified. The most effective compounds would be those that inhibit a broad spectrum of viruses, including DNA and RNA viruses as well as enveloped and naked viruses. One example of a series of compounds with potential as broad spectrum antiviral agents is 4-[6-(2-chloro-4-methoxyphenoxy)hexyl]-3, 5-heptanedione (arildone) (41). To date this compound has proven effective against some viruses which enter via the gastrointestinal tract such as polio and coxsackie virus as well as systemic viruses including herpes simplex and vesicular stomatitis. These compounds are effective both in vitro and in vivo (42,43). The fact that different derivatives of arildone can be synthesized with variable solubility and viral activity makes them attractive as broad-spectrum drugs. Many of the drugs described in the preceding section need not enter cells to exert their antiviral effect. Thus most of their effects are at the surface of cells therefore there is no need to worry about systemic absorption from the gastrointestinal tract.

The drugs that will be described in the following section inhibit virus infections after entry of the virus into the cell and therefore must actively enter the cell and act at intracellular sites of virus replication. We will not attempt to make an exhaustive list of all the drugs and all the viruses which can be inhibited by these drugs nor will we discuss all the different mechanisms by which each drug acts. In this section we will address classes of antiviral drugs that may have a potential in gastrointestinal infections.

Nucleoside analogs are one of the earliest antiviral drugs identified and such drugs continue to show promise for herpes viruses specifically, but newer analogs are showing activity against a wide variety of virus infections. Thus purine nucleosides such as 2-Amino-2'deoxy-9-β-D-ribofuranosyladenine, (S)-9-(2,3-Dihydroxyprophyl) adenine [(S)-DHPA], and (S)-9-(3-hydroxy-2-phosphonylmethoxyprophyl) adenine [S-HPMPA] have much broader antiviral activity than do the original ara-A derivatives whose activity was mainly restricted to herpes viruses. (44,45,46). Although the exact mechanism of action of these newer analogs is not fully understood the observation that they are active against both RNA viruses and DNA viruses (45,46) indicates their broad spectrum of activity. Specifically these drugs have been shown to be active against a number of adenovirus serotypes which cause infections of the respiratory tract. However, they have not yet been shown to be active against the human enteric adenoviruses nor have they been shown to be active against other enteric viruses such as coronavirus, astroviruses, etc.. If these drugs are found to be effective against these enteric viruses, the fact that they are active when given orally suggests that they should have potential as antiviral agents in vivo (46).

At present most of the pyrimidine nucleosides like the purine nucleosides are active against herpes viruses. However, drugs such as (1R)-(1α, 2β, 3β, 4α)-1(2,3-Dihydroxy-4-hydroxymethyl cyclopentyl) cytosine (carbodine) do have limited activity in vitro against influenza virus infections (47). Unfortunately in vivo activity has not been demonstrated to date. Whether other derivatives can be found that are active in vivo against a wider variety of viruses remains to be determined. The nucleoside ribavirin and its derivatives have been shown to have a broad spectrum antiviral activity (48,49). Once again in the majority of cases these drugs have been used systemically. Whether systemic administration will be affective in transfering antiviral states to the cells of the intestine remains to be determined. Furthermore, whether these drugs will be active if given orally also remains to be determined. The fact that they are active locally when given for respiratory infections indicates that some of these drugs may be effective if given by the oral route. This is further substantiated by the observation that ribavirin was at least partially active in vivo against rotavirus infections (50). Selenazole, (2β-D-Ribofuranosyl selenazole-4-carboxamide) an analog of ribavirin, appears superior to the parent compound both with respect to its antiviral activity as well as the maintenance of the antiviral state after the drugs is removed (51). Development of these types of drugs with sustained antiviral activity will ensure the antiviral state in the intestine between treatments. However, it must be indicated that drugs would probably still have to be given daily or every second day so that the new cells that are being continuously generated in the intestine continue to exhibit antiviral activity.

So far we have discussed antiviral drugs whose activity was mainly directed at DNA viruses but had some activity against RNA viruses. There are some drugs with in vitro activity against various RNA viruses but are inactive against DNA viruses. One example of such a compound is sodium-5 aminosulfonyl-2, 4-dichlorobenzoate (52). This compound is active against influenza, parainfluenza, respiratory syncytial, vesicular stomatitis, echo and rhinoviruses. Furthermore this drug has been shown to be active when given orally. The fact that this drug is tolerated at very high doses and has a very high therapeutic index indicates its potential for application in young infants and animals. Whether it will be active against the majority or at least the most serious gastrointestinal infection remains to be determined.

Inhibitors of viral maturation and assembly are also potential sites for directing antiviral chemotherapy. Thus in many cases viral specific processes associated with viral morphogenesis or maturation of specific proteins or polypeptides requires proteolytic cleavage of the precursor proteins to produce mature viral products. If the specific cleavage reactions are mediated by virus-coded proteases it is possible to direct antiviral compounds to these specific targets. Interference with these specific steps of viral replication should generally affect the formation of and release of infectious particles. Examples of such compounds include carbobenzoxyl leucylchloromethyl ketone, which inhibits cleavage of picornavirus precursor proteins without affecting cellular protein synthesis (53). 2-Amino-5 (2'-sulfamoylphenyl)-1,3,4-thiadiazole has been shown to act directly on viral structural proteins and prevent their assembly into virus particles of a variety of different RNA and DNA viruses (54). Although the mechanism of this inhibition remains unknown, studies on the effect of modifying various components of this molecule indicates that alterations in the activity against different viruses are possible. Thus it is possible as we learn more about the mechanisms of action of these specific drugs as well as the assembly and maturation of different virus infections it may be possible to engineer analogs of this compound which have the desired specific antiviral effect. In the case of rotaviruses, calcium is very important for stabilizing the virus and allowing infectious virus particle production. The growth of rotavirus in calcium-free media dramatically reduces virus infectivity (54a). Recent in vitro studies clearly indicate that chelators of calcium can dramatically reduce virus infectivity and replication in vivo as well as in vitro. Unfortunately, calcium chelators in the intestine have profound physiological effects which result in diarrhea (Ijaz, M.K. unpublished results). Thus, this approach does not appear at the present time to be feasible in controlling rotavirus infections.

One of the earliest antiviral mechanisms that the body has available for directly inhibiting virus replication is interferon. This compound was discovered in 1957 by Isaacs and Lindenmann (55) and interest has continued to be generated both in the direct application of interferon as an antiviral agent as well as induction of the bodies' own interferon system to develop an antiviral state. Since gastrointestinal virus infections are caused by a wide variety of viral agents the broad spectrum antiviral effects of interferon is very attractive for control of virus infections of the gastrointestinal tract. Another advantage of using interferon or interferon inducers as an antiviral compound is that interferons are generally considered to be most effective as prophylactic rather than therapeutic aids in infections. Since it is postulated that antiviral drugs in gastrointestinal infections will be applied prophylactically rather than therapeutically, this makes interferon a very attractive choice. One possible disadvantage of using interferon inducers is the possible induction of hyporesponsiveness after repeated treatment (56). As we learn more about the hyporeactive state and the mechanisms whereby various inducers stimulate interferon production it may be possible to overcome this hyporesponsive state. It is possible also to administer other compounds in combination with interferon inducers to reduce the hyporesponsive state and maybe even in some cases reduce the unwanted side effects of the interferon inducers (57). Polyinosinic: polycytidylic acid, [poly (I·C)], one of the initial synthetic inducers of interferon, was shown to be rapidly degraded by nucleases and generally had poor tolerability in animals. The tolerability was improved by altering the structural composition of poly I·C but maintaining the interferon inducing activity (58). Administration of poly I·C probably will require further modifications to increase the stability and prevent nuclease degradation of the compounds before they are active <u>in vivo</u>. Thus, although interferon inducers have some potential, they have not yet proven to be very effective.

The advent of recombinant DNA technology and availability of different cloned human and animal interferons may provide an effective economical method for future control of enteric infections of viral etiology. Recent studies have clearly indicated that animal interferons are capable of reducing gastrointestinal virus infections (59). However, before interferons can be employed routinely in the control of gastrointestinal infections, methods will have to be implemented to maintain these substances in the intestine for the duration of time that the animals are susceptible to the virus. Another impediment that will need to be overcome is the delivery of interferon to the cells of the gastrointestinal tract. Oral delivery creates problems due to the tremendous numbers and amounts of proteases that are present in the gastrointestinal tract. Since interferon is rapidly degraded by gastrointestinal proteases, methods of delivery and targeting of interferon will need to be developed to transport the compound to the target site and maintain sufficient levels of interferon in the intestine for a sufficient length of time to develop and maintain the antiviral state. Some of the methods developed for targeting interferon can also be very useful for targeting other antiviral drugs and viruses discussed in the preceding section. The development of enterocoating methods for oral delivery of these compounds will be a tremendous benefit to the oral delivery of specific antiviral drugs.

Combination chemotherapy has proven to be an effective means of dealing with drug toxicitiy and problems of drug resistance in both anti-microbial and cancer chemotherapy (60,61). Recent studies involving combination chemotherapy of viral infections clearly indicate that in some instances the use of two drugs, whose mechanism of action differs, greatly increases their efficacy (62,63). This approach has not been tried in gastrointestinal viral infections to date. In addition to acting synergistically some drugs can be antagonistic (63). Therefore, drug combinations will have to be selected judiciously.

SUMMARY

Presently there are a variety of antiviral drugs which have activity against a wide range of viruses whose mechanisms of replication are different. However few studies have focused on the application of antiviral drugs to specifically controlling gastrointestinal infections. The majority of effort in this area has been directed at controlling herpes virus infections with a considerable amount of success. It is anticipated that control of gastrointestinal infections by antiviral drugs will present a much greater challenge to molecular biologists and pharmacologists due to the variety of viral causative agents and the need to use them prophylactically rather than therapeutically. However, with the recent progress in molecular biology the identification of specific virus cell interactions, and identification of specific viral stages in replication within a cell should clearly provide important information required to develop broad-spectrum and selective antiviral drugs for specific viruses or virus groups. The combined activities of molecular virologists and pharmacologists to identify the modes of action of a number of the known antiviral compounds and their analogs should greatly enhance our ability to specifically synthesize more effective antiviral chemotherapeutic drugs.

Finally, new methods of drug delivery and/or drug targeting to improve potency and selectivity of antiviral compounds should greatly enhance the therapeutic index of a number of these drugs. In the case of gastrointestinal infections one of the major impediments will be to direct the drug to the epithelial cells and maintain the drug there for a considerable length of time to ensure the antiviral state is present at the time virus infection occurs. Whether this will be possible in cases where gastrointestinal infections can occur over a long period of time remains to be determined. Thus it will be important to be able to predict specific epidemics of gastrointestinal infection using more rapid diagnostic techniques and treat contacts of individuals suffering from these infections so as to greatly reduce the spread of virus infections within the community or hospital environment.

One of the best places for application of antiviral drugs may be in children as they enter hospitals for reasons other than gastrointestinal infections. In these situations it is possible to predict that gastrointestinal nosocomial infections will occur within a few days of hospitalization. Elimination of this added stress would increase the child's rate of recovery. Another application may be in day-care centers where contact with other children shedding gastrointestinal viruses occurs frequently.

REFERENCES

1. Snyder, J.D. and Merson, M.H. Bulletin of the World Health Organization. 60: 605-613, 1982.
2. Chen, L.C., Alauddin Chowdhury, A.K.M. and Huffman, S.L. Am. J. Clin. Nutr. 33: 1836-1845, 1980.
3. Black, R.E., Merson, M.H., Rahman, A.S.M.M., Yunus, M., Alim, A.R.M.A., Hug, I., Yolken, R.H. and Curlin, G.T. J. Infect. Dis. 142: 660-664, 1980.
4. Black, R.E., Merson, M.H., Imdadul, H., Alim, A.R. and Yumus, M.D. Lancet. 141-143, 1981.
5. Bishop, R.F. In: Virus Infection of the Gastrointestinal Tract. Ed. D.A.J. Tyrrell and A.Z. Kapikian, Dekker, New York, 1981, p. 195-210.
6. Kahn, D.E. J. Am. Vet. Med. Assoc. 173: 628-630, 1978.
7. Brown, E.H. Br. Med. J. 1: 169-171, 1973.

8. Nathanson, N. and Martin, J.R. Am. J. Epidemiol. 110: 672-692, 1979.
9. Dent, D.M., Days, P.J., Bind, A.R. and Birkenstock, W.E. S. Afr. Med. J. 49: 669-672, 1975.
10. Farmer, G.W., Vincent, M.M., Fucillo, L., Horta-Barbosa, L., Sever, J.L. and Gitnick, G.L. Gastroenterology. 65: 8-18, 1973.
11. Woode, G.N., Reed, D.E., Runnels, P.L., Herrig, M.A. and Hill, H.T. Vet. Microbiol. 7: 221-240, 1982.
12. Resta, S., Luby, J.P., Rosenfeld, C.R. and Siegel, J.D. Science. 229: 978-981, 1985.
13. Battaglia, M., Passarani, N., DiMatteo, A. and Gerna, G. J. Infect. Dis. 155: 140-143, 1987.
14. Schnagl, R.D., Greco, T. and Morey, F. Arch. Virol. 87: 331-337, 1986.
15. Horzinek, M.C., Ederveen, J. and Weiss, M. J. Gen. Virol. 66: 1287-1296, 1985.
16. Middelton, P.J., Szymanski, M.T. and Petric, M. Am. J. Dis. Child. 131: 733-737, 1977.
17. Coiro, J.R.R., Bendati, M.M.A., Neto, A.J.A., Heuser, M.C.F. and Basconcellos, V.L. Am. J. Trop. Med. Hyg. 32: 1186-1189, 1983.
18. Kaplan, J.E., Gary, G.W., Baron, R.C., Singh, N., Schonberger, L.B., Feldman, R. and Greenberg, H.B. Ann. Intern. Med. 96: 756-761,
19. Pearson, G.R. and McNulty, M.S., J. Comp. Pathol. 87: 363-373, 1977.
20. Leece, J.G., King, M.W. and Mock, R. Infect. Immun. 14: 816-825, 1976.
21. House, J.A. J. Am. Vet. Med. Assoc. 173: 573-576, 1978.
22. Snodgrass, D.R. and Wells, P.W. Arch. Virol. 52: 201-205, 1976.
23. Woode, G.N. and Bridger, J. Vet. Rec. 96: 85-88, 1975.
24. Saif, L.J., Redman, D.R., Smith, K.L. and Theil, K.W. Infect. Immun. 41: 1118-1131, 1983.
25. Hooper, B.E. and Haelterman, E.O. J. Am. Vet. Med. Assoc. 149: 1580-1586, 1966.
26. Bohl, E.H., Gupta, R.K.P., Olquin, F.M.W. and Saif, L.J. Infect. Immun. 6: 289-296, 1972.
27. Cotterill, A.M. and Walker-Smith, J.A. Br. Med. Bull. 42: 176-180, 1986.
28. Little, L.M. and Shadduck, J.A. Infect. Immun. 38: 755-763, 1982.
29. Crowell, R.L. and Lee Hsu,, K-H. In: Concepts of Viral Pathogenesis. Ed. Notkins, A.L. and Oldstone, M.B.A., Springer-Verlag. New York, 1986. pp. 117-125.
30. Noseworthy, J.H., Fields, B.N., Dichter, M.S., Sobotka, C., Pizer, E., Penry, L.L. Nepom, J.T. and Greene, M.I. J. Immunol. 131: 2533-2538, 1983.
31. Thiel, K.D., Helbig, B., Sprossig, M., Kocking, R. and Wutzler, P. Acta. Virol. 27: 200-209, 1983.
32. Walsh, D.S., Kappes, J.C., Duell, G.A., Tsuchiya, Y. and Nutini, L.G. IRCS Med. Sci. 13: 997-998, 1985.
33. Zhirnov, O.P., Ovcharenko, A.V. and Bukrinskaya, A.G. J. Gen. Virol. 63: 469-474, 1982.
34. Zhirnov, O.P., Ovcharenko, A.V. and Bukrinskaya, A.G. J. Gen. Virol. 65: 191-196, 1984.
35. Sabara, M.I., Frenchick, P.J. and Babiuk, L.A. J. Virol. (In press), 1987.
36. Clark, S.H., Roth, J.R., Clark, M.L., Barnett, B.B. and Spendlove, R.S. J. Virol. 39: 816-822, 1981.

37. Richardson, C.D., Scheid, A. and Choppin, P.W. Virol. 105: 205-222, 1980.
38. Dubovi, E.J., Geratz, J.D., Tidwell, R.R. Virol. 103: 502-504, 1980.
39. Dubovi, E.J., Geratz, J.D., Shver, S.R. and Tidwell, R.R. Antimicrob. Agents Chemother. 19: 649-659, 1981.
40. March, M., Wellstee, J., Kern, H., Harms, E. and Helenius, A. Proc. Natl. Acad. Sci. (USA) 79: 5297-5301, 1982.
41. Kim, K.S., Sapienza, V.J. and Carp, J.R. Antimicrobiol. Agents. Chemother. 18: 276-280, 1980.
42. McSharry, M.A. and Pancic, R. Handb. Exp. Pharmacol. 61: 419-444, 1983.
43. McKinely, M.A., MIralles, J.F., Brisson, C.T. and Pancic, F. Antimicrobiol. Agents Chemotherap. 22: 1022-1025, 1982.
44. Taguchi, F., Imeatani, Y., Nagaki, G., Nakagawa, A., Omura, S. J. Antibiot. 34: 313-318, 1981.
45. De Clercq, E., Descamps, J., DeSomer, P. and Holý, A. Science. 200: 563-565, 1978.
46. De Clercq, E., Holý, A., Rosenberg, I., Sakuma, T., Balzarini, J. and Maudgal, P.C. Nature, 323: 464-467, 1986.
47. Shannon, W.M., Arnett, G., Westbrook, L., Shealy, Y.F., O'Dell, C.A. and Brockman, R.W. Antimicrob. Agents Chemotherap. 20: 769-776, 1981.
48. Stephen, E.L., Jones, D.E., Peters, C.J., Eddy, G.A., Liozeaux, P.S. and Jahrling, P.B. In: Ribavirin: A broad spectrum antiviral agent. R.A. Smith and W. Kirkpatrick Ed. Academic Press, New York, pp. 169-183.
49. Koff, W.C., Pratt, R.D., Elm, J.C., Verkatershan, N.C. and Halstead, S.B. Antimicrob. Agents Chemotherap. 24: 134-136, 1983.
50. Smee, D.F., Sidwell, R.W., Clark, S.M., Barnett, B.B. and Spendlove, R.S. Antimicrob. Agents Chemotherap. 21: 66-73, 1982.
51. Kirsi, J.J., North, J.A., McKernan, P.A., Murray, B.K., Canonico, P.G., Huggins, J.W., Srivasta, P.C. and Rubbins, R.K. Antimicrob. Agents Chemotherap. 24: 353-361, 1983.
52. Ohnishi, H., Yamaguchi, K., Shimada, S., Himuro, S. and Suzuki, Y. Antimicrob. Agents Chemotherap. 22: 250-254, 1982.
53. Korant, B.D. In: Antiviral Chemotherapy: Design of inhibitors of viral functions. K.K. Gauri. Ed. Academic Press, New York, 1981, pp. 33-47.
54. Bonina, L., Ovzalesi, G., Merendino, R., Arena, A. and Mastroeni, P. Antimicrob. Agents Chemotherap. 22: 1067-1069, 1982.
54a. Shahrabadi, M.S. and Lee, P.W.K. Virology. 152: 298-307, 1986.
55. Isaac, A. and Lindenmann, J. Proc. Royal Soc. B147: 258-267, 1957.
56. DeClercq, E. Antibiot. Chemotherap. 27: 251-259, 1980.
57. Stringfellow, D.A. In: Methods of Enzymology. S. Pestka. Ed. 78A: 262-284, Acad. Press, New York. 1981.
58. Carter, W.A. and DeClercq, E. Science, 186: 1172-1178, 1974.
59. Wérenne, J. In: Adjuvants, Interferon and Non-Specific Immunity. Cancellotti, F.M. and Galassi, D. Ed. Comm. Eur. Commun. (EUR 8675 EN) pp. 12-30.
60. Jawetz, E. Ann. Rev. Pharmacol. 8: 151-170, 1968.
61. DeVita, V.T., Young, R.C. and Canellos, G.P. Cancer, 35: 98-110, 1975.
62. Fiala, M., Chow, A.W., Miuasaki, K. and Guze, L.B. J. Inf. Dis. 129: 82-85, 1974.
63. Ayisi, N.K., Gupta, V.S., Meldrum, J.B., Taneja, A.K. and Babiuk, L.A. Antimicrob. Agents Chemotherap. 17: 558-568, 1980.

21

PRINCIPLES OF ANTIRETROVIRAL THERAPY FOR AIDS AND RELATED DISEASES

J. BALZARINI[1] and S. BRODER

Clinical Oncology Program, National Cancer Institute, National Institutes of Health, Bethesda, Maryland 20892, U.S.A.

ABSTRACT

Every step in the life cycle of pathogenic human retroviruses could theoretically serve as a target for antiviral therapy. In particular, the process of reverse transcription appears to be an attractive target. A number of drugs and pharmacologic principles relevant to the administration of such drugs have been identified for application in the therapy of diseases mediated by pathogenic human retroviruses. One family of such drugs (2',3'-dideoxynucleoside analogues) shows special promise. These drugs are now being administered to patients with AIDS and its related disorders.

INTRODUCTION

The treatment of pathogenic human retroviral infections is a formidable challenge. The challenge has taken on a new urgency since the advent of the acquired immunodeficiency syndrome (AIDS), which has become a major public health problem after its recognition as a new clinical entity in 1981 (1-3). This disease is caused by the third known human T-lymphotropic virus (HTLV-III) (4) which replicates within critical cells of the immune system (particularly T cells and macrophages), leading to loss of CD4+ (helper-inducer) T cells and profound immunosuppression (5-8). This retrovirus has several other names including lymphadenopathy associated virus (LAV) (9) and AIDS-related virus (ARV) (10). It has recently been proposed by the human retroviral subcommittee of the International Committee on the Taxonomy of Viruses that the retrovirus which causes AIDS be called human immunodeficiency virus (HIV) (11), and in the future this could become the general designation for this category of retroviruses.

---

[1]On leave from the Rega Institute for Medical Research, Katholieke Universiteit Leuven, B-3000 Leuven, Belgium.

There have now been over 30,000 cases of AIDS reported in the United States, and the disease occurs throughout the world. It has become evident that there may be a long prodromal period before the development of fulminant AIDS (12,13). Over one million people in the USA are now infected with HTLV-III/LAV and this number is likely to increase. Perhaps 30 % or more of infected persons may develop AIDS in a 3 to 5 year period of time (14), so that even if the spread of HTLV-III/LAV infection slows down, the number of cases of AIDS is expected to continue to increase.

While AIDS is an exceedingly important priority in its own right, what we learn about treating the retrovirus which causes AIDS could have broad implications for other medical and veterinary illnesses.

RATIONALE FOR THE USE OF ANTIVIRAL THERAPY IN AIDS

The formal proof by Gallo and his co-workers in 1984 that HTLV-III/LAV was the causative agent of AIDS (4) permitted for the first time the consideration of specific antiretroviral therapy and, in partcular, made it possible to develop therapies aimed at inhibiting the replication of the retrovirus which causes this disorder. It is worth noting that antiretroviral therapy for the treatment of AIDS is based on the assumption that continued retroviral replication is involved in both the pathogenesis and progression of this disease. As we shall discuss, this assumption appears to be valid. In regard to the immunodeficiency, the depletion of CD4+ T cells can be mediated directly by HTLV-III/LAV infection of these cells (5,15) although it is possible that indirect mechanisms such as autoimmune reactions or the production of toxic lymphokines may also play a role (16). If the immune system of these patients retains some regenerative potential (either alone or with the addition of immunoreconstitutive therapy [17]), then inhibition of HTLV-III/LAV replication in a patient may permit some immune restoration to occur or, at the minimum, prevent further clinical deterioration, thereby prolonging survival.

Antiretroviral therapy may also address other disease manifestations of HTLV-III/LAV infection. Since the first observation that the brain is an important site of HTLV-III/LAV replication (18), there has been a growing recognition that serious neurologic dysfunction, ranging from peripheral neuropathy to profound dementia, may be caused by this virus (19-21). While it is still not known precisely how HTLV-III/LAV enters the nervous system (one likely mechanism is that it is transported there by infected macrophages [7]) or how it induces neurologic damage, it is clear that successful therapeutic strategies must somehow address the consequences of viral replication within the nervous system. As we will discuss later, improvements in the neurologic disease caused by HTLV-III/LAV are within the reach of the antiretroviral chemotherapy now being used.

STAGES IN THE LIFE CYCLE OF HTLV-III/LAV THAT MAY BE TARGETS FOR ANTIVIRAL AGENTS

Before turning our attention to a discussion of certain recent practical clinical developments in the antiretroviral therapy of AIDS, it is perhaps worthwhile to review briefly the replicative cycle of HTLV-III/LAV and consider how specific stages of viral replication might be targets for therapeutic strategies (Table 1). As noted above, HTLV-III/LAV belongs to a family of RNA viruses known as retroviruses. By definition, such viruses must replicate through a DNA intermediate (i.e., at one step in their cycle of replication, genetic information flows from RNA to DNA, thus in the "retro" direction) (22, 23). Many members of this family of viruses can induce neoplastic transformation in infected cells, and the terms "RNA tumor virus" or "leukemia virus" have often been used to describe these viruses. HTLV-III/LAV, however, has not been shown to be transforming *per se* although it is clearly linked to the causation of certain cancers through its ability to induce immunosuppression, and

Table 1. Stages in the replicative cycle of a pathogenic human retrovirus which may be targets for therapeutic intervention

Stage	Potential intervention
Binding to target cell	Antibodies to the virus or cell receptor
Early entry into target cell	Drugs that block fusion or interfere with retroviral uncoating
Transcription of RNA to DNA by reverse transcriptase	Reverse transcriptase inhibitors
Degradation of viral RNA of an RNA-DNA hybrid	Inhibitors of RNase H activity
Integration of DNA into host genome	Drugs which inhibit *pol* gene-mediated "integrase" function
Expression of viral genes	"Anti-sense" constructs; inhibitors of *tat*-III protein or *art/trs* protein
Viral component production and assembly	Myristylation, glycosylation and protease inhibitors or modifiers
Budding of virus	Interferons

this provides a powerful reminder of the relationship between abnormalities in the immune system and neoplasms (24). It is also worth stressing that HTLV-III/LAV is more complex than previously characterized retroviruses and has at least 8 genes.

The first step in the infection of a target cell by HTLV-III/LAV is its binding to the cell surface; in the case of helper-inducer T cells, the binding receptor is believed to be linked to the CD4 antigen (25,26). There is recent evidence that this binding between HTLV-III/LAV envelope glycoprotein and the CD4 antigen may play a major role in virally-induced T cell lysis through syncytia formation (27,28). This binding might possibly be blocked by antibodies either to the viral envelope glycoprotein or to the CD4 antigen; alternatively, synthetic ligands could be used to block viral attachment (29). There is substantial variation in the envelope of different isolates of HTLV-III/LAV (30,31), so that antibody directed against one isolate may theoretically not block binding of another. The envelope glycoprotein, however, is constrained by its need to bind to CD4, and it is likely that a relatively invariant region of the envelope will be shown to be crucial for viral infection.

After attachment, HTLV-III/LAV enters a target cell by an incompletely defined mechanism which may involve a fusion process. It then loses its envelope coat, and viral RNA (along with reverse transcriptase) is released into the cytoplasm. Compounds which block these steps have been found for other viruses (e.g., amantadine is believed to block the uncoating of influenza A virus [32]), and similar agents may be developed for HTLV-III/LAV.

At this point, an HTLV-III/LAV pol product, reverse transcriptase, makes a complementary (first strand) strand DNA copy of the RNA genome using a lysine transfer-RNA as the primer (30,33,34). This same enzyme then catalyzes the production of a positive strain DNA copy, so that, eventually, the genetic information is encoded in a double stranded DNA form. Reverse transcriptase activity is essential for viral replication and can for all practical purposes be considered as a unique viral function. For retroviruses as a class, it is thought that an RNase H activity is situated at the carboxyl terminus of the reverse transcriptase domain, and this RNase H eliminates the viral genomic RNA strand making it possible for the reverse transcriptase to generate the second DNA strand. If a DNA copy of the uncoated viral RNA is not made promptly, the RNA is susceptible to degradation by cellular enzymes. For these reasons, reverse transcriptase has been a major target for antiviral therapy in

AIDS, and, indeed, as will be discussed below, most of the agents now being investigated act at this step of viral replication (35-40). It should be noted that the progress in this area has been possible in part because of pioneering research directed at finding agents which inhibit the reverse transcriptase of murine and avian retroviruses (41-46).

The DNA copy of retroviral genetic material may become circularized during or soon after its formation and may remain in an unintegrated form or become integrated into the host cell genome through a viral endonuclease or "integrase" (thought to be a pol gene product) which mediates this step (47). It is possible that drugs which interfere with the "integrase" could be of value. At some later point, the DNA is transcribed to mRNA and to viral genomic RNA using host RNA polymerases, and the mRNA is translated to form viral proteins, again using to a large extent the biochemical apparatus of the host cell. One broad spectrum antiviral drug, ribavirin, is believed to interfere with a process called 5'-capping of virus-specific mRNA in other viral systems and it may possibly have a related activity in HTLV-III/LAV infection (32,48).

It has recently been shown that the production of HTLV-III/LAV proteins is under the regulatory control of at least two viral genes, one named tat-III (49) and the other named art or trs (50,51). The products of these unique viral genes are absolutely required for efficient HTLV-III/LAV replication, and they might be future targets for antiviral therapy. The product of the tat-III gene, for example, is believed to markedly increase production of viral proteins after binding to a region on what is called the 5'-long terminal repeat (LTR) of the viral mRNA (52). A cluster of positively charged amino acids in the protein encoded by tat-III has been postulated to be the site of its binding to mRNA (52). Drugs might be developed (perhaps designed on the basis of x-ray diffraction studies and computer-generated models of the protein [52-54]) with the goal of binding to and inhibiting the function of tat-III.

Another approach which may conceivably be used to inhibit the production of viral proteins may be the use of "anti-sense" oligodeoxynucleotides (55, 56). These are short sequences of DNA (or DNA which has been chemically modified to confer better cellular penetration and resistance to enzymatic degradation) with base pairs that are complimentary to a viral segment of the viral genome. Such oligonucleotides could theoretically block expression of viral proteins through hybridization arrest of mRNA (56,57) or interference with the binding of a regulatory protein (e.g., tat-III).

The final steps of viral replication involve secondary processing of certain viral proteins by proteases (thought to be a function of one of the viral pol gene products [33]) and myristylating and glycosylating enzymes (provided by the host cells) (8,59); these steps could be affected by inhibitors of these enzymatic functions. Finally, the virus is released by budding; alpha interferon, which has been shown to inhibit HTLV-III/LAV replication (60), is thought to possibly act at this step.

It has become evident that certain cells, including macrophages (6,7), EBV-infected B cells (61) (which are increased in patients with AIDS [62,63]), and even certain T cells (64,65), may be chronically infected with HTLV-III/LAV without necessarily being destroyed by the virus. Agents which act at the later steps of viral replication (beyond reverse transcription and integration) may theoretically address the problem of these chronically infected cells, while those agents which act at reverse transcription or earlier steps may be effective at protecting uninfected cells from HTLV-III/LAV infection. A combination of early and late acting agents may therefore be more effective than either used alone. Combined therapy in addition may reduce the chance of the virus developing resistance to any single agent.

2',3'-DIDEOXYNUCLEOSIDE ANALOGUES AS INHIBITORS OF HTLV-III/LAV INFECTION

We would now like to turn our attention to a group of compounds, 2',3'-dideoxynucleoside analogues, which can be metabolized in mammalian cells to become potent inhibitors of HTLV-III/LAV replication. 2',3'-Dideoxynucleosides have previously been studied in several pioneering laboratories (e.g., 66-68). Our group has observed that some of these compounds block the infection of target cells by HTLV-III/LAV at concentrations which are 10 to 20 fold lower than those at which they inhibit the proliferation and survival of the cells, even under conditions of high multiplicity of infection (39,40). As will be discussed further below, one drug, the 3'-azido derivative of 2',3'-dideoxythymidine (3'-azido-2',3'-dideoxythymidine, AzddThd, AZT), has recently been shown to confer a survival advantage when administered to certain patients with AIDS (69).

The 2',3'-dideoxynucleosides which have been shown to inhibit HTLV-III/LAV replication (39,40,70-75) differ from the normal substrates for nucleic acid synthesis (2'-deoxynucleosides) in that the 3'-hydroxy (-OH) group is replaced by an hydrogen (-H) or another group which cannot form phosphodiester linkages (Fig. 1). Individual compounds can be converted by mammalian kinases to 5'-triphosphate metabolites, but the degree of anabolic phosphorylation

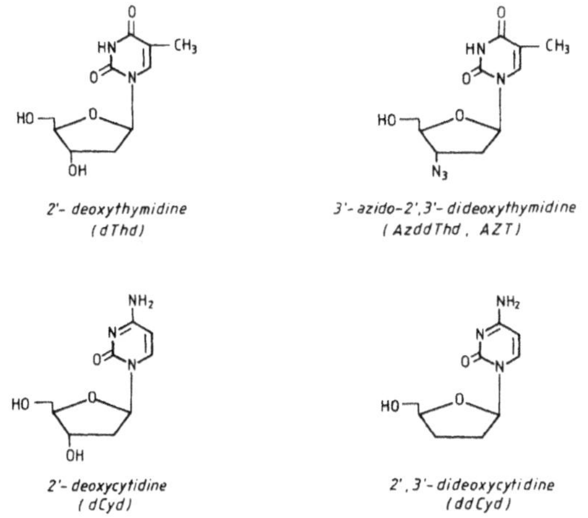

Fig. 1. Chemical structures of two 2',3'-dideoxynucleoside analogues now in clinical trials (right) compared to their respective physiologic nucleoside counterparts (left).

cannot be extrapolated from one drug to another and must be determined on a case by case basis (44,76,77). There is evidence that these phosphorylated nucleosides inhibit HTLV-III/LAV replication by acting as chain terminators (67,70,78); because of the 3' modification, once viral reverse transcriptase is fooled into adding them to the end of a growing chain of DNA, no subsequent 3',5'-phosphodiester linkages can be made (Fig. 2). The DNA polymerase of HTLV-III/LAV (reverse transcriptase) and other retroviruses is much more susceptible to the inhibitory effects of these compounds (as 5'-triphosphates) than is mammalian DNA polymerase alpha (44,70,71,76), and this is most likely one basis for their selective antiretroviral activity. Indeed, one might say that, based on the very low K_i values observed, reverse transcriptase prefers 2',3'-dideoxynucleotides to normal nucleotides as substrates.

As noted above, 2',3'-dideoxynucleosides must be phosphorylated to an active 5'-triphosphate moiety by host cellular kinases. Different species of animals (or even cell types within one species) differ in their efficiency of phosphorylation of these drugs (44,76,77), and for this reason, extreme caution must be used in extrapolating from experimental results obtained in one species (or a cell type within a species) to another. For example, 2',3'-dideoxythymidine is poorly phosphorylated in human Molt/4F cells (K_i/K_m for thy-

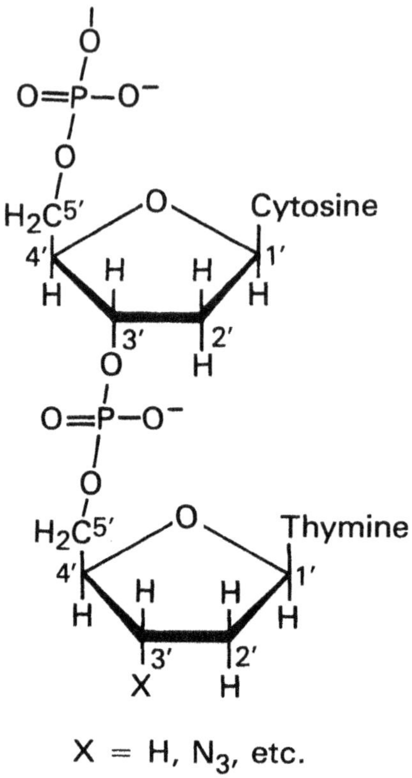

Fig. 2. One mechanism to explain the antiretroviral activity of 2',3'-dideoxynucleoside analogues. After anabolic phosphorylation within a target cell, 2',3'-dideoxynucleotides can be incorporated into a growing chain of DNA as the genomic RNA of HTLV-III is transcribed to DNA; this will elongate the DNA by one residue and then terminate DNA synthesis because the 3'-carbon is not available for further phosphodiester linkage.

midine kinase ∿ 130; comparing drug to normal substrate (74) (Fig. 3)), and is among the least potent antiviral agents of any 2',3'-dideoxynucleoside tested in ATH8 cells (40,74). Substitution of the 3'-hydrogen by an azido (N_3) group in the erythro configuration, however, produces a compound mentioned above, 3'-azido-2',3'-dideoxythymidine (AzddThd, AZT) (Fig. 1), that is an excellent substrate for human thymidine kinase (K_m : 3 µM; K_i/K_m ∿ 1) (71,76, and Balzarini and Broder, unpublished), and very efficient at inhibiting HTLV-III/LAV replication *in vitro* (39,40,79). The *threo* isomer is very much less potent as

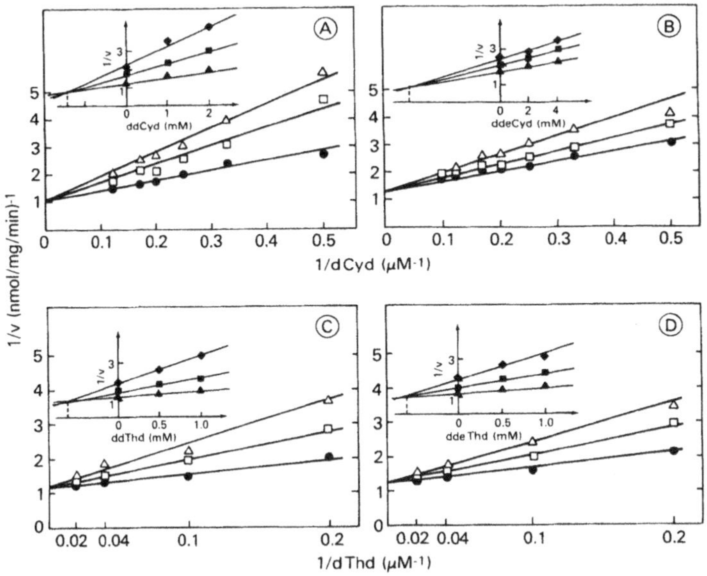

Fig. 3. Double-reciprocal plots for inhibition of Molt/4F dCyd kinase by ddCyd (panel A) and ddeCyd (panel B), and Molt/4F dThd kinase by ddThd (panel C) and ddeThd (panel D). Inhibitor concentrations: none (●), 0.5 mM (□) and 1 mM (Δ) for ddThd; none (●), 0.5 mM (□) and 1 mM (Δ) for ddeThd; none (●), 1 mM (□) and 2 mM (Δ) for ddCyd; none (●), 2 mM (□), and 4 mM (Δ) for ddeCyd. In every panel are also given the Dixon plots of the data at the following concentrations of dCyd in panel A: 2 μM (◆); 3 μM (■) and 6 μM (▲), and panel B: 3 μM (◆); 4 μM (■) and 8 μM (▲); and dThd in panels C and D: 5 μM (◆); 10 μM (■) and 50 μM (▲).

an antiretroviral agent (unpublished data).

Of all the nucleoside analogues that have been tested so far in our laboratory for the ability to inhibit the in vitro replication and cytopathic effect of HTLV-III/LAV against ATH8 cells [a normal helper-inducer T cell line immortalized after transformation with HTLV-I and highly sensitive to the cytopathic effect of HTLV-III/LAV (77),] the pyrimidine analogue 2',3'-dideoxycytidine (ddCyd) is the most potent on a molar basis. At very high multiplicity of infection, essentially complete inhibition is obtained at a concentration of 0.5 μM and the virus-inhibiting effect is durable over prolonged culture periods (40,71). At a low multiplicity of infection, viral suppression may be seen at concentrations of 10 nanomolar or less in vitro (Mitsuya and Broder, unpublished).

The metabolism of ddCyd to its 5'-mono-, 5'-di- and 5'-triphosphate metabolites is similar in uninfected and HTLV-III/LAV-infected ATH8 cells. This observation suggests that no specific viral enzyme is involved in the activation and metabolism of the drug (Fig. 4). The following observations led us to conclude that ddCyd is phosphorylated by the cellular deoxycytidine (dCyd) kinase to its 5'-monophosphate derivative ddCMP : (i) the cytostatic effects of ddCyd against human Molt/4F cells are dramatically reversed by dCyd (> 150 fold) (72,74); (ii) the anti-HTLV-III/LAV effects of ddCyd in ATH8 cells are reversed by dCyd (70); (iii) dCyd severely depresses intracellular phosphorylation of ddCyd (72,77); (iv) ddCyd lacks any appreciable cytostatic effect against a dCyd kinase-deficient cell line (ID_{50} : > 2400 μM compared to 320 μM for the wild-type cell line (72,74); (v) phosphorylated metabolites of ddCyd could not be detected in a dCyd kinase-deficient cell line which contains normal levels of Cyd/Urd kinase (77); (vi) ddCyd shows competitive inhibition of partially purified Molt/4F dCyd kinase with respect to dCyd (74) (Fig. 3). However, dCyd kinase has a much lower affinity for ddCyd (K_m = 201 μM) than for its physiological substrate dCyd (K_m = 3.5 μM) (72,74,80). Since ddCTP is thought to be the active intracellular metabolite for the antiretroviral effect of ddCyd, limited intracellular levels of ddCTP or high intracellular ratios of dCTP to ddCTP could prevent ddCyd from achieving its optimal antiviral effect. Nevertheless we could demonstrate that ATH8 cells as well as a number of other human cell lines, including PHA-stimulated normal lymphocytes, deve-

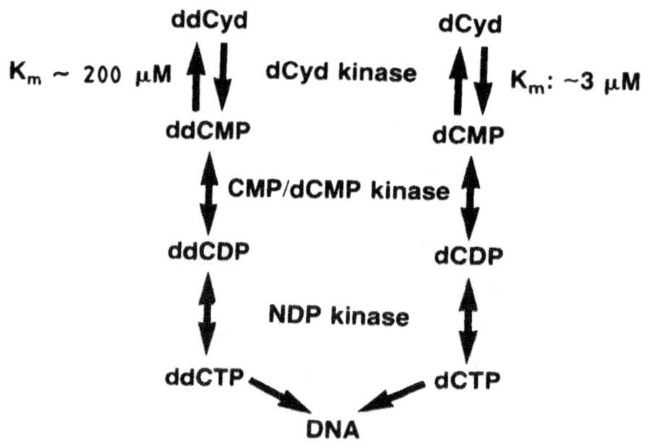

Fig. 4. Pathways of anabolic phosphorylation for 2',3'-dideoxycytidine (ddCyd) and its normal counterpart 2'-deoxycytidine (dCyd). The K_m for the first step of anabolic phosphorylation (mediated by 2'-deoxycytidine kinase) is shown for both drug and the normal substrate.

loped intracellular ddCTP levels of about 0.5 μM when incubated with 1 μM ddCyd for 24 hours (77). These ddCTP levels exceed the Ki-value of ddCTP for HTLV-III/LAV reverse transcriptase ($K_i \sim 0.2$ μM, determined by Hao Z, Johns DG, Broder S et al., unpublished). Moreover, Mitsuya et al. (81) showed that ddCTP could bring about a chain termination of DNA synthesis catalyzed by HTLV-III reverse transcriptase at a concentration which is 40 fold below the dCTP concentration present in the reaction mixture. Taking into account that the levels of ddCTP attainable upon incubation of human lymphocytes with 1 μM ddCyd are $\geq$ 1/40 the normal intracellular dCTP levels present in the cells [e.g., 20 μM in human H9 cells (76)], we may conclude that these ddCTP levels are high enough to account for an antiretroviral (chain terminating) effect of ddCyd.

We are exploring ways to manipulate the activity of certain mammalian enzymes involved in pyrimidine nucleoside metabolism, in particular dCyd kinase and CDP reductase, in order to increase the anabolic phosphorylation of ddCyd. One potential approach is to decrease the intracellular dCTP pools. Since dCTP is a potent feedback inhibitor of dCK (82,83), a reduction of the dCTP levels will stimulate dCyd phosphorylation and at the same time enhance ddCyd phosphorylation.

In Fig. 5, potential target enzymes for achieving decreased dCTP pools are indicated. All inhibitors shown in Fig. 5, have previously been demonstrated to decrease dCTP pools (84-98). We found that a 12-hour preincubation period of murine leukemia L1210 cells with hydroxyurea [an inhibitor of CDP reductase (84,85)], 3-deazauridine [an inhibitor of CTP synthetase as its 5'-triphosphate derivative (86,87)], and dThd [an allosteric inhibitor of CDP reductase as its 5'-triphosphate derivative (88,89) as well as an activator of dCyd kinase under certain conditions (99)], led to a 4-6 fold increase in phosphorylated ddCyd metabolites. Pyrazofurin [an inhibitor of OMP decarboxylase (90,91)] increased ddCTP levels only 2-3 fold while PALA [an inhibitor of aspartate transcarbamoylase (92-94)] showed only marginal stimulation under our experimental conditions. A 4-quinoline carboxylic acid derivative, an inhibitor of dihydroorotate dehydrogenase (95,96) and 6-azauridine, an inhibitor of OMP decarboxylase (97,98), were without any effect. However, it should be emphasized that optimal exposure time and concentration of these antimetabolites remain to be determined for each particular compound.

We then focused on dThd in more detail. When the effect of dThd on the stimulation of ddCyd phosphorylation was examined more closely, we found that

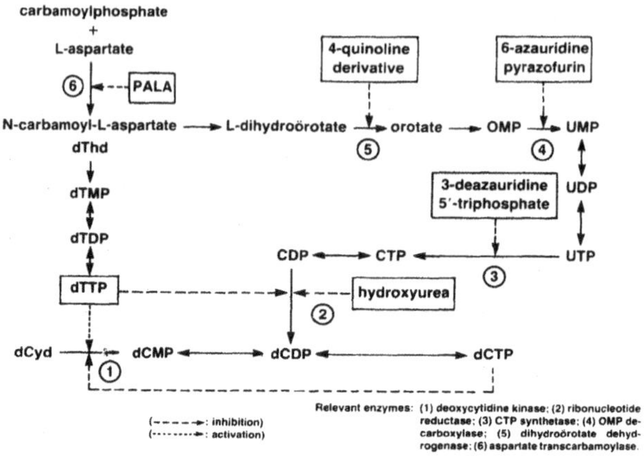

Fig. 5. Potential target enzymes for achieving a decreased intracellular concentration of dCTP.

the degree of stimulation of ddCyd phosphorylation was highly dependent on preincubation time as well as concentration of dThd. We observed a correlation between the increased ddCTP levels and decreased dCTP pools. However, we also found that the drop in dCTP pools seemed necessary but probably not sufficient to obtain efficiently increased ddCTP levels.

Finally, we demonstrated that under defined conditions, dThd enhanced the protective effect of ddCyd against HTLV-III-infected ATH8 cells. Moreover, the dThd-induced increase in the concentration of phosphorylated ddCyd metabolites did not parallel the slight increase in apparent toxicity. Therefore, toxic and antiviral effects may be modulated by different mechanisms, an observation of potential chemotherapeutic importance.

The possible mechanisms for the stimulation of ddCyd phosphorylation are as follows (Fig. 6) : (i) Conversion of dThd to dTTP in the cells, the latter metabolite being capable of enhancing the activity of dCyd kinase under constant dCTP levels (99); (ii) inhibition of CDP reductase by dTTP which would result in decreased amounts of dCTP, a potent feedback inhibitor of dCyd kinase (82,83); consequently, dCyd kinase would be stimulated and thus enhance ddCyd phosphorylation. (iii) The drop in dCTP pools would increase the competition of ddCTP with dCTP for reverse transcriptase, the putative target enzyme for the antiviral effects of ddCyd. Thus, a combination of ddCyd with dThd should potentially lead to increased intracellular ddCTP pools. Moreover,

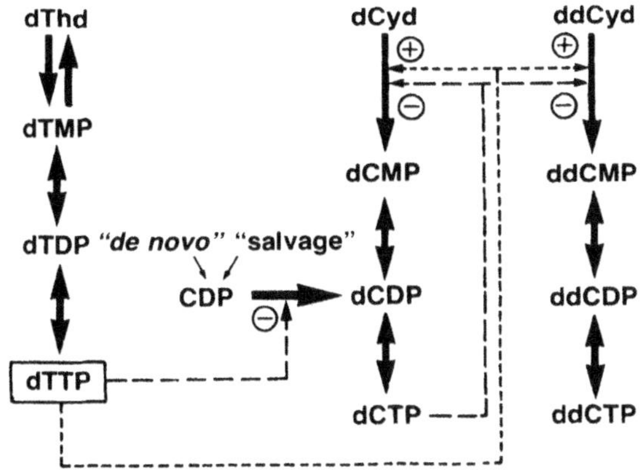

Fig. 6. Possible mechanism for stimulating ddCyd phosphorylation and reducing dCTP formation using thymidine (dThd). The + sign indicates catalytic steps that can be enhanced. The - sign indicates catalytic steps that can be inhibited.

dThd has already been used in phase I clinical trials (100) and should be a good candidate to be included in treatment schedules for AIDS patients in the near future. Indeed, as will be discussed later, the toxic effect of AzddThd is probably caused by a severe drop in dTTP levels; therefore, a schedule for drug administration in AIDS patients may be considered in which several weeks treatment with high concentrations of AzddThd is followed by several weeks treatment of a combination of ddCyd with dThd. This treatment scheme may result in an increased chemotherapeutic index by enhancing the antiviral effect and/or lowering (or rescuing) toxicity.

Additional tests have demonstrated that 2',3'-dideoxycytidine has other properties desirable as a potential therapeutic agent for treatment of HTLV-III/LAV infection: it is relatively resistant to cytidine deaminase (a major catabolic enzyme for cytidine analogues) (101); it is well absorbed when administered by the oral route in animal tests; it has straightforward pharmacokinetic clearance by the kidney; it does not appear to induce pyrimidine starvation; and finally, it has comparatively little toxicity in test animals. Entry to the central nervous system is also rapid, but the cerebrospinal fluid to plasma concentration ratio after i.v. administration is low (101). However, it was observed that drug levels which are effective in completely inhibiting the virus in vitro at high multiplicity of infection can be achieved after a single i.v. bolus administration at 27 mg/kg in the monkey (101). We have initiated a Phase I (feasibility and toxicity) study of 2',3'-dideoxycytidine in patients with AIDS as the next step in the development of this drug (Yarchoan R and Broder S, unpublished). A comparable study for a purine analogue, 2',3'-dideoxyadenosine, is planned for the future. The preliminary data using 2',3'-dideoxycytidine in patients look promising.

Another 2',3'-dideoxynucleoside analogue already touched upon, 3'-azido-2',3'-dideoxythymidine (AzddThd, AZT), was synthesized more than 20 years ago by Horwitz et al. (66) and was shown to inhibit C-type murine retrovirus replication in vitro by Ostertag more than 12 years ago (45). In the beginning of 1985, our group, in collaboration with the Wellcome Research Laboratories, observed that AzddThd at concentrations of 1 to 3 μM inhibited HTLV-III/LAV replication even under test conditions that utilize a high multiplicity of infection (39). Under less stringent test conditions AzddThd is inhibitory at substantially lower concentrations. However, it should be emphasized that AzddThd, upon longer incubation with HTLV-III-infected ATH8 cells, partially loses its antiretroviral activity. The fact that ddThd but not ddCyd and ddAdo also loses its antiretroviral potency upon longer incubation times with HTLV-III-infected ATH8 cells suggests that this phenomenon is not an unusual property of some ddThd analogues, an observation which might be important from a therapeutic viewpoint (74).

AzddThd metabolism to its 5'-triphosphate metabolites is similar in uninfected and HTLV-III-infected H9 cells, indicating that no specific viral enzyme is involved in the activation and metabolism of the drug (71,76) (Fig. 7). AzddThd is phosphorylated by the cellular dThd kinase (TK) to AzddThd 5'-monophosphate (AzddTMP). Its K_m for this enzyme nearly equals that of the physiological substrate dThd ($K_m \sim 3$ μM); the K_i/K_m ratio of AzddThd for dThd kinase (as measured with radiolabeled dThd as the substrate) is close to 1, inhibition of TK is competitive with respect to dThd, and its V_{max} is 60 % of the V_{max} for dTMP formation (71 and Balzarini, Johns and Broder, unpublished). These data suggest that AzddThd is an excellent substrate for TK and readily converted to its 5'-monophosphate derivative.

When evaluated for its substrate properties for thymidylate kinase (dTMP-K), AzddTMP shows a K_m of 8 μM, that is a 2-fold higher value than the K_m for the physiological substrate dTMP (Fig. 7). However, V_{max} is only 0.3 % of V_{max} for dTMP formation (71). Thus, AzddTMP has to be considered as a potent alternative substrate-inhibitor of dTMP kinase.

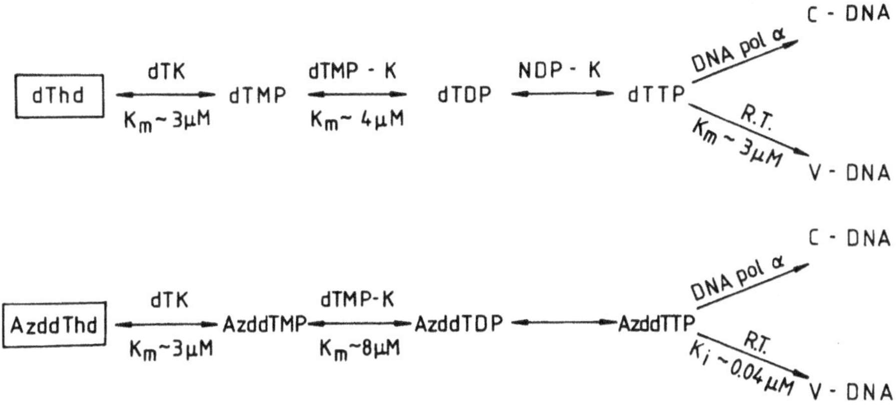

Fig. 7. Anabolic phosphorylation of thymidine (dThd) and the 3'-azido analogue of 2',3'-dideoxythymidine (AzddThd). The final normal nucleotide product (dTTP) and drug product (AzddTTP) can be possibly incorporated into both cellular DNA (cDNA) or retroviral DNA (vDNA). However, the viral DNA polymerase (R.T.) has a high affinity for AzddTTP, and this at least partially accounts for the selectivity of AzddThd as an antiretroviral agent. Also, see Fig. 2. Some of the toxicity of AzddThd appears to be related to the reduction of dTTP caused by the drug in inhibiting thymidylate kinase (dTMP-K) (see text).

Although the substrate properties of AzddTDP for nucleoside-5'-diphosphate kinase (NDP-K) have not yet been determined or reported, one may consider dTMP-K as the most likely rate-limiting step in the conversion of AzddThd to its active metabolite AzddThd-5'-triphosphate (AzddTTP).

Once converted to its 5'-triphosphate derivative, AzddThd may theoretically interfere with all cellular DNA polymerases (i.e., DNA polymerase α, β, γ; TdT transferase) and the viral DNA polymerase (reverse transcriptase). The observation that HTLV-III reverse transcriptase has a very high affinity for AzddTTP (K_i = 0.04 μM), this affinity being several orders of magnitude higher than the affinity for cellular DNA polymerase α and β (71,76, and Fig. 7), may explain the selectivity and potency of AzddThd as an antiretroviral chemotherapeutic agent in human beings. Thus, assuming that inhibition of reverse transcriptase is the main target for the observed antiviral effects of AzddThd, low intracellular ratios of dTTP <u>versus</u> AzddTTP will increase the potency of AzddThd as an antiretroviral agent, but may also enhance the cytostatic and/or cell toxic effects of the drug. Therefore, determination of intracellular nucleotide pool levels are necessary to obtain more information concerning intracellular balances between phosphorylated AzddThd metabolites and the natural nucleotide pools and to provide a rationale in our attempt to manipulate the antiviral and/or cytostatic action of the drug.

When H9 cells are incubated with AzddThd, dramatic changes in the dTTP pools occur: they drop at least 20-fold (71). How can the change of the dTTP pools be explained and what is their relevance to the toxic effects of AzddThd ? Most likely, the decrease of intracellular dTTP levels is due to the inhibition of dTMP-K by AzddTMP. Since the entire flow to dTTP from dTMP, either formed by salvage or de novo synthesis, has to go through this enzyme, it is obvious that a blockage at this level results in a decrease of dTDP and dTTP pools (consumed during the remaining DNA synthesis and/or temporary accumulation of dTMP). Indeed, we observed increased dTMP levels and decreased dTTP levels in AzddThd-treated ATH8 and Molt/4F cells. Thus, starvation of dTTP by AzddThd treatment will shut down DNA synthesis and may lead to the equivalent of a thymineless death of the cells. However, as a consequence of the inhibition of dTMP-K by AzddTMP, a dramatic accumulation of AzddTMP would also occur. This phenomenon has been demonstrated by Furman et al. in H9 cells (76) and by Balzarini and co-workers in human ATH8 and Molt/4F cells and in caprine Tahr cells (unpublished). Under some experimental conditions, AzddTMP pools exceed AzddTTP pools by 50-100 fold and intracellular levels of AzddTMP as high as 800 μM could be recorded in H9 cells treated with 50 μM AzddThd (71). The accumulation of AzddTMP into cells makes this metabolite also a potential candidate for thymidylate synthase (TS) inhibition. Indeed, TS represents a key enzyme in DNA synthesis since it is the only de novo source for the cells to synthesize dTMP, and it has been demonstrated that inhibition of this enzyme may lead to potent inhibition of cell growth and DNA synthesis (102,103). However, when AzddTMP was evaluated for its inhibitory effect against partially purified TS from Molt/4F cells, no significant inhibition could be demonstrated at an AzddTMP concentration as high as 250 μM. Also when evaluated in intact cells by measuring the effect of AzddThd on tritium release from [5-^{3}H]dUrd and [5-^{3}H]dCyd, no inhibition by 10 μM of AzddThd was recorded even when tritium release was measured at different time points (up to 72 hours) during AzddThd incubation. Deoxycytidylate (dCMP) deaminase, another important source of the substrate dUMP for the thymidylate synthase reaction and thus a potential target for toxicity, is also not affected by AzddThd and its phosphorylated metabolites. Thus, it is unlikely that inhibition of TS or dCMP deaminase by AzddThd or its phosphorylated metabolites accounts for the observed drop in dTTP pools. Moreover, since addition of dThd could easily overcome the inhibitory effect of AzddThd against Molt/4F cell proliferation, it seems likely that the severe decrease in dTTP levels and the toxic effects achieved by AzddThd must be ascribed to an inhibition of the dThd flow to dTTP at the dTMP-K level by AzddTMP. A similar observation has been made by us for 2',3'-ddThd which also accumulates as its 5'-monophosphate metabolite upon administration to Molt/4F and ATH8 cells although phosphorylation occurs considerably less extensively than for AzddThd. These data suggest that the phenomenon of inhibition of dTMP-K by AzddTMP and ddTMP is not an unusual property of 2',3'-dideoxythymidine analogues and one may speculate that toxicity exerted by this subset of compounds will be primarily due to dTTP starvation in human target cells.

However, inhibition of dTMP-K by AzddTMP is not a general phenomenon when cells from different species are compared. Indeed, we found dramatic differences in metabolism of AzddThd in murine leukemia (L1210), caprine ovary (Tahr) and human lymphoblast (Molt/4F) and lymphocyte (ATH8) cells. While accumulation of AzddTMP was noted in the human lymphoid and caprine ovary cells (ratio AzddTMP/AzddTTP : 50-200), extensive formation of AzddTTP versus AzddTMP was noted in murine leukemia L1210 cells (ratio AzddTMP/AzddTTP: $\sim$ 0.5) (Fig. 8). These data clearly indicate that metabolism of AzddThd is highly dependent on the species and cell type. This may also explain the observation that AzddThd is a potent inhibitor of mouse viremia and retroviral disease (e.g., splenomegaly) in Rauscher murine leukemia virus infected mice (104). It also explains our observation that in Moloney murine sarcoma virus infected newborn NMRI mice, AzddThd treatment severely depresses tumor formation and dramatically prolongs survival rate of the test mice (105). These protective effects against murine retroviruses are probably due to the extremely high levels of AzddTTP found in the murine cells (versus the human cells), rather than the species origin of the virus because Dahlberg, Aaronson, Mitsuya and Broder found that AzddThd dramatically inhibits the replication of an amphotropic murine leukemia virus in cells of the mouse and rat but is only moderately effective in human cells (unpublished data). Thus, one should not automatically extrapolate metabolic data from murine or other animal cell lines to human cell metabolism.

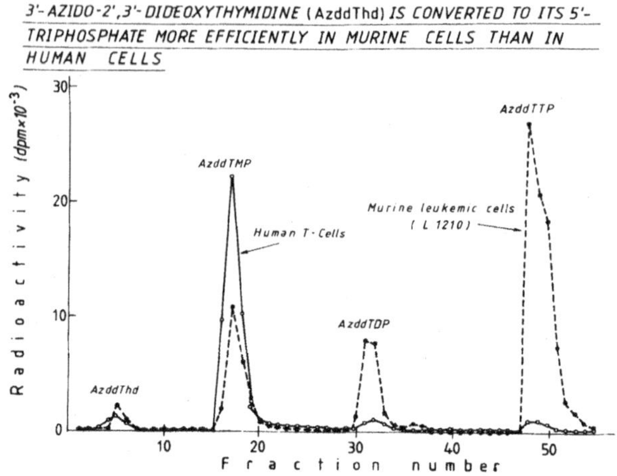

Fig. 8. Different prolifes of anabolic phosphorylation of AzddThd in human T cells versus mouse L1210 cells. The antiretroviral effects and possibly some toxic effects of the drug are mediated by the 5'-triphosphate derivative (AzddTTP). Note that in human cells, only a small fraction of the drug is metabolized to a triphosphate form.

EVALUATION OF 3'-AZIDO-2',3'-DIDEOXYTHYMIDINE IN PATIENTS WITH ADVANCED HTLV-III DISEASES

In the middle of 1985, we began administering AzddThd to patients with AIDS and related conditions in a phase I trial conducted in collaboration with Duke University Medical Center and Wellcome Research Laboratories. One of the purposes of this trial was to study the pharmacokinetics of AzddThd. We found that the drug has good oral bioavailability (approximately 60 %) and that levels of AzddThd that were inhibitory for HTLV-III/LAV replication in vitro were achievable in patients (106,107). The half-life of AzddThd was found to be approximately 1 hour with much of the drug clearance being via glucuronidation (106-108). At least one drug which is known to interfere with hepatic glucuronidation, probenecid, increases the half life of AzddThd [108], and it is possible that other inhibitors of glucuronidation such as acetaminophen, morphine, and sulfonamides could have similar effects. Finally, AzddThd was found to have excellent penetration across the blood-brain barrier (106,107).

In this trial and extensions of this trial, we also observed that patients administered intermediate doses of AzddThd (15 to 30 mg/kg/day orally) over a 6 week period had partial restoration of immune function including 2- to 3-fold increases in the absolute number of helper-inducer (CD4+) T cells and the development of cutaneous delayed type hypersensitivity reactions in previously anergic individuals. These changes were often detectable after the second week of therapy (106), but they were not always dramatic. Indeed, one of the principles we now know is that small increments of T4 counts can confer a significant survival advantage to patients. The majority of patients given AzddThd also gained weight, reported less fatigue, and had improvement in other clinical abnormalities associated with AIDS. Finally, we observed that some patients with HTLV-III/LAV-induced neurologic disease (particularly those with dementia) had substantial improvements in neurologic function upon being given AzddThd (69,109 and Yarchoan and Broder, unpublished observation). AzddThd, however, would not a priori be expected to affect secondary neurological manifestations of AIDS such as toxoplasmosis or central nervous system lymphoma, and these conditions should be excluded in evaluating the effects of AzddThd on HTLV-III-induced neurologic disease.

Bone marrow toxicity (particularly anemia and leukopenia, and, occasionally, decline in the number of CD4+ lymphocytes) was observed in some patients receiving high doses of AzddThd (90 mg/kg/day orally) for 4 to 6 weeks and in some patients receiving lower doses of AzddThd (25 to 30 mg/kg/day orally) for more than 8 weeks (107). Increase in the red blood cell mean corpuscular volume (reflecting megaloblastic changes) is often an early sign of this toxicity (110). Interestingly, thrombocytopenia is much less common and indeed, patients often have increases in the number of platelets upon being given short courses of AzddThd (106, and Yarchoan and Broder, unpublished observation). Some patients receiving AzddThd reported mild headaches, and two patients receiving high doses (90 mg/kg/day orally) developed transient agitation. Otherwise, non-hematologic toxicity was rarely observed.

We have now administered AzddThd to patients with AIDS for as long as 18 months, and many patients appear to tolerate the drug over this period of time, although some patients required dose reductions. Seven of the 8 patients who entered the phase I trial with a prior episode of Pneumocystis carinii pneumonia are alive after a median time of 54 weeks on AzddThd (Yarchoan, Bolognesi, Durack and Broder, unpublished); in the absence of therapy, only half would be expected to survive past 35 weeks (111). Formal proof that AzddThd could prolong survival, at least in AIDS patients who had recently recovered from *Pneumocystis carinii* pneumonia, was provided by a multicenter double-blind placebo-controlled trial in which there was a dramatic difference in mortality between the treated and untreated patients (Dept. Health and Human Services announcement, Sept. 19, 1986).

The trial was begun in February 1986. By September 1986, there were 16 deaths in the placebo arm and one death in the AzddThd arm. At the same time, patients in the drug arm showed other evidence of clinical and immunological improvements. Interestingly, while patients in the AzddThd arm, in addition to an improved survival, had a clear reduction in the number of opportunistic infections or other AIDS-related events, a major difference between the two arms of the study did not occur until after six weeks, suggesting that the benefits conferred by AzddThd did not fully materialize until at least six weeks of administration.) At that point, an independent data-monitoring board recommended that the placebo group begin to receive the drug. The study is still underway (with both arms now receiving drug) to determine long-term safety and efficacy.

AzddThd is now being made widely available to physicians for the treatment of patients with AIDS who have had *Pneumocystis carinii* pneumonia and who fit certain other criteria. It is worth stressing that the clinical experience with AzddThd is at this time limited, and the availability of AzddThd represents only a first step. The optimal dose for long term therapy is still being explored. We also need to learn the best schedule of administration and to learn whether the development of AIDS may be slowed or prevented by administering AzddThd early in the course of HTLV-III/LAV infection. We are just beginning to explore strategies which may permit a greater antiviral activity without a concomitant increase in bone marrow toxicity. In this regard, our group has recently observed that acyclovir (a guanosine analog with potent activity against *herpes* viruses but little or no antiretroviral activity *per se*) can potentiate the antiretroviral effect of AzddThd *in vitro* through an as yet unknown mechanism (70, and Mitsuya and Broder, unpublished observation). Acyclovir does not characteristically cause bone marrow suppression and it is possible that the combination of these two agents may be of benefit. We are presently initiating a clinical trial to test this hypothesis.

Thus, AzddThd is a drug selected on the basis of its *in vitro* antiviral effect against HTLV-III and shown to confer at least a short-term clinical benefit in patients with advanced disease. In this respect, it represents a first step in developing practical chemotherapy against pathogenic human retroviruses. While the clinical results may be important in their own right, they are also important because they appear to validate some of the strategies of therapy discussed here, and they reinforce the idea that therapy for established retroviral infections will eventually be an attainable goal.

CONCLUSION

AIDS and its related disorders are caused by the third known pathogenic human retrovirus (HTLV-III), which infects and destroys helper/inducer T cells, thereby irreparably damaging the immune system. The genome of this virus contains the standard replicative genes found in other retroviruses; however, it contains several additional genes not previously known. The product(s) of each gene represents a potential target of opportunity in developing new experimental therapies for diseases caused by this virus, and we have discussed several issues related to the viral life cycle which may be relevant to future therapeutic strategies. We have also discussed some basic pharmacologic principles for designing antiretroviral treatment. Curative therapy for diseases caused by pathogenic retroviruses will probably not be possible until the molecular biology of the virus and the structural chemistry of key viral products are defined through further basic research. Nevertheless, HTLV-III has revealed enough of its life history for us to initiate new therapeutic strategies, and it is not necessary to await new breakthroughs before implementing some strategies that we already know have clinical applications in patients with advanced disease.

REFERENCES

1. Gottlieb, M.S., Schroff, R., Schanker, H.M., Weisman, J.D., Fan, P.T., Wolf, F.A. and Saxon, A. New Engl. J. Med. 305 : 1425-1431, 1981.
2. Masur, H., Michelis, M.A., Greene, J.B., Onorato, I., Vande Stouwe, R.A., Holzman, R.S., Wormser, G., Brettman, L., Lange, M., Murray, H.W. and Cunningham-Rundles, S. New Engl. J. Med. 305 : 1431-1438, 1981.
3. Siegal, F.P., Lopez, C., Hammer, G.S., Brown, A.E., Kornfeld, S.J., Gold, J., Hassell, J., Hirschman, S.Z., Cunningham-Rundles, C., Adelsberg, B.R., Parham, D.M. Siegal, M., Cunningham-Rundles, S. and Armstrong, D. New Engl. J. Med. 305 : 1439-1444, 1981.
4. Gallo, R.C., Salahuddin, S.Z., Popovic, M., Sherer, G.M., Kaplan, M., Haynes, B.F., Palker, T.J., Redfield, R., Oleske, J., Safai, B., White, G., Foster, P. and Markham, P.D. Science 224 : 500-503, 1984.
5. Popovic, M., Sarngadharan, M.G., Reed, E. and Gallo, R.C. Science 224 : 497-500, 1984.
6. Ho, D.D., Rota, T.R. and Hirsch, M.S. J. Clin. Invest. 77 : 1712-1715, 1986.
7. Gartner, S., Markovits, P., Markovitz, D.M., Kaplan, M.H., Gallo, R.C. and Popovic, M. Science 233 : 215-219, 1986.

8. Fahey, J.L., Prince, H., Weaver, M., Groopman, J., Visscher, B., Schwartz, K. and Detels, R. Am. J. Med. 76 : 95-100, 1984.
9. Barré-Sinoussi, F., Chermann, J.C., Rey, F., Nugeyre, M.T., Chamaret, S., Gruest, J., Duaguet, C., Axler-Blin, C., Vézinet-Brun, F., Rouxioux, C., Rosenbaum, W. and Montagnier, L. Science 220 : 868-871, 1983.
10. Levy, J.A., Hoffman, A.D., Kramer, S.M., Landis, J.A., Shimabukuro, J.M. and Oshiro, J.S. Science 225 : 840-842, 1984.
11. Coffin, J., Haase, A., Levy, J.A., Montagnier, L., Oroszlan, S., Teich, N., Temin, H., Toyoshima, K., Varmus, H., Vogt, P. and Weiss, R. Science 232 : 697, 1986.
12. Groopman, J.E., Salahuddin, S.Z., Sarngadharan, M.G., Markham, P.D., Gonda, M., Sliski, A. and Gallo, R.C. Science 226 : 447-449, 1984.
13. Ho, D.D., Schooley, R.T., Rota, T.R., Kaplan, J.C., Flynn, T., Salahuddin, S.Z., Gonda, M. and Hirsch, M.S. Science 226 : 451-453, 1984.
14. Goedert, J.J., Bigger, R.J., Weiss, S.H., Eyster, M.E., Melbye, M., Wilson, S., Ginzburg, H.M., Grossman, R.J., Digioia, R.A., Sanchez, W.C., Giron, J.A., Ebbesen, P., Gallo, R.C. and Battner, W.A. Science 231 : 992-995, 1986.
15. Klatzmann, D., Barré-Sinoussi, F., Nugeyre, M.T., Dauguet, C., Vilmer, E., Griscelli, C., Brun-Vézinet, F., Rouzioux, C., Gluckman, J.C., Chermann, J.C. and Montagnier, L. Science 225 : 59-63, 1984.
16. Lawrence, J., Gottlieb, A.B. and Kunkel, H.G. J. Clin. Invest. 72 : 2072-2081, 1983.
17. Lane, H.C., Masur, H.M., Longo, D.L., Klein, M.G., Rook, A.H., Quinnam, G.V., Steis, R.G., Macher, A., Whalen, G., Edgar, L.C. and Fanci, A.S. New Engl. J. Med. 311 : 1099-1103, 1984.
18. Shaw, G.M., Harper, M.E., Hahn, B.H., Epstein, L.G., Gajdusek, D.C., Price, R.W., Navia, B.A., Petito, C.K., O'Hara, C.J., Groopman, J.E., Cho, E.-S., Oleske, J.M., Wong-Staal, F. and Gallo, R.C. Science 227 : 177-182, 1985.
19. Ho, D.D., Rota, T.R., Schooley, R.T., Kaplan, J.C., Allan, J.D., Resnick, L., Felsenstein, D., Andrews, C.A. and Hirsch, M.S. New Engl. J. Med. 313 : 1493-1497, 1985.
20. Resnick, L., DiMarzo-Veronese, F., Schüpbach, J., Tourtellote, W.W., Ho, D.D., Muller, F., Shapshak, P., Vogt, M., Groopman, J., Markham, P.D. and Gallo, R.C. New Engl. J. Med. 313 : 1498-1504, 1985.
21. Navis, B.A., Jordan, B.D. and Price, R.W. Ann. Neurol. 19 : 517-524, 1986.
22. Baltimore, D. Nature 226 : 1209-1211, 1970.
23. Temin, H.M. and Mizutani, S. Nature 226 : 1211-1213, 1970.
24. Klein, G. Cancer 45 : 2486-2499, 1980.
25. Dalgleish, A.G., Beverly, P.C.L., Clapham, P.R., Crawford, D.H., Greaves, M.F. and Weiss, R.A. Nature 312 : 763-767, 1984.
26. Klatzmann, D., Champagne, E., Chamerat, S., Gruest, J., Guetard, D., Hercend, T., Gluckman, J.-C. and Montagnier, L. Nature 312 : 767-768, 1984.
27. Sodroski, J., Goh, W.C., Rosen, C., Campbell, K. and Haseltine, W.A. Nature 322 : 470-474, 1986.
28. Lifson, J.D., Feinberg, M.B., Reyes, G.R., Rabin, L., Banapour, B., Chakrabarti, S., Moss, B., Wong-Staal, F., Steiner, K.S. and Engleman, E.G. Nature 323 : 725-728, 1986.
29. Pert, C.B., Hill, J.M., Ruff, M.R., Berman, R.M., Robey, W.G., Arthur, L.O. et al. Proc. Natl. Acad. Sci. USA, in press.
30. Wong-Staal, F. and Gallo, R.C. Nature 317 : 395-403, 1985.
31. Hahn, B., Shaw, G.M., Taylor, M.E., Redfield, R.R., Markham, P.D., Salahuddin, S.Z., Wong-Staal, F., Gallo, R.C., Parks, E.S. and Parks, W.P. Science 232 : 1548-1553, 1986.

32. Dolin, R. Science 227 : 1296-1303, 1985.
33. Wain-Hobson, S., Sonigo, P., Danos, O., Cole, S. and Alizon, M. Cell 40 : 9-17, 1985.
34. Tanese, N., Sodroski, J., Haseltine, W.A. and Goff, S.P. J. Virol. 59 : 743-745, 1986.
35. Mitsuya, H., Popovic, M., Yarchoan, R., Matsushita, S., Gallo, R.C. and Broder, S. Science 226 : 172-174, 1984.
36. Rozenbaum, W., Dormont, D., Spire, B., Vilmer, E., Gentilini, M., Griscelli, C., Montagnier, L., Barré-Sinoussi, F. and Chermann, J.C. Lancet i : 450-451, 1985.
37. Sandstrom, E.G., Kaplan, J.C., Byington, R.E. and Hirsch, M.S. Lancet i : 1480-1482, 1985.
38. Anand, R., Moore, J., Feorino, P., Curran, J. and Srinivasan, A., Lancet i : 97-98, 1986.
39. Mitsuya, H., Weinhold, K.J., Furman, P.A., St. Clair, M.H., Nusinoff-Lehrman, S., Gallo, R.C., Bolognesi, D., Barry, D.W. and Broder, S. Proc. Natl. Acad. Sci. USA 82 : 7096-7100, 1985.
40. Mitsuya, H. and Broder, S. Proc. Natl. Acad. Sci. USA 83 : 1911-1915, 1986.
41. Ting, R.C., Yang, S.S. and Gallo, R.C. Nature New Biol. 236 : 163-166, 1972.
42. Furmanski, P., Bourguignon, G.L., Bolles, C.S., Corombos, J.D. and Das, M.R. Cancer Lett. 8 : 307-315, 1980.
43. De Clercq, E. Cancer Lett. 8 : 9-22, 1979.
44. Waqar, M.A., Evans, M.J., Manly, K.F., Hughes, R.G., Huberman, J.A. J. Cell. Physiol. 121 : 402-408, 1984.
45. Ostertag, W., Roesler, G., Krieg, C.J., Kind, J., Cole, T., Crozier, T., Gaedicke, G., Steinheider, G., Kluge, N. and Dube, S. Proc. Natl. Acad. Sci. USA 71 : 4980-4985, 1974.
46. Chandra, P., Demirhan, I. and De Clercq, E. Cancer Lett. 12 : 181-193, 1981.
47. Panganibau, A.T. and Temin, H.M. Proc. Natl. Acad. Sci. USA 81 : 7885-7889, 1984.
48. McCormick, J.B., Getchell, J.P., Mitchell, S.W. and Hicks, D.R. Lancet ii : 1367-1369, 1984.
49. Sodroski, J.G., Rosen, C.A., Wong-Staal, F., Salahuddin, S.Z., Popovic, M., Arya, S., Gallo, R.C. and Haseltine, W.A. Science 221 : 171-173, 1985.
50. Sodroski, J., Goh, W.C., Rosen, C., Dayton, A., Terwilliger, E. and Haseltine, W. Nature 321 : 412-417, 1986.
51. Feinberg, M.B., Jarrett, R.F., Aldovini, A., Gallo, R.C. and Wong-Staal, F. Cell 46 : 807-817, 1986.
52. Fisher, A.G., Feinberg, M.B., Josephs, S.F., Harper, M.E., Marselle, L.M., Reyes, G., Gonda, M.A., Aldovini, A., Debouk, C., Gallo, R.C. and Wong-Staal, F. Nature 320 : 367-371, 1986.
53. Rosen, C.A. Sodroski, J.G., Goh, W.C., Dayton, A.I., Lippke, J. and Haseltine, W.A. Nature 319 : 555-559, 1986.
54. Smith, T.J., Kremer, M.J., Luo, M., Vriend, G., Arnold, E., Kamer, G., Rossmann, M.G., McKinlay, M.A., Diana, G.D. and Otto, M.J. Science 233 : 1286-1293, 1986.
55. Izant, J.G., Weinbraub, H. Cell 36 : 1007-1015, 1984.
56. Pestka, S., Daugherty, B.L., Jung, V., Hotta, K. and Pestka, R.K. Proc. Natl. Acad. Sci. USA 81 : 7525-7528, 1984.
57. Zamenick, P.C., Goodchild, J., Taguchi, Y. and Sarin, P. Proc. Natl. Acad. Sci. USA 83 : 4143-4146, 1986.

58. Schultz, A.M. and Oroszlan, S. J. Virol. 46 : 355-361, 1983.
59. Henderson, L.E., Krutzch, H.C. and Oroszlan, S. Proc. Natl. Acad. Sci. USA 80 : 339-343, 1983.
60. Ho, D.D., Hartshorn, K.L., Rota, T.R., Andrews, C.A., Kaplan, J.C., Schooley, R.T. and Hirsch, M.S. Lancet i : 602-604, 1985.
61. Montagnier, L., Gruest, J., Chamaret, S., Dauguet, C., Axler, C., Guétard, D., Nugeyre, M.T., Barré-Sinoussi, F., Chermann, J.-C., Brunet, J.B., Klatzmann, D. and Gluckman, J.C. Science 225 : 63-66, 1984.
62. Birx, D.L., Redfield, R.R. and Tosato, G. New Engl. J. Med. 314 : 874-879, 1986.
63. Yarchoan, R., Redfield, R.R. and Broder, S. J. Clin. Invest. 78 : 439-447, 1986.
64. Hoxie, J., Haggarty, B.S., Rackowski, J.L., Pillsbury, N. and Levy, J.A. Science 229 : 1400-1402, 1985.
65. Folks, T., Powell, D.M., Lightfoote, M.M., Benn, S., Martin, M.A. and Fauci, A.S. Science 231: 600-602, 1986.
66. Horwitz, J.P., Chua, J. and Noel, M. J. Org. Chem. 29 : 2076-2078, 1964.
67. Toji, L. and Cohen, S.S. Proc. Natl. Acad. Sci. USA 63 : 871-877, 1969.
68. Lin, T.S. and Prusoff, W.H. J. Med. Chem. 21 : 109-112, 1978.
69. Yarchoan, R. and Broder, S. In: Vaccine 87. (Eds. R.M. Channock, R.A. Lerner, F. Brown and H.S. Ginsberg), Cold Spring Harbor Press, Cold Spring Harbor, 1987, in press.
70. Mitsuya, H., Matsukura, M. and Broder, S. In AIDS: Modern Concepts and Therapeutic Challenges (Ed. S. Broder), Marcel Dekker Inc., New York, 1987, pp 303-333.
71. Mitsuya, H. and Broder, S. Nature 325 : 773-778, 1987.
72. Balzarini, J., Pauwels, R., Herdewijn, P., De Clercq, E., Cooney, D.A., Kang, G.-J., Dalal, M., Johns, D.G. and Broder, S. Biochem. Biophys. Res. Commun. 140 : 735-742, 1986.
73. Baba, M., Pauwels, R., Herdewijn, P., De Clercq, E., Desmyter, J. and Vandeputte, M. Biochem. Biophys. Res. Commun. 142 : 128-134 (1987).
74. Balzarini, J., Kang, G.-J., Dalal, M., Herdewijn, P., De Clercq, E., Broder, S. and Johns, D.G. Mol. Pharmacol., in press (1987).
75. Balzarini, J., Pauwels, R., Baba, M., Robins, M.J., Zou, R., Herdewijn, P. and De Clercq, E. Biochem. Biophys. Res. Commun., in press (1987).
76. Furman, P.A., Fyfe, J.S., St. Clair, M.H., Weinhold, K., Rideout, J.L., Freeman, G.A., Lehrman, S.N., Bolognesi, D.P., Broder, S., Mitsuya, H. and Barry, D.W. Proc. Natl. Acad. Sci. USA 83 : 8333-8337, 1986.
77. Cooney, D.A., Dalal, M., Mitsuya, H., McMahon, J., Nadkarni, M., Balzarini, J., Broder, S. and Johns, D.G. Biochem. Pharmacol. 35 : 2065-2068, 1986.
78. Sanger, F., Nicklen, S. and Coulson, A.R. Proc. Natl. Acad. Sci. USA 74 : 5463-5467, 1977.
79. Pauwels, R., De Clercq, E., Desmyter, J., Balzarini, J., Goubau, P., Herdewijn, P., Vanderhaeghe, H. and Vandeputte, M. J. Virol. Meth. : in press, 1987.
80. Starnes, M.C. and Cheng, Y.C. J. Biol. Chem. 262 : 988-991, 1987.
81. Mitsuya, H., Jarrett, R.F., Matsukura, M., Veronese, F., De Vico, A.L., Sarngadharan, M.S., Johns, D.G., Reitz, M.S. and Broder, S. Proc. Natl. Acad. Sci. USA, in press, 1987.
82. Maley, F. and Maley, G.F. Biochemistry 1: 847-851, 1962.
83. Durham, J.P. and Ives, D.H. Fed. Proc. 26 : 808a, 1967.
84. Skoog, L. and Nordenskjöld, B. Eur. J. Biochem. 19 : 81-89, 1971.
85. Theiss, J.C. and Fisher, G.A. Biochem. Pharmacol. 25 : 73-79, 1976.

86. McPartland, R.P., Wang, M.C., Bloch, A. and Weinfeld, H. Cancer Res. 34 : 3107-3111, 1974.
87. Brockman, R.W. Shaddix, S.C., Williams, M., Nelson, J.A., Rose, L.M. and Schabel, F.M. Jr. Ann. N.Y. Acad. Sci .USA 255 : 501-521, 1975.
88. Morris, N.R. and Fisher, G.A. Biochim. Biophys. Acta 68 : 84-92, 1963.
89. Ives, D.H., Morse, P.A. Jr. and Potter, V.R. J. Biol. Chem. 238 : 1467-1474, 1963.
90. Sweeney, M.J., Davis, F.A., Gutowski, G.E., Hamill, R.L., Hoffman, D.H. and Poore, G.A. Cancer Res. 33 : 2619-2623, 1973.
91. Plagemann, P.G.W. and Behrens, M. Cancer Res. 36 : 3807-3812, 1976.
92. Collins, K.D. and Stark, G.R. J. Biol. Chem. 246 : 6599-6605, 1971.
93. Swyryd, E.A., Seaver, S.S. and Stark, G.R. J. Biol. Chem. 249 : 6945-6950, 1974.
94. Yoshida, T., Stark, G.R. and Hoogenraad, N.J. J. Biol. Chem. 249 : 6951-6955, 1974.
95. Dexter, D.L., Hesson, D.P., Ardecky, R.J., Rao, G.V., Tippet, D.L., Tusak, B.A., Paull, K.D., Plowman, J., Delarco, B.M., Narayanan, V.L. and Forbes, M. Cancer Res. 45 : 5563-5568, 1985.
96. Chen, S.F., Ruben, R.L. and Dexter, D.L. In: Proc. 77th Annual Meeting of the American Association for Cancer Research, Los Angeles, 1986, p. 1183.
97. Handschumacher, R.E. and Pasternak, C.A. Biochim. Biophys. Acta 30 : 451-452, 1958.
98. Handschumacher, R.E. J. Biol. Chem. 235 : 2917-2919, 1960.
99. Durham, J.P. and Ives, D.H. J. Biol. Chem. 245 : 2276-2284, 1970.
100. Woodcock, T., Damin, L., O'Hehir, M., Andreef, M. and Young, C. Proc. Am. Assoc. Cancer Res. 20 : 462, 1979.
101. Kelley, J.A., Litterst, C.L., Roth, J.S., Vistica, D.T., Poplack, D.G., Cooney, D.A., Nadkarni, M., Balis, F.M., Broder, S. and Johns, D.G. Drug Met. Disp., in press, 1987.
102. Cohen, S.S. Ann. N.Y. Acad. Sci. USA 186 : 292-301, 1971.
103. Bjursell, G. and Reichard, P. J. Biol. Chem. 248 : 3904-3909, 1973.
104. Ruprecht, R.M., O'Brien, L.G., Rossoni, L.D. and Nusinoff-Lehrman, S. Nature 323 : 467-469, 1986.
105. Balzarini, J., Pauwels, R., Baba, M., De Clercq, E., Broder, S. and Johns, D.G. III International Conference on Acquired Immunodeficiency Syndrome (AIDS), June 1-5, 1987, Washington DC, USA, submitted, 1987.
106. Yarchoan, R., Klecker, R.W., Weinhold, K.J., Markham, P.D., Lyerly, H.K., Durack, D.T., Gelmann, E., Lehrmann, S.N., Blum, R.M., Barry, D., Shearer, G., Fischl, M.A., Mitsuya, H., Gallo, R.C., Collins, J.M., Bolognesi, D.P., Myers, C.E. and Broder, S. Lancet i : 575-580, 1986.
107. Klecker, R.W., Collins, J.M., Yarchoan, R., Thomas, R., Jenkins, J.F., Broder, S. and Myers, C.E. Clin. Pharmacol. Therap., in press, 1987.
108. de Miranda, P., Good, S.S., Blum, M.R., Thomas, R.V., Yarchoan, R. and Broder, S. Program of the International Conference on Acquired Immunodeficiency Syndrome (AIDS), Paris, France, June 23-25, 1986.
109. Yarchoan, R., Berg, G., Brouwers, P., Fischl, M.A., Spitzer, A.R., Wichman, A., Grafman, J., Thomas, R., Safai, B., Brunetti, A., Perno, C., Schmidt, P.J., Larson, S.M., Myers, C.E. and Broder, S. Lancet i : 132-135, 1987.
110. Yarchoan, R. and Broder, S. In: AIDS: Modern Concepts and Therapeutic Challenges (Ed. S. Broder), Marcel Dekker Inc., New York, pp. 335-360, 1987.
111. Broder, S. and Gallo, R.C. New Engl. J. Med. 311 : 1292-1297, 1984.

22

PROMISES TO KEEP: CLINICAL USE OF ANTIVIRAL DRUGS

G.J. GALASSO

Office of the Director, National Institutes of Health,
Bethesda, Maryland, 20892, USA

....I have promises to keep,
And miles to go before I sleep,
And miles to go before I sleep.
Robert Frost

ABSTRACT

Research on the development and clinical use of antiviral agents is progressing at an impressive rate after a very slow start. The need, problems, and current status of antiviral agents in clinical usage are discussed.

INTRODUCTION

Like Robert Frost, musing in the woods on a snowy evening, so too do antiviral agents hold great promise for the control of infectious diseases, but have far to go to realize this promise. Research on antiviral agents is a relatively new discipline with a very slow beginning, only recently gaining the recognition it properly deserves. A few years ago, research in this field was considered both unrewarding and unproductive.

Viral inhibition was considered a helpful tool in the study of viral replication, but there was no expectation that this would lead to useful therapeutics. Clinical efficacy, free of drug toxicity, was not believed attainable. So entrenched was the concept that antivirals would not be clinically useful, that medical practitioners, and even virologists, were not prepared to accept amantadine, an extremely effective and beneficial agent for the prevention and treatment of influenza A infections. It took more than 10 years before this proven drug was accepted, and it is still not used as much as it deserves. However, the potential role and need of antiviral agents now is generally recognized, and each year brings increased research, new agents, and clinical benefits.

VACCINES

To put antivirals in proper perspective, one must first consider vaccines. In the control of any viral infection, the best approach is prevention. As long as there is a vaccine that is safe, effective, reasonable in cost, and its risks are greatly outweighed by the potential morbidity and mortality of the diseases, its use should be advocated strongly. Smallpox vaccination provides an excellent example. This vaccine was a very crude preparation with several undesirable side effects. While smallpox remained a threat to public health, vaccination was encouraged and practiced throughout the world, and was eventually instrumental in the eradication of the disease. When, however, it was shown that 15 importations of smallpox cases would be required each year to achieve the same level of mortality that resulted from vaccination, the program was discontinued in the U.S. prior to eradication. A similar situation occurred in other developed nations. Vaccines continue to be important in the control of viral diseases such as polio, mumps, and rubella, but they cannot meet all our needs for the control of every important viral disease.

Despite the impossibility of a vaccination program against every virus, there remain three major non-scientific obstacles to vaccine solutions for viral disease prevention--resistance to preventive medicine, complacency, and litigation. Unlike pediatricians who have championed childhood immunization programs, internists and general practitioners have not been strong advocates of existing adult vaccination programs for pneumococcal pneumonia, hepatitis, or influenza, reflecting, perhaps, their traditional inexperience in the use of vaccines. For example, influenza vaccine is underused, despite the considerable mortality and morbidity of the disease among the elderly, other high risk populations, and essential personnel. Even in well-organized programs, acceptance in the target population is only 10-30%. The reason is largely due to the resistance of health care personnel. In a study of 206 health care personnel, 86% were aware of the complication of influenza, and 84% believed the vaccine was protective, but only 16% received the vaccine and only 47% recommended it to their patients. There appeared to be no effort to avoid the considerable morbidity of the disease or to follow the recommendations to decrease nosocomial infection through prophylaxis (1). While the medical community and the general public are coming to an appreciation of disease prevention in other health areas through diet and exercise, the beneficial use of vaccines has not been promoted effectively.

Complacency is generally defined as self-satisfaction accompanied by unawareness of actual dangers or deficiencies. Paradoxically, as diseases such as diphtheria, polio, pertussis, mumps, and measles have become less common because of the success of immunization programs, concerns about side effects have increased among patients and physicians. The result has been a decline in the advocacy of immunization programs. But these diseases are neither trivial nor near eradication. Continued complacency based upon the false conception of the near eradication of these diseases could predictably lead to their recrudescence. This has already been the case with pertussis in Europe and Japan.

The third major factor in the disinclination to advocate vaccines is litigation. The tort system in the U.S. has already contributed to several pharmaceutical companies withdrawing from the manufacture of vaccines. Costs of current law suits against vaccine manufacturers exceed their annual earnings from the sale of vaccines (2). Supplies of diphtheria-pertussis-tetanus (DPT) vaccines remain precarious; there is currently only one U.S. manufacturer of this vaccine. Liability is a serious problem and may only be resolved, in the U.S., by federal legislation which would standarize compensation for unavoidable adverse reaction, medical expenses, and loss of income, but limit legal fees and eliminate awards for pain and suffering, as well as punitive damages. Without some relief from the threat of litigation and foreseeable legal judgments, the vaccine industry faces stagnation and eventual decline.

All this argues more forcibly for the need for expanded research in the development and improvement of antiviral agents. Viral infectious diseases remain one of the major causes of morbidity and mortality, and, rather than coming under control, they are increasing in number and severity. In the past few years, there have been new infectious diseases and increased incidence of others, such as: Legionnaires' disease, toxic shock syndrome, genital herpes, genital warts, and acquired immunodeficiency syndrome (AIDS). If vaccines cannot provide prevention, there must be control through treatment.

DEVELOPMENT OF ANTIVIRAL AGENTS

The progress in development of antivirals to date has been painstakingly slow. The birth of antiviral chemotherapy can be traced to 1962 when Kaufman demonstrated the effectiveness of idoxuridine against herpes keratitis (3). Although impressive, idoxuridine did not generate much stimulus for antivirals because it was a topical drug without wide applicability. Shortly thereafter, amantadine was shown to be effective against Asian influenza (4). However, unwarranted concerns for its side effects, lack of experience in the use of antivirals, and several other peripheral factors contributed to its poor acceptance. Only the threat of swine flu led to renewed interest, further clinical trials, and wider use. However, as mentioned earlier, medical practitioners still hesitate to prescribe amantadine. The initial lethargy of the field was due to the disbelief that agents could be developed that could enter cells, inhibit cellular functions used in the replication of viruses, and not be toxic to the uninfected cells. Therefore, all the agents currently available have been developed through serendipity. Screening programs seeking anti-cancer agents or studying viral replication have yielded viral inhibitors. Once inhibition is established, stepwise chemical manipulation of the agent is attempted in order to achieve maximal viral inhibition. The most promising agents, identified by _in vitro_ studies, are tested _in vivo_ for toxicity and pharmacology and eventually in animal models that mimic the viral disease in man. Finally, clinical efficacy studies are performed. This process is unavoidably slow, taking 5-10 years and an expenditure measured in millions of dollars. Still, the system has been relatively effective.

The real impetus for research on antivirals was the finding, in 1977, that vidarabine could be used successfully in treatment of a serious ongoing viral infection, herpes encephalitis, with a resultant reduction in mortality (5). Vidarabine also proved effective against neonatal herpes and varicella-zoster in immunocompromised patients. The success of acyclovir against herpes infections promoted further interest in antiviral agents (6,7). Most recently, science, politics, and public apprehension about AIDS have fueled further the engine of antiviral development. The urgent need to treat this tragic disease has stimulated a burst of antiviral research with promising results. In a short time, the potential of suramin, ribavirin, azidothymidine, HPA-23, foscarnet, and interferon have been, and are being, evaluated clinically with several other drugs under development. The extraordinary efforts in this area have resulted in demonstration of efficacy for azidothymidine in a remarkably short 19 months which attests to what can be accomplished. The momentum established in antiviral research must not only be nurtured and maintained, but accelerated.

The stage of relying on chance identification of new antivirals in screening programs designed for other purposes has passed. These programs are important and must continue. But remarkable capabilities have also developed in molecular biology to dissect the various components of a virus particle and to identify virus-specific enzymes, structural proteins, and other viral components which may serve as targets for development of new agents. Crystallography, computer-assisted molecular modeling, and so on, are yielding data on the structure of viruses, including the spatial configuration of their component proteins (8). These data curently are being used for identification of immunogenic sites (epitopes), and may also prove useful for the identification of targets for antiviral agents (9).

The feasibility of a targeted approach is demonstrated by the success of agents directed against the thymidine kinases of herpes viruses and the reverse transcriptase of Human Immunodeficiency Virus (HIV). Other similar targets can be identified for HIV, such as the genome of the <u>trans</u>-activator protein. This protein is a stimulator of replication and, therefore, may prove an excellent target for viral inhibition.

The need for treatment of viral diseases remains great because the available agents barely scratch the surface. Important candidate viruses for the development of antivirals are listed in Table 1. Of particular importance are those viruses with rapidly changing surface antigens, such as influenza, that limit the effective life of vaccines, and viruses with animal reservoirs, that complicate control measures. An additional area for antiviral development are those diseases for which there is some indication that a viral etiological agent may be involved, such as diabetes, arthritis, multiple sclerosis, cancer, heart disease, and many other chronic and degenerative diseases.

Viruses warranting increased efforts for antiviral agent development

HTLV-III/LAV (HIV)	Parainfluenza
Herpes	Hepatitis
Papilloma	Enteroviruses
Influenza	Caliciviruses
Rhinovirus	Arenaviruses
Coronavirus	Arboviruses
Respiratory Syncytial	Rabies

The ideal antiviral would have a broad spectrum of activity, although an agent directed at a specific viral structure or enzyme foreign to the host cell is likely to be less toxic. The existence of broad spectrum antivirals such as interferon and ribavirin are, nonetheless, encouraging. The list of potential viral targets includes the protein, lipid and polysaccharide receptors of host cells, as well as enzymes required for synthesis or assembly of nucleic acids and proteins, and inhibitors of glycosylation and phosphorylation.

Perhaps the greatest obstacle to the treatment of viral diseases is viral latency. Following infection, the nucleic acid of some viruses such as herpes and HIV can be incorporated into the genome of the host cell and may replicate indefinitely with the host cell without apparent viral progeny. Periodically, the viral genome can be expressed, destroying the cell and causing disease. Antiviral agents, thus far, have been capable of inhibiting viral replication during the acute stage of disease, but have not been effective in eradicating the latent virus. It is important to realize, therefore, that patients with these diseases may have to receive life-long prophylactic drug treatment for the rest of their lives.

An important factor in the development of viral therapeutics is the ability of the drug to reach the target site. For viruses such as HIV, the causative agent of AIDS, effective agents must be able to penetrate the blood-brain barrier, since the virus infects the brain as well as the immune cells. In addition to the crossing of the natural barriers, systems must be developed that will selectively deliver the agent to the target organ or cell. Combining the drug with a matrix that will permit slow release of a constant inhibitory concentration would permit a steady state of drug availability without the need of continuous infusion or repeated injections. Other mechanisms to enhance the effective delivery of an antiviral agent to a specific target organ could theoretically include complexing with cell-specific monoclonal antibodies. Drugs have been incorporated into liposomes to assure slow release and targeting to a specific organ (10,11). Liposome-encapsulated ribavirin was concentrated in the liver of mice, the primary site of Rift Valley fever virus proliferation. This led to a fivefold greater beneficial effect than administration of the same concentration of free drug. Liposomes are cleared from the circulation by phagocytic cells of the reticuloendothelial system. Liposomes in the lung capillaries are engulfed by circulating blood monocytes which may subsequently become alveolar macrophages. Liposomes thus offer a means of targeting antivirals to viral infections of the monocyte-macrophage series of cells. Other slow release strategies are also under study.

Topical agents, effective in treating ophthalmic infections, have not been very successful in treating viral infections elsewhere in the body, probably because of difficulty in penetrating the skin. However, topical agents can be formulated with effective penetrating compounds such as dimethylsulfoxide (DMSO), ozone, and propylene glycol. Trifluorothymidine (TFT) combined with ozone and propylene glycol was tested against herpes simplex type 1 infections in a dorsal cutaneous guinea pig model (12). These studies demonstrated that the penetration-enhancing agents permitted the effective delivery of the drug and the expression of its antiviral effect.

The development of resistance to an antiviral agent is also a concern; an agent that acts on more than one site of viral replication reduces the likelihood of viral resistance. Combination drug therapy also would reduce the development of drug resistance by concomitantly attacking multiple viral target sites. If the combination acts synergistically, the lower dosage of each drug might also reduce toxicity with greater efficacy. Thus, combination drugs could have the multiple effect of reducing toxicity, avoiding resistance, improving efficacy, and overcoming other single-compound shortcomings. Several examples of such effects can be cited using interferon and anti-herpes drugs in herpes keratoconjunctivitis (13) as well as ribavirin and closely related C-glycosyde nucleoside analogs (14,15). Ideally, antivirals should not suppress normal immunological responses so that the patient will be protected from future infection by the virus.

Successful antiviral therapy is dependent on rapid viral diagnosis. If the former is to be successful, equal effort must be expended on the development of the latter. Vidarabine and acyclovir are both effective in treatment of herpes encephalitis, but their beneficial effect is dependent on the stage of the disease at which treatment is initiated. Individuals can be saved without sequelae if treated early in the disease process, but cannot be helped in the comatose stage. Antivirals probably will be effective only in those diseases in which progression is linked to continued viral replication and where diagnosis can be made early.

RESPIRATORY DISEASES

1) The Common Cold: The common cold remains one of the major causes of morbidity and work absenteeism. Considerable attempts have been made at a cure. The major problem is the wide range of viral agents and strains. A specific problem in combating the common cold is that it is a relatively mild infection of short duration. Therefore, for an antiviral to be acceptable, it should be almost free of toxicity and side effects. Ideally, it should be an over-the-counter drug. In addition to Eli Lilly, which developed enviroxime, other firms that have developed antirhinovirus compounds include: Wellcome Research, Smith Kline Beckman, Pfizer, Glaxo, Rhone-Poulenc, Sterling-Winthrop, Merrill Dow, and Imperial Chemical. None of their candidates have demonstrated adequate clinical efficacy. Other means of controlling the common cold are exemplified by the work of Elliot Dick in controlling transmission. Using virucidal facial tissues (paper tissues treated with 9.1% citric acid, 4.5% malic acid, and 1.8% sodium lauryl sulfate) he and his associates have demonstrated that viral transmission can be interrupted (16). The citric acid was the major component against rhinoviruses, malic acid was added for synergy, and sodium lauryl sulfate, a surfactant, provided added activity against paramyxoviruses. Volunteers experimentally infected with rhinovirus were confined for 12 hours in a room with healthy poker-playing partners who served as controls. In two experiments, all of the participants were given plain cotton handkerchiefs to use whenever they wished. In two other experiments, the participants were given virucidal tissues and instructed to use them carefully to smother all coughs and sneezes. In the handkerchief experiments, 42%

and 75% of the normals became infected, whereas none of the virucidal tissue experiment normals caught colds; a total of 14/24 versus 0/24. Although these results were quite striking, the experiments would have been more conclusive if the handkerchief volunteers had received the same instructions for use as the virucidal tissue experiment participants.

Another approach to the treatment of the common cold has involved the use of zinc lozenges (17). A double blind, placebo-controlled, clinical trial was performed using zinc gluconate lozenges every two wakeful hours in patients with common colds. After seven days, 89% of 37 treated patients were asymptomatic compared to 47% of 28 patients in the placebo group. The lozenges shortened the average duration of common colds by about seven days. Although the treatment group did fare better than the control group, there were several shortcomings in the study, including unpalatability and distortion of taste in many subjects. If these disadvantages can be overcome, zinc lozenges may prove to be a beneficial over-the-counter medication.

The common cold was one of the first infections to be tested with interferon. It now appears, after many years of study, that this substance may have an effect against this disease. Recombinant alpha interferon is effective in postexposure of rhinovirus colds in a family setting. If family members used an interferon nasal spray within 48 hours of onset of illness in an index case, respiratory infections could be prevented in family contacts (18,19,20). However, nasal congestion and blood tinged mucus or nasal mucosal bleeding were observed more frequently in the interferon group compared to placebo. Further work is warranted to make this a practical regimen for the prevention of colds. The current side effects and prescription status of the drug dictate against wide usage.

2) Influenza: Amantadine remains perhaps the most effective and least toxic of all the antiviral agents. It has both prophylactic and therapeutic effects against influenza A infections. Its efficacy as a prophylactic agent was demonstrated in the mid-60s, but, for a variety of reasons, was not generally accepted by the medical community. Its importance was again demonstrated in the late 70s when clinical studies again proved its value not only as a prophylactic agent but also as a

therapeutic one (21). Another study showed 78% reduction in influenza-like illness and a 91% reduction in laboratory-confirmed influenza (22). Rimantadine, an amantadine analog, was introduced because of continuing concern about the effects of amantadine on the central nervous system (CNS). Rimantadine not only compared favorably as a therapeutic and prophylactic drug, it also had the advantage of no CNS effects. This compound was shown to be 65% effective in preventing influenza-like illness and to reduce laboratory-confirmed influenza by 85% (22).

A recent placebo-controlled, double-blind trial demonstrated that one dose of rimantadine, 200 mg/day, for five days was sufficient to produce a significant therapeutic effect against uncomplicated influenza A infection when treatment was initiated within 48 hours of the onset of symptoms (23). Although amantadine is an excellent agent against influenza A both prophylactically and therapeutically, rimantadine may evolve as the drug of choice. Once influenza A breaks out in a given community, the drug should be administered for the duration of the outbreak (generally 6-12 weeks) to those for whom influenzal illness may be a problem. Amantadine has been approved by the FDA for influenza A, and rimantadine will be submitted shortly for FDA review.

Ribavirin has activity against both influenza A and B. However, the difficulty of administration, relative ineffectiveness against uncomplicated influenza, and relative toxicity of ribavirin make rimantadine the drug of choice. The relative merits of these drugs against influenzal pneumonia remain to be demonstrated.

3) Respiratory Syncytial Virus: Ribavirin, a highly water-soluble nucleoside, is a broad spectrum antiviral agent recently approved by the FDA for aerosol treatment of respiratory syncyctial virus (RSV) infection in neonates (24). RSV is the major respiratory pathogen of infants and young children, and is responsible for a significant number of pediatric hospitalizations and intensive-care admissions each year. Infected infants were treated in a double-blind, placebo-controlled study with a 12-20 hour/day aerosol of either drug or water for approximately five days. The infants treated with ribavirin improved within three days with significant improvement by the fourth day compared to the controls. The treatment group also displayed less viral

shedding (25). However, a more effective drug or treatment regimen is still needed.

HERPES VIRUS INFECTIONS

The greatest success of antiviral agents has been against herpes viruses. Perhaps this is due to the presence of a virus-specific enzyme, thymidine kinase, which offers an excellent target for antiviral agents. Of the six antiviral agents currently approved for indicated use by the FDA, four are directed against herpes virus.

1) Herpes Infections of the Eye: Herpes eye infections are relatively common and quite debilitating. Since this infection lends itself to topical treatment, drugs that cannot be used systemically can be of value. The first antiviral drug approved by the FDA was idoxuridine for herpetic keratitis. Subsequently, vidarabine, trifluridine, acyclovir, bromovinyldeoxyuridine, and interferon all proved effective in treating herpes infections of the eye. The drug of choice appears to be trifluridine, but it is also one of the more expensive. The eye afforded the first opportunity for antiviral agents, for this same reason (an isolated organ), it may provide the first opportunity for combined therapy. Several studies have already demonstrated that trifluridine in combination with interferon and acyclovir combined with interferon are more effective than any agent by itself.

2) Herpes Encephalitis: Vidarabine was the first drug that proved effective when administered systemically against this ongoing, life-threatening viral disease (5). But, subsequently, direct comparison studies indicated that acyclovir was the drug of choice (26). A major reason is the ease of administration of the drug. The necessity of rapid viral diagnosis is clearly evident. Patients treated early in the disease respond more favorably with fewer sequelae. Since the only accurate method for diagnosis is brain biopsy, a more suitable noninvasive rapid method is needed.

3) Varicella-Zoster: This is a serious infection in normal patients, but it is quite severe and life-threatening in immunocompromised patients. Again, at first vidarabine proved to be effective in treating these patients, but more recent studies comparing acyclovir to vidarabine showed that acyclovir was more effective in preventing

cutaneous dissemination, in reducing viral shedding and the appearance of new lesions, in healing existing lesions, and in shortening the median interval until the first decrease in pain (27). 2'-Fluoro-5-iodoarabinosylcytosine (FIAC) was also compared to vidarabine in a randomized, double-blind study and reported to be superior to vidarabine in varicella-zoster patients (28). Analogs of FIAC (FIAU and FMAU) are also being studied.

Bromovinyldeoxyuridine is a potent viral inhibitor _in vitro_ against herpes simplex 1 and varicella-zoster. The drug seems promising based on initial clinical studies and these are now expanded to double-blind controlled trials.

4) _Neonatal Herpes_: Since acyclovir proved more effective than vidarabine in the treatment of herpes encephalitis and varicella-zoster, a comparison was made in neonatal herpes (29). A study in 182 infants with herpes simplex infections was recently completed; 87 received vidarabine (V) and 95 acyclovir (A). They were divided into three categories: 64 with CNS involvement, 43 with disseminated disease (DIS), and 75 with infection localized to the skin, eye or mouth (SEM). None of the SEM babies died at one year; 85% V and 93% A were developing normally one year post-infection. The mortality in neonates with CNS infection was 13% V and 8% A, and 37% V and 34% A were developing normally at one year. The highest mortality was in DIS disease: 50% V and 65% A, with only 23% V and 29% A developing normally at one year. Therefore, unlike with the previous two indications, there is no advantage for either drug; both drugs appeared equally effective.

5) _Oral/Genital Herpes_: Acyclovir has been shown to prevent effectively reactivation of herpes in seropositive immunocompromised patients (30,31). It has also proven effective in speeding the resolution of symptoms and lesions of oral and genital herpes, but its effectiveness in recurrent episodes is limited. It has no effect on recurrence rate in those patients effectively treated for first episodes (32,33). Studies have been done on patients with severe and frequent recurrences where prophylactic dosages of two to five 200mg capsules of oral acyclovir were administered daily. Although these studies showed prophylactic efficacy, long-term usage must be approached with great caution. Further data are needed on long-term toxicity, development of resistant strains, and transmission of disease (34).

6) Cytomegalovirus: The compounds vidarabine and acyclovir have little or no effect against cytomegalovirus (CMV). However, the acyclic nucleoside structurally related to acyclovir, 9 - [2- hydroxy -1- (hydroxymethyl) ethoxymethyl] guanine (DHPG), does. Ten marrow transplant recipients with biopsy-proven CMV pneumonia were treated with DHPG (35). Viruria and viremia ceased in four days, and CMV was eliminated from respiratory secretions after a median of eight days, but only one patient survived the pneumonia. DHPG is proving more effective in other CMV infections, particularly in AIDS patients. These patients with CMV retinitis and gastrointestinal disease seem to improve with therapy, and the virus does recur after termination of treatment (36,37).

Foscarnet has a broad anti-herpes spectrum, by inhibition of DNA polymerases. Preliminary data indicate that it has some beneficial clinical results with CMV infections. It has an affinity for the inorganic matrix of bone, with 10% being bound in adult animals and 30% in growing animals. Prospects for this drug against herpes infections do not appear high, but more information is needed.

7) Epstein-Barr Virus: Because of the effectiveness of acyclovir against herpes simplex virus as well as varicella-zoster and its antiviral activity against Epstein-Barr virus in vitro, 31 patients with clinical and laboratory diagnosis of infectious mononucleosis were treated with the drug. A placebo-controlled, double-blind trial was performed in patients with symptoms for seven or fewer days. Although the acyclovir-treated group showed reduced viral shedding, there was no significant efficacy shown (38).

ARENAVIRUS INFECTIONS

Ribavirin is a broad spectrum antiviral agent against a wide range of DNA and RNA viruses including retroviruses. This activity is due to its resemblance to nucleosides. As stated above, it is approved by the FDA for use against respiratory syncytial virus in infants. But one of the more exciting uses of ribavirin is against arenaviruses. A study comparing the effectiveness of intravenous ribavirin versus convalescent plasma in the treatment of Lassa fever in Sierra Leone was reported recently (39). Lassa fever is a serious, often fatal, arenavirus infection with insidious onset and a wide variety of signs and symptoms. The acute phase lasts one to four weeks, with death in 15-20% of hospitalized patients. Patients selected for this study were those with a much higher risk of death. A serum aspartate aminotransferase level equal to, or greater than, 150 units per liter had a case-fatality rate of 55%. Patients in this category, whose treatment with ribavirin was started within the first six days of onset of fever, had a case-fatality rate of 5%; if treatment was initiated seven days or more after the onset of fever, mortality rose to 26%. Viremia with levels equal to, or greater than, $10^{3.6}$ TCID per ml on admission had a case-fatality rate of 76%. These patients, treated within the first six days of onset of fever, had a case-fatality rate of 9% compared to 47% if treated seven days or more post onset of fever. Oral ribavirin was also effective, but convalescent serum did not significantly reduce mortality nor increase the benefit of ribavirin when given concomitantly.

Ribavirin has demonstrated some potential against Rift Valley fever virus, Hantaan virus (the etiological agent of Korean hemorrhagic fever), as well as Junin, Machupo, and Pichinde viruses. Ribavirin is also effective against viruses of the toga and bunya families. These insect- and rodent-transmitted viruses cause considerable morbidity and mortality worldwide. Clinical studies are currently under way in Korea against Hantaan virus.

RETROVIRUS INFECTIONS--AIDS

Retroviruses have a role in leukemia and other diseases, but the one of major current interest is AIDS, first reported in 1981. Scientists have risen to the challenge; within two years the etiological agent (HIV) was identified, and a year later a serological test was developed which could prevent transmission through blood products. Shortly thereafter, an antiviral agent, azidothymidine (AZT), was shown to have an *in vitro* inhibitory effect against HIV, and, as stated above, within 19 months clinical efficacy was shown and it was being distributed to specified AIDS patients. Although AZT has been demonstrated to prolong the life of AIDS patients, the rapidity of the studies have not allowed the opportunity to understand all the benefits and the problems of the drug. AZT is not without toxicity (hematopoietic effect); this must be more fully understood. The role this drug may have against AIDS-related complex (ARC) still needs to be elucidated, and what effects, other than prolongation of life, this drug affords are currently under study. AZT represents a major step in the control of the disease and the advancement of antiviral agents, but more potent and/or selective compounds are needed.

There are a number of other antiviral agents against AIDS currently under clinical study (40). The final answer is not yet available, but in descending order of potential are: interferon, ribavirin, foscarnet, suramin and HPA 23. In addition, there are several other compounds awaiting clinical trials, including immune modulators. Two of the more promising compounds just beginning clinical trials are dideoxycytidine, and dideoxyadenosine.

CONCLUSION

The ideal antiviral agent is one that, in addition to being effective, is non-toxic or at least minimally toxic (high therapeutic index), is highly viral-specific, is able to reach the targeted organ, has a low frequency of induced drug resistance, does not inhibit immunological response, and preferably is able to overcome viral latency.

Antiviral agents are becoming more widely accepted both by clinicians and by investigators as a field worthy of their endeavors. The number of compounds with clinical efficacy has grown slowly, but, with recent successes and with the increased interest caused by AIDS, it should grow more rapidly over the next few years.

REFERENCES

1. Pachucki, C.T., Lentino, J.R., Jackson, G.G. J.Inf.Dis. 151:1170-1171, 1985.
2. Georges, P. J.Inf.Dis. 151:981-983, 1985.
3. Kaufman, H.E. Proc. Soc. Exptl. Biol. Med. 109:251-252, 1962.
4. Jackson, G.G., Muldoon, R.L., Ackers, L.W. Antimicrob. Ag. Chemother. 3:703-707, 1963.
5. Whitley, R.J., Soong, S-J., Dolin, R., Galasso, G.J. NEJM 279:289-294, 1977.
6. Corey, L., Nahmias, A.J., Guinan, M.E., Benedetti, J.K., Critchlow, C.W., Holmes, K.K. NEJM 306:1313-1319, 1982.
7. King, D., Galasso, G.J. Amer.J.Med. 73A:Supplement, 1982.

8. Rossmann, M.D., Arnold, E., Erickson, J.W., Frankenberger, E.A., Griffith, J.P., Hecht, H-J., Johnson, J.E., Kramer, G., Luo, M., Mosser, A.G., Rueckert, R.R., Sherry, B., Vriend, G. Nature 317:145-153, 1985.
9. Smith, T.J., Kremer, M.J., Luo, M., Vriend, G., Arnold, E., Kamer, G., Rossmann, M.G., McKinlay, M.A., Diana, G.D., Otto, M.J. Science 233:1286-1293, 1986.
10. Kende, M., Alving, C.R., Rill, W.L., Schwartz, G.M., Jr., Canonico, P.G. Antimicrob. Ag. Chemother. 27:903-907, 1985.
11. Koff, W.C., Fidler, I.J. Antiviral Research 5:179-190, 1985.
12. Spruance, S.L., McKeough, M., Sugibayashi, K., Robertson, F., Golde, P., Clark, D.S. Antimicrob.Ag.Chemother. 26:819-823, 1984.
13. DeKoning, E.W.J., van Bijsterveld, O.P., Cantell, K. Brit. J. Ophthal. 66:509-512, 1982.
14. Huggins, J.M., Robins, R.K., Canonico, P.G. Antimicrob. Ag. Chemother. 26:476-480, 1984.
15. Kirsi, J.J., McKernan, P.A., Burns III, N.J., North, J.A., Murray, B.K., Robins, R.K. Antimicrob. Ag. Chemother. 26:466-475, 1984.
16. Dick, E.C., Houssain, S.U., Mink, K.A., Meschievitz, C.K., Schultz, S.B., Raynor, W.J., Inhorn, S.L. J. Inf. Dis. 153:352-356, 1986.
17. Eby, G.A., Davis, D.R., Halcomb, W.W. Antimicrob. Ag. Chemother. 25:20-24, 1984.
18. Monto, A.S., Shope, T.C., Schwartz, S.A., Albrecht, J.K. J. Infect. Dis. 154:128-133, 1986.
19. Hayden, F.G., Albrecht, J.K., Kaiser, D.L., Gwaltney, J.M., Jr. NEJM 314:71-75, 1986.
20. Douglas, R.M., Moore, B.W., Miles, H.B., Davies, L.M., Graham, N.M., Ryan, P., Worswick, D.A., Albrecht, J.K. NEJM 314:5-70, 1986.
21. Monto, A.S., Gunn, R.A., Bandyk, M.G., King, C. JAMA 241:1003-1007, 1979.
22. Dolin, R., Reichman, R.C., Madore, H.P., Maynard, R., Linton, P.N., Webber-Jones, J. NEJM 307:580-584, 1982.
23. Hayden, F.G., Monto, A.S. Antimicrob. Ag. Chemother. 29:339-341, 1986.
24. Fernandez, H., Banks, G., Smith, R.A. European J. Epidemiol. 2:1-14, 1986.
25. Hall, C.B. In: The Antimicrobial Agents Annual 1 (Eds. Peterson, P.K., Verhoef, J.), Elsevier, Amsterdam, 1986, pp. 358-369.
26. Whitley, R.J., Alford, C.A., Jr., Hirsch, M.S., Schooley, R.T., Luby, J.P., Aoki, F.Y., Hanley, D., Nahmias, A.J., Soong, S-J., and the NIAID Collaborative Antiviral Study Group NEJM 314:144-149, 1986.
27. Shepp, D.H., Dandliker, P.S., Meyers, J.D. NEJM 314:208-212, 1986.
28. Leyland-Jones, B., Donnelly, H., Groshen, S., Myskowski, P., Donner, A.L., Fanucchi, M., Fox, J., and the Memorial Sloan-Kettering Antiviral Working Group. J. Inf. Dis. 154:430-436, 1986.
29. Whitley, R.J., Arvin, A., Corey, L., Powell, D., Plotkin, S., Starr, S., Alford, C.A., Jr., Connor, J., Nahmias, A.J., Soong, S-J., and the NIAID Collaborative Study Group. Society for Pediatric Research, Washington, D.C., 1986.
30. Saral, R., Burns, W.H., Laskin, O.C., Santos, G.W., Lietman, P.S. NEJM 305:63-67, 1981.
31. Wade, J.C., Newton, B., Flournoy, N., Meyers, J.D. Ann. Intern. Med. 100:823-828, 1984.

32. Mertz, G.J., Critchloro, C., Benedetti, J., Reichman, R.C., Dolin, R., Connor, J., Redfield, D.C., Savoia, M.C., Richman, D.D., Tyrell, D.L., Miedzinski, L., Portnoy, J., Keeney, R.E., Corey, L. JAMA 252:1147-1151, 1984.
33. Reichman, R.C., Badger, G.J., Mertz, G.J., Corey, L., Richman, D.D., Connor, J.D., Redfield, D., Savoia, M.C., Oxman, M.N., Bryson, Y., Tyrell, D.L., Portnoy, J., Creigh-Kirk, T., Keeney, R.E., Ashikaga, T., Dolin, R. JAMA 251:2103-2107, 1983.
34. Corey, L., Spear, P.G. NEJM 314:749-757, 1986.
35. Shepp, D.H., Dandliker, P.S., de Miranda, P., Burnette, T.C., Cederberg, D.M., Kirk, E., Meyers, J.D. Ann. Intern. Med. 103:368-373, 1985.
36. Collaborative DHPG Treatment Study Group. NEJM 314:801-805, 1986.
37. Felsenstein, D., D'Amico, D., Hirsch, M., Neumeyer, D.A., Cederberg, D.M., de Miranda, P., Schooley, R.T. Ann. Int. Med. 103:377-380, 1985.
38. Anderson, J., Britton, S., Ernberg, I., Anderson, U., Henle, W., Skoldenberg, B., Tisell, A. J. Inf. Dis. 153:283-290, 1986.
39. McCormick, J.B., King, I.J., Webb, P.A., Scribner, C.L., Craven, R.B., Johnson, K.M., Elliott, L.H., Belmont-Williams, R. NEJM 314:20-26, 1986.
40. DeClercq, E. J. Med. Chem. 29:1561-1569, 1986.

SUBJECT INDEX

GPSR Compliance

The European Union's (EU) General Product Safety Regulation (GPSR) is a set of rules that requires consumer products to be safe and our obligations to ensure this.

If you have any concerns about our products, you can contact us on ProductSafety@springernature.com

In case Publisher is established outside the EU, the EU authorized representative is:

Springer Nature Customer Service Center GmbH
Europaplatz 3
69115 Heidelberg, Germany

Batch number: 10761599

Printed by Printforce, the Netherlands